Contents

	Introduction to GNVQs in engineering	iv
	Acknowledgements	vii

PART ONE   ENGINEERING MATERIALS AND PROCESSES

1	Materials and components for engineered products	3
2	Engineering processes for electromechanical products	44
3	Producing electromechanical products to specification	109
	Sample unit test for Unit 1	122

PART TWO   GRAPHICAL COMMUNICATION IN ENGINEERING

4	Graphical methods for communicating engineering information	129
5	Scale and schematic drawings in engineering	172
6	Interpreting engineering drawings	243
	Sample unit test for Unit 2	275

PART THREE   SCIENCE AND MATHEMATICS FOR ENGINEERING

7	Scientific laws and principles applied to engineering	285
8	The measurement of physical quantities	331
9	Mathematical techniques	350
	Sample unit test for Unit 3	371

PART FOUR   ENGINEERING IN SOCIETY AND THE ENVIRONMENT

10	The application of engineering technology in society	379
11	Careers in engineering	417
	Sample unit test for Unit 4	457

Answers to the sample unit tests	460
Index	461

Introduction to GNVQs in engineering

GNVQs (General National Vocational Qualifications) are a new type of qualification available for students who want to follow a course linking traditional areas of study with the world of work. Those who qualify will be ideally placed to make the choice between progressing to higher education and applying for employment.

GNVQs are available at three levels:

Foundation normally studied full time for 1 year
Intermediate normally studied full time for 1 year
Advanced normally studied full time for 2 years, and often referred to as 'vocational A levels'.

Each GNVQ is divided into various types of units:

Mandatory units which everyone must study
Optional units from which a student must choose to study
Additional units offered by the various awarding bodies in order to allow students to top up their skills and knowledge
Key skills units covering the essential skills related to numeracy, IT and communication.

Units are clearly structured. The principal components are:

Elements which focus on specific aspects of the unit
Performance criteria which tell you what you must be able to do
Range which tell you what areas you must be able to apply your knowledge to.

Intermediate GNVQs in Engineering

This book is intended primarily for students of the Intermediate GNVQ in Engineering. This course is divided as follows:

Four mandatory units covering all key aspects of engineering as a discipline, namely:

Unit 1: Engineering materials and processes
Unit 2: Graphical communication in engineering
Unit 3: Science and mathematics for engineering
Unit 4: Engineering in society and the environment

from which the GNVQ student must choose and study two.

Additional units chosen from those offered by the various awarding bodies.
Key skills units in the three areas at level two.

Assessment

GNVQs are assessed using two methods:
Unit tests — taken by everybody
Portfolio of evidence — developed individually by each student to demonstrate competence and understanding.

The unit tests can be sat on a number of occasions throughout the year, and can be sat again should a student not pass. Any number of attempts can be made to pass, but a pass must be achieved before a student's portfolio will be assessed. A GNVQ cannot be awarded until the student has passed the unit test for each of the mandatory units.

Each unit test consists of a number of multiple choice questions, and the pass mark is 70%. A practice paper is included at the end of each unit in this book, to give students an idea of what is expected of them.

Portfolio of evidence

Each student must develop a portfolio (i.e. organised collection) of evidence that they have achieved the required level of skill and understanding for each of the elements of the units they take. The unit specifications say exactly what type of evidence will be acceptable for assessment, but do not dictate the details of any project to be undertaken, nor exactly how the evidence is to be compiled. Written, photographic, video or audio tape records of achievement are all acceptable, as are testimonials from work placements.

Regardless of the media employed to deliver evidence, it is absolutely essential that the content is organised and referenced so that it can be easily and effectively assessed. There is no point in assembling masses of evidence if its presentation makes it impossible to assess or understand. An index should be included, and evidence relevant to more than one element cross-referenced, or duplicated if necessary.

Grading

A GNVQ can be awarded as a pass, a merit or a distinction. To achieve a merit or distinction a student must demonstrate notable ability in a range of skills related to the development of high-quality portfolio evidence. Your lecturer or teacher will be able to give you detailed guidance in how to develop and demonstrate these skills.

Additional titles developed to support GNVQs in Engineering are listed on the back cover of this book, and further details are available from Stanley Thornes Publishers Customer Services Department on 01242 228888.

The following table shows how the assignments offer the opportunities to provide evidence of the key skills.

Users of this book should note that the many activities inserted through the text provide numerous opportunities to develop evidence of competence in the key skills. Each activity has been left sufficiently flexible to allow students and lecturers the option of adapting them in order to demonstrate these skills as required.

Key skill coverage in the assignments

Key skills	Assignments
Communication	
2.1 Take part in discussions	1, 2, 3, 11
2.2 Produce written materials	1, 2, 3, 4, 6, 7, 10, 11
2.3 Use images	1, 2, 3, 4, 5, 6, 10, 11
2.4 Read and respond to written material	4, 6, 8, 10, 11
Information technology	
2.1 Prepare information	4, 5, 6, 10, 11
2.2 Process information	4, 5, 6, 10, 11
2.3 Present information	4, 5, 6, 10, 11
2.4 Evaluate the use of information technology	5
Application of number	
2.1 Collect and record data	8
2.2 Tackle problems	3, 9, 10
2.3 Interpret and present data	3, 7, 10

Acknowledgements

The authors and publishers are grateful to the following individuals or organisations for permission to reproduce photographs and other material:

British Aerospace Defence Ltd; British Gas plc; British Steel plc; Colour Unlimited; Desoutter Ltd; The Engineering Council; Esso UK plc; The Ford Motor Company Ltd; Maplin Electronics Group; The National Radiological Protection Board; The Office for National Statistics; The Oil Spill Recovery Centre; Silverscreen Solid Modelling (distributed in the UK by Leonardo Computer Systems); Staubli Unimation; WaterAid; Quantel Electronics plc; Gerald Dunne, Alan Rudge and Pamela Wilson; Pictor International Ltd and Alan Rutherford for the cover illustration.

Every effort has been made to contact copyright holders, and we apologise if any have been overlooked.

PART ONE: ENGINEERING MATERIALS AND PROCESSES

Chapter 1: Materials and components for engineered products
Chapter 2: Engineering processes for electromechanical products
Chapter 3: Producing electromechanical products to specification

Sample unit test for Unit 1

Engineered products are many and varied. A paper clip is an engineered product and so is Concorde. They are all important to Britain's economy. Their manufacture provides employment and their export brings wealth into the country. An engineered product should do the job for which it was designed, sell at a competitive price, require a minimum of maintenance and have a reasonably long service life. A product that fulfils these conditions is said to have 'quality'.

The design and manufacture of a quality product requires teamwork. Design and production engineers and purchasing and scheduling personnel must work together to make sure that a product is not only fit for its purpose but also can be produced and sold economically.

The engineers must select the most appropriate design and production methods. In order to do this they need a wide knowledge of the available materials, components and processing methods. The purchasing and scheduling personnel must ensure that the materials are available in the quantities and at the times required so that production runs smoothly.

The team must also include the technicians and operators who process and assemble the product. The quality of the finished product depends heavily on their skills and attention to detail. Many successful engineering companies operate 'quality circles', in which members of the workforce meet to discuss ideas for design changes and the more efficient use of materials and processes.

Chapter 1: Materials and components for engineered products

> **This chapter covers:**
> Element 1.1: Select materials and components for engineered products
> **... and is divided into the following sections:**
> - Materials for mechanical products
> - Components for mechanical products
> - Selecting materials and components for a mechanical product
> - Electrical and electronic components for electrical products
> - Selecting electrical and electronic components for an electrical product
> - Selecting suitable materials and components for an electromechanical product.

Materials for mechanical products

The solid materials used to make mechanical products can be divided into:
- ferrous metals
- non-ferrous metals
- thermoplastics
- thermosetting plastics
- rubbers
- ceramics
- semiconductors
- composite materials.

Ferrous metals

Ferrous metals
Metals in which the main constituent is iron

Ferrous metals are those in which the main constituent is iron. Pure iron has a dull grey colour. It is very soft and does not machine to a good finish. When molten, pure iron is not very fluid and this makes it difficult to cast. For these reasons, it is not used in its pure state as an engineering material.

Plain-carbon steels

The addition of small amounts of carbon greatly improves the properties of iron, making it much stronger and tougher. The resulting alloy is called plain-carbon steel. It should be noted that an alloy is a mixture of a metal and at least one other element. Plain-carbon steels can contain up to 1.4% carbon. Its different grades are known as dead mild steel, mild steel, medium-carbon steel and high-carbon steel. Their carbon content and some of their properties and uses are shown in Table 1.1.

The effect of increasing the carbon content in plain-carbon steel is shown in Figure 1.1.

Table 1.1 Plain-carbon steels

Name	Carbon content (%)	Properties	Uses
Dead mild steel	0.10–0.15	Fairly soft, suitable for cold forming	Wire, drawn tube, nails, rivets, sheet steel for pressings
Mild steel	0.15–0.30	Strong, general-purpose material	Girders, boiler plate, barstock, nuts and bolts
Medium-carbon steel	0.30–0.80	Strong and tough, can be hardened by heat treatment	Crankshafts, axles, couplings, gears, wire ropes, hammer heads, cold chisels
High-carbon steel	0.80–1.40	Strong and tough, can be made very hard by heat treatment	Springs, knives, screwcutting taps and dies, saw blades

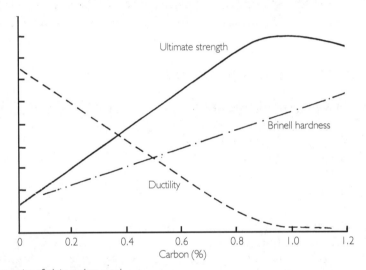

Figure 1.1 Properties of plain-carbon steel

As can be seen, the hardness of the steel increases uniformly with carbon content, as does its strength up to a composition of around 1% carbon, after which it starts to fall. The ductility of the steel, which is its ability to be drawn out in tension without breaking, falls with carbon content.

Grey cast iron

If more carbon is added to steel, it is found that at a content above about 1.7% flakes of graphite appear among the grains when it is cooled down. The alloy is then known as grey cast iron, which has quite different properties to plain-carbon steel. For practical purposes, a carbon content of 3.2–3.5% is found to be the most suitable. This makes the material very fluid when molten and enables it to be cast into intricate shapes.

After cooling, grey cast iron is strong in compression but rather brittle and weak in tension. It has the ability to absorb vibrations and is widely used for the beds and frames of machines, decorative columns and arches in buildings, manhole covers and a host of other uses in which the loading is compressive rather than tensile. Carbon in the form of graphite is a good lubricant and the graphite flakes in grey cast iron make the material self-lubricating and easy to machine without a cutting fluid. The only problem is that black carbon dust is given off and dust extraction equipment is necessary where cast iron is machined regularly.

Alloy steels

In addition to carbon, other elements are added to steel to give it particular properties. Two common examples are stainless steel and high-speed steel, which are referred to as alloy steels. Stainless steel contains nickel and chromium to give it corrosion resistance. High-speed steel is used for cutting tools. It contains tungsten, chromium and vanadium to help it keep its sharp cutting edge when cutting at high speeds.

Case study

The hip joints of elderly people often become painful owing to wear and tear. At one time there was little that could be done and the condition often became crippling. Today the situation is much improved. Artificial hip joints have been developed and hip joint replacement is a common operation.

The improvements that have taken place over the last 40 years are due to advances in surgery and cooperation between surgeons and material technologists. The hip joint consists of a ball and socket and both parts are replaced. The materials used must be strong, tough, non-corrosive and wear resistant. They must also be compatible with the body so that they do not react with the blood or trigger rejection by the body.

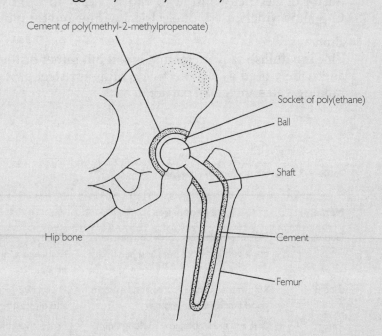

Figure 1.2 Total hip joint replacement

The worn head of the femur (the thigh bone) is removed and a metal ball on a shaft is cemented into it. Stainless steel is widely used for this, but new alloys that can move and flex with the bone are being developed. The socket in the pelvis is enlarged and an artificial plastic socket is cemented in. High-density polyethene is widely used for this but here, again, new materials that are more wear resistant are under investigation. It is hoped that eventually the cement will no longer be necessary by using materials that the bone will grow into and so form a natural bond.

Non-ferrous metals

Non-ferrous metals and alloys
Those that contain no iron, or in which iron is present only in very small amounts

Non-ferrous metals and alloys are those that contain no iron, or in which iron is present in only very small amounts. Some of the most common non-ferrous base metals used in engineering are:
- copper
- zinc
- tin
- aluminium.

Some of the most common non-ferrous alloys are:
- brasses
- bronzes
- cupro-nickels
- aluminium alloys.

Copper

Copper has a distinctive reddish colour and was one of the earliest metals to be used by man. It is a malleable and ductile material, which means that it is relatively soft and easy to deform. It is also resistant to corrosion, a very good conductor of electricity and heat, and it is easy to join by soldering and brazing. Copper is widely used for electrical wiring and for plumbing.

Zinc

Zinc is a dullish grey metal that is soft but rather brittle. It is corrosion resistant and widely used as a protective coating for steel products such as waste bins, railings, gates and crash barriers.

Table 1.2 Pure non-ferrous metals

Name	Properties	Uses
Copper	Ductile, malleable, corrosion resistant, good conductor of heat and electricity	Electrical wire and cable, water pipes, heat exchangers, alloying to make brasses and bronzes
Zinc	Soft, rather brittle, corrosion resistant, good fluidity when molten	Protective coatings, alloying to make brasses, and die casting alloys
Tin	Soft and malleable, low melting point, good corrosion resistance	Protective coatings, alloying to make tin bronzes and soft solders
Aluminium	Malleable and ductile, corrosion resistant, good fluidity when molten, good conductor of heat and electricity	A variety of engineering and domestic products, alloying to make aluminium alloys and bronzes

Tin

Tin is a shiny silvery metal that is very soft and malleable. It is corrosion resistant and used as a protective coating for the thin mild-steel sheet used for food cans. This material is called 'tinplate'. Tin is also used for coating copper wire to make it easier to join by soldering.

Aluminium

Aluminium is a whitish grey metal. It is very light, having less than half the density of iron. It is very malleable and ductile and has good corrosion resistance. Compared with steel, aluminium is not very strong in its pure form, but it is a good conductor of electricity and heat. It is widely used as the conducting core in overhead transmission lines surrounded by high-tensile steel, which carries the load. It is also widely used for cooking utensils.

Progress check

1. What grade of plain-carbon steel is used for general engineering purposes such as for making nuts and bolts?
2. What are the typical uses of high-carbon steel?
3. What effect do the flakes of graphite in grey cast iron have on its properties?
4. What are the alloying elements that give stainless steel its corrosion resistance?
5. What are the properties of copper that make it suitable for use in electrical wiring?
6. What is 'tinplate' and what is it widely used for?
7. What are the properties of aluminium that make it suitable for use in saucepans and other kitchen utensils?
8. What are the main uses of zinc in engineering?

Brass

Brass is an alloy of copper and zinc that sometimes also contains small amounts of tin and lead, and has a distinguishing yellow colour. It is stronger and tougher than either of its main constituents and has good corrosion resistance. Brasses with a high copper content are more ductile and malleable. They are used for

Table 1.3 Common brasses

Name	Composition (%)	Properties	Uses
Cartridge brass	Copper 70 Zinc 30	Very ductile	Cold-formed deep-drawn components, e.g. cartridge cases, condenser tubes
Admiralty brass	Copper 70 Zinc 29 Tin 1	Very ductile and corrosion resistant	Cold-formed deep-drawn components for marine and other uses
Standard brass	Copper 65 Zinc 35	Ductile and tough	Cold pressings
Muntz metal	Copper 60 Zinc 40	Strong and tough	Hot forming, e.g. hot-rolled plate, forgings and castings
Naval brass	Copper 62 Zinc 37 Tin 1	Strong, tough and corrosion resistant	Hot-formed components for marine and other uses

cold-forming processes such as rolling, drawing and extrusion. Brasses with a high zinc content are more suitable for hot-forming processes such as forging, hot rolling and casting.

Bronze

Bronze, or to be more precise tin bronze, was the first alloy to be widely used by man. Its discovery about 3000 years ago gave rise to the Bronze Age, when it was used for tools and weapons. Tin bronze is an alloy of copper and tin to which a little phosphorus and zinc are sometimes added. As with brass, the alloy is stronger and tougher than either of its main constituents. Tin bronzes with a high copper content take on a reddish colour. They are malleable and ductile

Table 1.4 Common tin bronzes

Name	Composition (%)	Properties	Uses
Low-tin bronze	Copper 96 Tin 3.9 Phosphorus 0.1	Very malleable and ductile when annealed, elastic when cold worked	Springs, electrical contacts, instrument parts
Cast phosphor-bronze	Copper 90 Tin 9.5 Phosphorus 0.5	Tough with good anti-friction properties	Bearings and worm gears
Admiralty gun metal	Copper 88 Tin 10 Zinc 2	Tough with good fluidity and corrosion resistance	Miscellaneous castings, e.g. valve and pump components
Bell metal	Copper 78 Tin 22	Sonorous and tough with good fluidity	Bells and other castings

Table 1.5 Common aluminium bronzes and cupro-nickel alloys

Name	Composition (%)	Properties	Uses
Wrought aluminium bronze	Copper 91 Aluminium 5 Nickel and manganese 4	Ductile and malleable with good corrosion resistance	Boiler and condenser tubes, chemical plant components
Cast aluminium bronze	Copper 86 Aluminium 9.5 Nickel 1 Manganese 1	Tough, good fluidity, good corrosion resistance	Sand and die cast, e.g. valve and pump parts, gears, propellers
Coinage cupro-nickel	Copper 74.75 Nickel 25 Manganese 0.25	Tough, strong and corrosion resistant	'Silver' coinage
Monel metal (cupro-nickel)	Copper 29.5 Nickel 68 Iron 1.25 Manganese 0.25	Tough, strong and corrosion resistant	Chemical plant and marine components
Wrought aluminium bronze	Copper 91 Aluminium 5 Nickel and manganese 4	Ductile and malleable with good corrosion resistance	Boiler and condenser tubes, chemical plant components

and used for cold-forming operations. Tin bronzes with a higher zinc content take on a more yellow appearance and can sometimes be mistaken for brass. They are less ductile and used for hot-forming operations.

Aluminium bronze and cupro-nickel
These are engineering alloys that also have copper as their main constituent. They are tough, strong and highly corrosion resistant. Some of the more common types and their uses are described in Table 1.5.

Aluminium alloys
These have been developed over the years for a wide range of applications in which low weight, corrosion resistance and good strength are needed. As can be seen from Table 1.6, small amounts of copper, silicon, manganese, magnesium and nickel are added to aluminium to give a range of casting and cold-forming alloys.

Table 1.6 Common aluminium alloys

Name	Composition (%)	Properties	Uses
Casting alloy	Aluminium 92 Silicon 5 Copper 3	Good fluidity and moderate strength	Sand and die castings for light-duty applications
Casting alloy	Aluminium 88 Silicon 12	Very good fluidity and moderate strength	Sand and die castings for motor vehicle and marine components
'Y' alloy	Aluminium 92 Nickel 2 Manganese 1.5	Good fluidity, can have its hardness modified by heat treatment	Motor vehicle engine parts, e.g. pistons and cylinder heads
Duralumin	Aluminium 94 Copper 4 Magnesium 0.8 Manganese 0.7 Silicon 0.5	Ductile and malleable in soft condition, can be heat treated	Structural uses, e.g. motor vehicle and aircraft panels
Wrought alloy	Aluminium 97.3 Magnesium 1 Silicon 1 Manganese 0.7	Ductile, with good strength and electrical conductivity, can be heat treated	Structural uses, e.g. ladders, scaffold tubes, overhead power lines

Progress check

1. What is the name given to brass that is composed of 70% copper and 30% zinc, and what are its typical uses?
2. What kind of bronze is used for gearwheels and bearings?
3. What is the alloy used to make 'silver' coinage?
4. What are the general properties of aluminium alloys that make them important engineering materials?
5. What is 'duralumin' and what is it used for?

Engineering materials and processes

Activity 1.1

The plumbing and central heating system in your home, school or college uses a number of the metals and alloys that have been described. Take a close look at the following items and find out what they are made from: (a) a kitchen sink; (b) hot and cold taps; (c) central heating radiators; (d) hot and cold water pipes.

Activity 1.2

There are usually three different types of metal saucepan or frying pan on sale in department stores and hardware shops. Find out what these metals are and see if you can add a fourth one to your list.

Thermoplastics

Thermo-plastics
Polymer materials that can be softened and remoulded by being reheated

Thermoplastics are those polymer materials that can be softened and remoulded by being reheated. Some of the more common thermoplastics are:
- polythene
- polypropene
- PVC (polyvinyl chloride)
- polystyrene
- Perspex
- PTFE (polytetrafluoroethylene)
- nylon
- Terylene.

Like all plastic materials, thermoplastics are made up of chains of carbon atoms called polymers, to the sides of which are attached hydrogen atoms, and sometimes also nitrogen, chlorine, fluorine and silicon atoms. They are good electrical insulators and are used for a wide range of industrial and domestic products.

Polythene

Low-density polythene is tough, durable and resistant to water and chemical solvents. It is used for plastic bags and wrappings, waterproof covers and cable insulation. High-density polythene has longer polymers that are packed more closely together. It is a useful engineering material, being harder, stiffer and stronger than low-density polythene. It is moulded into bottles, pipes, tubs, crates, tanks and other containers.

Polypropene

This is also widely known as 'polypropylene'. It has similar properties to high-density polythene but is stronger, harder and has a higher melting point. It has similar uses to high-density polythene and can also be produced as a fibre for use in ropes and nets. Its higher melting point makes it suitable for hospital equipment, such as bowls and trays, that need to be sterilised frequently.

PVC

This plastic is made from polychloroethene, which used to be called polyvinyl chloride, hence the initials PVC. In its flexible form this plastic is known as pPVC, which is tough and has good solvent resistance. It is used for the insulation on electric wiring and cables, for wellington boots and for items of clothing. In its rigid form it is known as uPVC, which is tough and rigid. It is used for plastic window frames and doors, and also for safety helmets.

Polystyrene

Polystyrene can be made tough and dense for household and industrial use or, in expanded cellular form, it can be used as a light packaging material. The linings for refrigerators, disposable coffee cups and cartons for margarine are all familiar items made from polystyrene. A disadvantage is that it tends to be brittle and is easily broken.

Table 1.7 Common thermoplastic materials

Name	Polymer	Properties	Uses
Low-density polythene	Polyethene	Tough, flexible, easily moulded, solvent resistant, gradual deterioration if exposed to light	Packaging, piping, squeeze containers, cable and wire insulation
High-density polythene	Polyethene	Similar to low-density polythene but harder, stiffer and with good tensile strength	Pipes, mouldings, tubs, crates, food containers, medical equipment
Polypropene	Polypropene	High strength, hard, high melting temperature, can be produced as a fibre	Tubes, pipes, fibres, ropes, electronic components, kitchen utensils, medical equipment
PVC	Polychloroethene	Can be made tough and hard or soft and flexible, solvent resistant, soft form tends to harden with time	When hard, window frames, piping and guttering; when soft, cable and wire insulation, upholstery
Polystyrene	Polyphenyl ethene	Tough, hard, rigid but somewhat brittle, can be made into a light cellular foam, can be attacked by petrol-based solvents	Mouldings for refrigerators and other appliances, moulded foam used for packaging
Perspex	Methyl-2-methylpropenoate	Strong, rigid, transparent but easily scratched, easily softened and moulded, can be attacked by petrol-based solvents	Lenses, protective shields, aircraft windows, light fittings, corrugated sheets for roof lights
PTFE	Polytetrafluoroethylene	Tough, flexible, heat resistant, highly solvent resistant, has a waxy low-friction surface	Bearings, seals, gaskets, non-stick coatings, tape
Nylon	Polyamide	Tough, flexible and very strong, good solvent resistance but does absorb water and deteriorates with outdoor exposure	Bearings, gears, cams, bristles for brushes, textiles
Terylene	Thermoplastic polyester	Strong, flexible and solvent resistant, can be made as a fibre, tape or sheet	Textile fibres, recording tape, electrical insulation tape

Perspex
Perspex is strong, rigid and transparent, and can be softened and remoulded using boiling water. It is used for protective shields, lenses and corrugated sheets for roof lights. Unfortunately, it is easily scratched and is attacked by petrol.

PTFE
This name is derived from the polymer polytetrafluoroethylene. It is a tough, flexible, heat-resistant thermoplastic that is not attacked by any solvent. It also has very low frictional resistance. PTFE is widely used as a bearing and gasket material in industry and as a non-stick coating for domestic utensils under the trade name of Teflon.

Nylon
Nylon is strong, tough and flexible and has good solvent resistance. It is used in fibre form for fishing line, ropes and nets. Nylon is also hard wearing, which makes it suitable for use as a material for bearings and gears. A disadvantage that it shares with some other thermoplastics is that it deteriorates after long exposure to sunlight.

Terylene
Terylene and Dacron are trade names for materials produced from polymers known as polyesters. In fibre form they are strong and stable with good solvent resistance but they can be attacked by strong acids and alkalis. Polyester materials are also used for recording tapes and for electrical insulation.

Thermosetting plastics

> **Thermo-setting plastics**
> *Polymer materials that cannot be softened and remoulded by reheating*

Thermosetting plastics cannot be softened by heat after they have been moulded to shape. Their raw material comes in the form of a powder or a resin. The powder is heated under pressure in a mould and the resin is moulded after mixing it with a chemical hardener. During the moulding process, the polymers become joined together by chemical bonds known as 'cross-links'. Once formed, these cannot be broken by heat and they hold the material rigidly in shape. Some common thermosetting plastic materials used in engineering are:
- Bakelite
- formica
- melamine
- epoxy resins
- polyester resin.

Thermosetting plastics are resistant to most solvents and excellent insulators of both heat and electricity. They tend to be more rigid and brittle than thermoplastics, and very often filler materials are added to improve their properties. These include very fine sawdust called wood flour, powdered chalk or limestone, mica granules, glass fibres and carbon fibres.

Bakelite
Bakelite was developed by the Belgian chemist Dr Leo Baekland early in this century and was one of the first thermosetting plastics to become widely used. Wood flour is commonly used as a filler material to give it strength and it is usually black or brown in colour. It is still used for insulated handles, knobs and electrical fittings. Bakelite is resistant to most solvents but does absorb water. It does not soften when heated but will decompose at temperatures above 200°C and should not be subjected to working temperatures above about 80°C.

Materials and components for engineered products

Formica
This can be made in a variety of colours and with different filler materials. Depending on the filler, it can have different degrees of hardness, rigidity and toughness. Like Bakelite, it has good solvent resistance, but it will absorb water and has the same recommended working temperature limit of 80°C. It is now more widely used than Bakelite for control knobs, handles and electrical fittings such as plugs and switches.

Melamine
This material has properties similar to those of Bakelite and formica, but it is tougher and has better resistance to heat. It is also highly water resistant, tasteless, odourless and available in a wide range of colours. It is widely used for plastic tableware, work surfaces, instrument panels and electrical fittings.

Epoxy resin
Epoxy resin is mixed with a chemical hardener before moulding and is often reinforced with glass fibres, carbon fibres, cloth and paper. Its strength and toughness vary with the filler, but epoxy resin-based materials are generally resistant to solvents and water and can be used at service temperatures up to 200°C. With different fillers and reinforcements it has many applications, ranging from boat hulls to printed circuit boards. Epoxy resin and its hardener can also be used as an adhesive. It bonds well to metals and is widely used in DIY repair kits.

Polyester resin
This is also mixed with a hardener and reinforced with the same materials as epoxy resin. It is also added to paints and enamels to improve their wear resistance. Polyester resin with different reinforcing materials is used for car bodies, boat hulls, crash helmets and mouldings for electrical equipment. As in the case of epoxy resin products, strength and toughness vary with the filler. Polyester products have good resistance to heat, water and most solvents but can be attacked by concentrated acids and alkalis.

Rubbers

Rubbers
Polymer materials that can return to their original shape after being deformed

Rubbers are polymer materials that are sometimes called 'elastomers'. Their polymers are long, intertwined and able to return to their original shape after being deformed.

Rubber is an important material for applications in which flexibility is required and it is often reinforced with fabric or wire mesh. Natural rubber is obtained from the sap of the rubber tree, which is a native of South America and also grown in Far Eastern countries. It is not used alone because it is readily attacked by solvents and perishes when exposed to the atmosphere. The following are a number of common synthetic rubbers that have been developed to overcome these deficiencies:
- styrene rubber
- Neoprene
- butyl rubber
- silicone rubber.

Styrene rubber
This type of synthetic rubber was developed during the Second World War in the USA because of a shortage of natural rubber for vehicle tyres. It is sometimes blended with natural rubber and has good resistance to oils and petrol, and to attack from the atmosphere and sunlight. Styrene rubber has been developed

and improved over the last 50 years and is still widely used for vehicle tyres, rubber boots, conveyer belts and as electrical insulation.

Neoprene
This is a synthetic rubber that is resistant to attack from the atmosphere and mineral oils, and has a working temperature range of –10°C to 90°C. It is more expensive than styrene rubber but more suitable for applications such as oil seals, gaskets, hoses and wet suits.

Butyl rubber
This rubber has good resistance to chemicals and heat and is impermeable to air and other gases. It is relatively cheap to produce and is sometimes blended with small amounts of natural rubber. Butyl rubber is used for tyre inner tubes, moulded diaphragms, hoses and tank linings.

Silicone rubber
This type of rubber has the widest working temperature range. It retains its flexibility down to –80°C and can withstand temperatures up to 300°C. It is expensive but widely used in applications in which wide variations in temperature occur, such as in aircraft.

Progress check

1. Which thermoplastic material is widely used for packaging and for squeeze containers?
2. What are the main properties of Teflon and what is it widely used for in engineering?
3. Which thermoplastic material is used for making safety helmets?
4. Which thermoplastic materials are used in the manufacture of textiles?
5. Name three filler materials that are used with thermosetting plastics.
6. Which white-coloured thermosetting plastic is widely used for electrical switches, control knobs and plug tops?
7. Name three things that are made from reinforced polyester resin.
8. Why is natural rubber not used by itself as an engineering material?
9. Name two typical uses of butyl rubber.
10. Which synthetic rubber has the widest working temperature range?

Ceramics

Ceramics
Ceramics contain clays and oxides of silicon, aluminium and magnesium

The main ingredients of ceramics are clays and mixtures containing oxides of silicon, aluminium and magnesium. They are used to make a variety of products, ranging from building bricks and cements to grinding wheels and cutting tools.

Ceramics are good electrical insulators, and some of them can resist extremely high temperatures. They can be made very hard and wear resistant but they tend to be brittle and should not be subjected to shock loading. Some of the main types are:
- amorphous ceramics
- crystalline ceramics
- bonded ceramics
- cements.

Amorphous ceramics
These include the different forms of glass used for window panes, lenses, containers and glass fibre matting. When amorphous ceramics are cooled down, they do not produce crystals and do not have a definite melting or solidification

point. The more common forms of glass gradually become molten at temperatures around 400°C. Glass is highly resistant to chemicals and a good electrical insulator.

Crystalline ceramics

These ceramics include the crystals of magnesium oxide and aluminium oxide. Magnesium oxide, or 'magnesia', is an excellent electrical insulator, which is used in compressed-powder form to insulate the core of copper-sheathed, mineral-insulated cables. Aluminium oxide, or 'alumina', is better known as 'emery', which is used as the abrasive grit in lapping paste and in emery paper and cloth.

Bonded ceramics

These are a combination of the above two types of ceramic. They include grinding wheels, bricks, tiles and the porcelain used for pottery and as an electrical insulator. Bonded ceramics contain crystals of aluminium oxide, silicon oxide and other similar materials that are bonded together at high temperatures in an amorphous glassy matrix.

Cements

Cements can also be part crystalline and part amorphous after solidifying. They include the cement used to make mortar and concrete in the construction industries and the fire clays used in furnace linings. Portland cement, which solidifies as a result of chemical action when mixed with water, and resembles Portland stone, is the type most commonly used for building.

Case study

Beer and soft drink containers need be made from a material that does not corrode, is non-toxic and does not react chemically in any way with the liquids. The containers must also be strong and shock resistant. Low material and manufacturing costs are another consideration, and the containers must be light in weight to keep handling and transport costs to a minimum. Until about 40 years ago returnable glass bottles were the only type of container used. Glass is cheap, hygienic and recyclable, but it also has its disadvantages. Glass bottles are heavier than alternative metal and plastic containers and they are fragile. Coated steel and aluminium cans are unbreakable and lighter than glass. They are more expensive to produce but they can be recycled. Plastics are unbreakable, cheap and light. However, they are not as easy to recycle and some plastics allow gas to escape through them. In recent years, the use of metal cans has gradually overtaken the use of glass bottles. The main reason is that the cheaper handling and transport costs of cans outweigh the cheaper material costs of glass. The invention of the improved ring-pull, which stays with the can, and the 'widget', which allows the canned product to resemble draught beer, have further increased the popularity of cans with the public.

Semiconductor materials

Semiconductor materials, such as the elements silicon and germanium, are poor electrical conductors at low temperatures, but their conductivity improves as the temperature rises. The reason for this behaviour lies in the structure of their atoms. The electrons in the outer shell of conducting metal atoms are loosely attached and can break free easily to become current carriers when a potential difference is applied to the material. With semiconductors, the outer electrons are securely bound to the atoms at low temperatures, but as the temperature rises more electrons break free and quite large increases in current occur.

Engineering materials and processes

Semiconductor materials

Semiconductor materials are poor conductors at low temperatures but their conductivity improves with temperature rise

Silicon and germanium both have four electrons in the outermost shell of their atoms, and in the pure form they are known as intrinsic semiconductors. For use in electronic components such as transistors, they are 'doped' by the addition of impurity atoms and are then called extrinsic semiconductors. There are two types of extrinsic semiconductor:
- n-type semiconductors
- p-type semiconductors.

n-type semiconductors
These are doped with small amounts of phosphorus, antimony and arsenic, which have five electrons in the outer shell, i.e. one more than silicon and germanium.

p-type semiconductors
These are doped with small amounts of aluminium, indium and gallium, which have three electrons in the outer shell, i.e. one fewer than silicon and germanium.

Both kinds of doping improve the conductivity of the material. Junction diodes, transistors and integrated circuits are made using different combinations of p-type and n-type material, often in very thin layers, to give them their particular electrical properties.

Composite materials

Composites are made up of two or more materials. They may be grouped into:
- reinforced plastic composites
- ceramic composites
- wood composites.

Reinforced plastic composites
These have already been described. They are the thermosetting plastics with their different filler and reinforcing materials. Common examples are glass fibre- and carbon fibre-reinforced epoxy and polyester resins used to make boat hulls, car bodies, fishing rods and archery bows. Sometimes additional metal reinforcement in the form of rods or mesh is used to give improved strength and rigidity.

Ceramic composites
These include bonded ceramics, which consist of hard abrasive crystals held together in a glass-like matrix, and are used to make grinding wheels. In recent years, composites known as 'cermets' have been developed, which consist of ceramic grains held in a matrix of metal. They are hard and rigid and able to

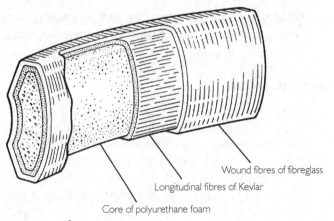

Figure 1.3 Composite tennis racquet frame

retain their properties at high working temperatures. Cemented carbide cutting tools are cermets, of which tungsten carbide is perhaps the best known. Cermets are also used for high-temperature applications such as nuclear reactor and jet engine components.

The concrete used by the building and construction industry is also a ceramic composite. It consists of hard granite chippings held in a cement matrix. Reinforced concrete contains steel rods to give it additional strength.

Wood composites

These include the different forms of plywood, blockboard and laminated chipboard used in building and in furniture manufacture. Plywood consists of thin layers of wood bonded together with their grain directions alternately at right angles. Blockboard consists of strips of wood bonded together and sandwiched between two thin outer sheets. Laminated chipboard is made up of compressed wood shavings and sawdust, which are bonded together and sandwiched between thin outer sheets. Thermosetting resins are used as the bonding adhesives. The wood composites are flexible, resistant to warping and do not have the weaknesses caused by grain direction that are present in timber.

Progress check

1. What is the most common amorphous ceramic?
2. Under what names is the crystalline ceramic aluminium oxide also known?
3. Describe the structure of a bonded ceramic product such as a grinding wheel.
4. How does the electrical conductivity of silicon change with temperature?
5. What is meant by the 'doping' of semiconductor materials?
6. What are the reinforcing materials that are used with thermosetting plastic resins to make boat hulls and fishing rods?
7. What is laminated chipboard composed of?

Activity 1.3

Obtain samples of polythene, nylon, PVC, Bakelite and melamine that are roughly the same shape and mass (about 20–30 g will be enough) Carefully record the mass of each specimen on a chemical balance and then immerse them in water for 24 hours. Remove them, dry them off with kitchen roll and again record the mass of each specimen. Put the specimens back into the water for another 24 hours and then dry them off and record their masses for a third time.

Calculate the percentage of water that each material has absorbed and present your results in the form of a table. Which of these materials do you think would be most suitable for (a) the insulation for an outdoor electric cable and (b) the case for a voltmeter that will be used on outdoor electrical installations?

Components for mechanical products

Having identified possible materials for engineered products it is now necessary to identify some of the more common components that are used to keep the different parts of a product together. Products such as bicycles, power tools and domestic appliances are assembled from many different parts. The way in which the parts are joined together depends on whether they will need to be removed

or replaced during servicing and repair. The most common mechanical fixing components are:
- screwed fastenings
- locking devices
- rivets.

Screwed fastenings

Screwed fastenings may be classed as a semipermanent method of joining that allows components to be removed for repair and servicing. Most countries now use the International Organization for Standardization (ISO) metric screw thread for general engineering applications. This has the thread profile shown in Figure 1.4a.

The British Association (BA) thread is also employed internationally for small-sized screw fastenings, particularly those used in electrical and electronic equipment. Its thread profile is different to the ISO metric, as shown in Figure 1.4b.

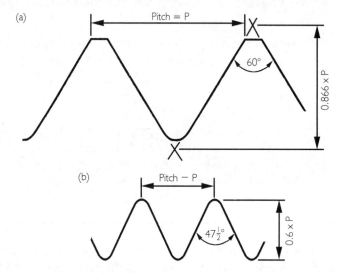

Figure 1.4 Screw thread profiles

The most common screwed fastenings are:
- nuts and bolts
- setscrews
- nuts and studs
- self-tapping screws.

Metric nuts and bolts

These are specified in a particular way on drawings and in suppliers' catalogues to describe their shape, diameter, pitch and bolt length. For example, a nut and bolt may be specified as:

$$\text{Steel, hex hd bolt} - M\ 12 \times 1.25 \times 75$$

$$\text{Steel, hex hd nut} - M\ 12 \times 1.25$$

where M specifies the ISO metric system, 12 specifies a thread diameter of 12 mm, 1.25 specifies a thread pitch of 1.25 mm and 75 specifies a bolt length of 75 mm.

Materials and components for engineered products

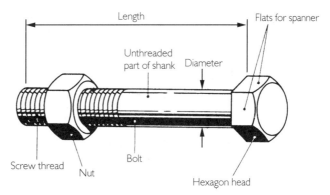

Figure 1.5 Hexagonal head nut and bolt

Bolts of a given diameter and length are available with different lengths of thread. The bolts used for general applications have their head and shank forged to shape and are known as 'black' bolts. For applications in which accurate positioning is required, fitted bolts are used. These have their shank diameters machined accurately to size. For applications in which high strength is required, high-tensile bolts are available. These are made from high-tensile alloy steels and have their threads formed by a process known as thread rolling to give them added strength. Steel nuts and bolts are often plated with cadmium or zinc to improve their corrosion resistance.

When two components are joined by a nut and bolt, the plain, unthreaded part of the shank should extend through the joint face. This ensures that the threaded part is not carrying any shearing load. It is also good practice to fit a washer under a nut, as shown in Figure 1.6. This has the effect of spreading the load and avoids damaging the component surface.

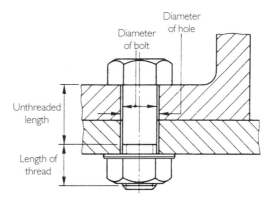

Figure 1.6 Nut, bolt and washer

Metric screws

These are also sometimes called setscrews or machine screws. They are similar to bolts but they are not used with a nut.

Screws are used for applications in which one of the components being joined is too large for a bolt to pass through it. The larger component is drilled and tapped with a suitable size of thread and the smaller, or thinner, component is

Engineering materials and processes

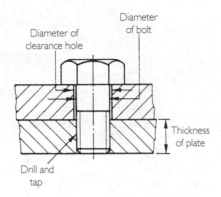

Figure 1.7 Metric screw

drilled to a clearance diameter. Screws need to be selected with the threaded part of the shank slightly shorter than the thickness of component that it passes through. When joining very thin plate or sheet-metal components, screws are usually threaded over the whole length of the shank.

Metric screws are available with different shapes of head for different applications. Some of the more common ones are shown in Figure 1.8.

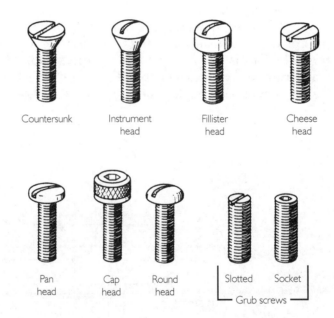

Figure 1.8 Metric screws

Studs and nuts

These are often used as an alternative to setscrews for applications in which a component may need to be removed regularly. Inspection covers are a typical example. If setscrews were to be used, it is possible that the threaded holes in the larger and usually more expensive component could become worn or damaged. A stud is a length of bar that is threaded at both ends.

20

Materials and components for engineered products

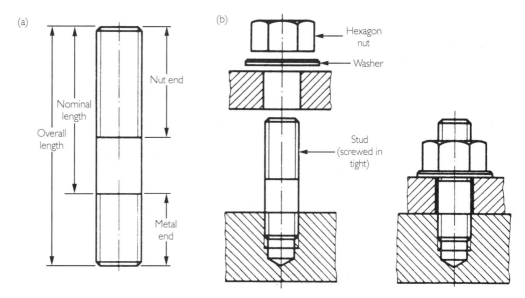

Figure 1.9 Studs: (a) stud form; (b) use of studs

The longer threaded part is screwed into the large component, sometimes using a locking adhesive to prevent it from becoming easily unscrewed. The unthreaded part in the middle of the stud should be a little shorter than the component through which it passes and the shorter threaded part should be slightly longer than the nut. As with nuts and bolts, it is good practice to include a washer to prevent damage to the component surface and to spread the load.

Self-tapping screws
These are widely used with sheet-metal and plastic components and may be of a thread-forming or a thread-cutting type. They screw into a pilot hole that is drilled to the root diameter of the thread.

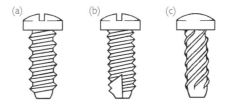

Figure 1.10 Self-tapping screws: (a) thread-forming type; (b) thread-cutting type; (c) drive type

The thread-forming type displaces material to form a thread and is used with soft metals and plastics. The thread-cutting type cuts a thread in the pilot hole and is used with harder materials.

Locking devices

Locking devices are necessary with screwed fastenings that are subjected to vibration. This is especially true in the case of motor vehicles, aircraft and production machinery, in which the loss of a nut or setscrew might have devastating effects. There are two basic types of locking device: those that depend on friction and those that provide a positive mechanical locking action.

Engineering materials and processes

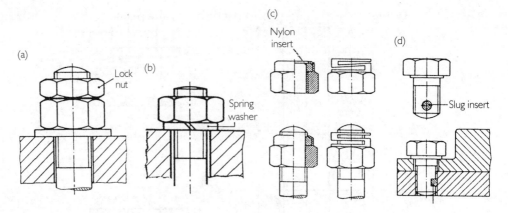

Figure 1.11 Friction locking devices: (a) lock nut; (b) spring washer; (c) friction nuts; (d) friction screw

The insert in the friction nuts may be nylon, as shown in Figure 1.11, in which case they are referred to as 'nylock' nuts, or compressed fibre, in which case they are generally referred to as 'Simmonds' nuts. Spring washers may be of the coiled type or the 'shakeproof' type.

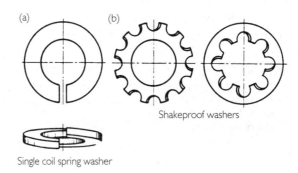

Figure 1.12 Spring and shakeproof washers: (a) single-coil spring washer; (b) shakeproof washers

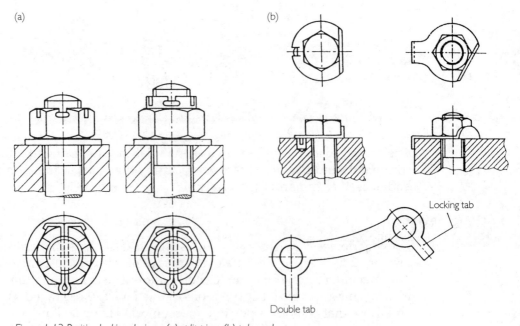

Figure 1.13 Positive locking devices: (a) split pins; (b) tab washers

Materials and components for engineered products

When spring washers and friction nuts have been removed during maintenance and servicing they should be discarded and new ones fitted.

Split pins and tab washers must always be replaced with new ones once they have been removed.

Rivets

Rivets are used to join materials and components permanently. Many of their previous applications have been taken over by welding and the use of adhesives. They are, however, still widely used in applications such as aircraft assembly to join aluminium and titanium alloy components that are difficult to weld. An advantage of a riveted joint is that it may not be as rigid as one that has been welded, allowing a little flexing to take place under load. This is sometimes desirable in load-bearing structures.

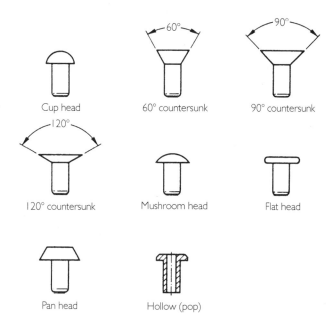

Figure 1.14 Types of rivet

Rivets are made with different-shaped heads for different applications, and from a variety of materials. The composition of the rivets to be used should always be as close as possible to that of the components being joined. This is to reduce the risk of electrolytic corrosion, which can take place where two different metals are in contact and moisture is present. The rivet material must of course be malleable to allow the head to be 'closed' when making a joint.

Rivets should not be loaded in tension. They are intended to withstand shearing forces acting across the shanks. Pan-head and cup-head rivets have the strongest heads. Mushroom and flat heads have less strength but do not protrude as much. Countersunk heads are used when a flush joint surface is required.

Pop rivets are widely used to join thin plate material when access is only possible from one side of the joint. The rivet is formed by pulling the head of a central pin through the rivet using a special tool. As the rivet head is formed, the pin breaks off, as shown in Figure 1.16.

Engineering materials and processes

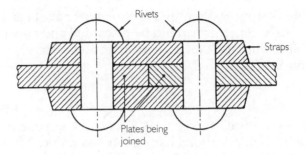

Figure 1.15 Section through a joint with cup-head rivets

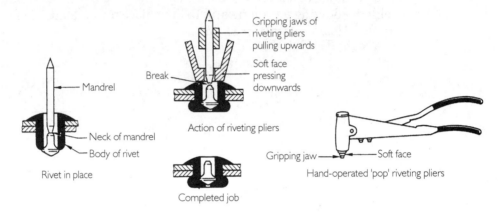

Figure 1.16 Pop riveting

Progress check

1. What kind of applications are screws with British Association (BA) threads used for?
2. What is the diameter, thread pitch and length of a bolt specified as M 10 × 1.25 × 50?
3. How do fitted bolts differ from black bolts?
4. In what kind of application would it be better to use studs rather than setscrews?
5. What are the two basic categories of locking device?
6. What should be done with spring washers and split pins that have been removed during maintenance work?
7. In what way do riveted joints sometimes have an advantage over welded joints?
8. Why should rivets be of the same material composition as the materials being joined?
9. What type of rivet head gives a smooth joint surface?

Activity 1.4

BA screws and nuts are widely used for small-diameter screwed fastenings in electrical and electronic equipment. Find out how BA screws are listed in suppliers' catalogues and draw up a table that shows the diameter of each available size.

Selecting materials and components for a mechanical product

The materials and components used in mechanical products must be fit for their purpose and readily available at an economic price. When comparing the different materials and components available, a design engineer must consider their:
- mechanical properties
- thermal properties
- resistance to attack
- cost
- availability.

The choice might also be affected by the availability of the plant, equipment and skilled labour needed to process and assemble the materials. Here the design engineer may need to seek the advice of the production engineer before making a final choice.

Mechanical properties

Mechanical properties
The mechanical properties of a material are a measure of how it behaves under load

The mechanical properties of a material are a measure of how it behaves under load. The mechanical properties of engineering materials include:
- ultimate tensile strength
- hardness
- ductility
- malleability
- toughness.

Ultimate tensile strength (UTS)

This is a measure of how much tensile load a material can carry before breaking. It is determined from tests in which specimens are loaded until they break. The UTS is the tensile stress that causes the material to fracture. It is calculated using the following formula.

$$\text{UTS} = \frac{\text{tensile load at fracture}}{\text{original cross-sectional area}}$$

The original cross-sectional area is always used, even though the material may be much thinner by the time it fractures. This is because it is sometimes difficult to measure the cross-sectional area at the point where it breaks. The units most often used for the UTS are newtons per square millimetre (N/mm^2), i.e. the load that can be carried by each square millimetre of cross-section before the material breaks.

Hardness

This is a measure of the resistance of a material to wear and abrasion. The most common way of measuring hardness is to carry out an indentation test. This involves pressing an indentor into the surface of the material or component. The Brinell test, in which the indentor is a hardened steel ball, is used for softer materials. The Vickers pyramid test, in which the indentor is a pointed diamond, is used for harder materials.

The size of the indentation can then be used to calculate a hardness number for the material. This is written after the letters BHN or VPN to denote whether it is a Brinell hardness number or a Vickers pyramid hardness number. Unlike tensile tests, indentation tests are non-destructive and can be carried out on raw materials and finished components alike. They leave only a small indentation on the surface of a component that does not damage it or affect its strength.

Ductility
This is a measure of how much a material can be pulled or stretched by tensile forces before it breaks. Materials that are drawn through a die to make small-diameter rod and wire must be ductile. There are different ways of measuring ductility. Two of the most common are the measurement of the percentage elongation and the percentage reduction in area of a material that has undergone a tensile test. They are calculated by measuring the original and final length and cross-sectional area of a specimen and using the following formulae.

$$\text{Percentage elongation} = \frac{\text{increase in length}}{\text{original length}} \times 100$$

$$\text{Percentage reduction in area} = \frac{\text{decrease in cross-sectional area}}{\text{original cross-sectional area}} \times 100$$

Alternatively, the broken pieces of a standard tensile test specimen can be placed on special gauges that enable these values to be read off directly. It must be noted that the values are not the same for a particular material and the method of calculation should always be stated.

Malleability
This is the ability of a material to be deformed by compressive forces. Materials that are rolled or forged to shape must be malleable, and so must fixing components such as rivets. Malleability cannot be measured as easily as the above properties. This is because compressive failure is not as easy to detect as tensile failure in malleable materials. Design engineers know from experience which are the most malleable materials. Table 1.8 gives some common engineering materials in order of decreasing malleability and ductility. Note that the order is not the same. Lead, for instance, is very malleable but not as ductile.

Toughness
This is a measure of the resistance of a material or component to impact forces and shock loading. The materials used for hammers and forging dies must be tough. The toughness of a material is measured by carrying out an impact test in which a standard specimen is broken by a heavy swinging pendulum. The amount of energy, in joules (J), lost by the pendulum as it breaks the specimen is recorded on a scale and gives a measure of the material toughness.

Table 1.8 Decreasing order of malleability and ductility

Malleability	Ductility
Lead	Silver
Silver	Copper
Copper	Aluminium
Aluminium	Mild steel
Tin	Zinc
Mild steel	Tin
Zinc	Lead

Materials and components for engineered products

Thermal properties

> **Thermal properties**
>
> *The thermal properties of a material are a measure of how it behaves when subjected to temperature change*

The thermal properties of a material are a measure of how it behaves when subjected to temperature change. The thermal properties of engineering materials include:
- thermal conductivity
- thermal expansivity.

Thermal conductivity

This is a measure of the ability of a material to allow heat energy to flow through it. Metals are good conductors of heat energy and have a high thermal conductivity. Copper is a particularly good thermal conductor, which is why it is used for soldering iron bits and in car radiators. Plastics and ceramics are poor conductors of heat energy and have a low thermal conductivity. They are used as thermal insulators to prevent heat loss from equipment and for personal protection.

Thermal expansivity

This is a measure of the effect of temperature change on the dimensions of a material. When their temperature rises, most materials expand, and when cooled down they contract. The effect is greater in metals than in non-metals, and certain thermoplastics, such as those used for shrink insulation sleeving, will in fact contract when heated.

Resistance to attack

There is a tendency for all engineering materials to deteriorate over a period of time as a result of attack from substances in their environment. This can take the form of:
- corrosion
- solvent attack.

Corrosion

Corrosion occurs mainly in metals. It is due to a chemical reaction taking place between the metal and some other substance, usually oxygen, in its service environment. Iron and steel are particularly affected by the oxygen in moisture, which causes rusting. At high temperatures, iron and steel also combine readily with atmospheric oxygen, which causes the formation of the black flaky oxide known as 'millscale'.

The non-ferrous metals used in engineering have a much higher resistance to corrosion. Although some of them combine more readily with oxygen, the oxide film that forms on their surface is very dense and protects the metal against further attack. This is not the case with iron and steel, on which the rust and millscale are very loose and porous. The corrosion resistance of components can be increased by painting, coating with plastics or plating with a more corrosion-resistant metal. Tin, zinc, cadmium and chromium are all widely used as plating materials.

Polymer and ceramic materials have a high resistance to oxidation but can deteriorate in other ways. Some plastics become brittle if exposed to the ultra-violet radiation in sunlight and some ceramics deteriorate at high temperatures.

Solvent attack

This is a more likely cause of degradation in polymer materials. Solvents are chemicals that cause plastics to dissolve or become mechanically weaker. Some plastics are highly resistant to solvents, but care should be taken when selecting polymer materials that will come into contact with petrochemical products and strong acids and alkalis.

Costs

Material and component costs often form a large part of the total cost of a product, and they should be kept as low as possible. When selecting materials and components, it may be found that those with the best properties are too expensive. Cheaper ones of lower quality may then need to be found. Here, however, the advice of the production engineer should again be sought because the cheaper material might be more costly to process.

Wherever possible, materials and components that can be purchased in standard forms and sizes and which require a minimum of processing should be selected. Discount can often be obtained by purchasing large quantities or by placing regular repeat orders with a supplier. Here, however, there is a danger that the cost of storing the material or components will be more than the discount and there is also a danger of relying too heavily on a single supplier.

Availability

Engineering materials are supplied in a variety of forms. These include:
- ingots
- granules
- liquids
- barstock
- pipe and tube
- rolled sections
- extruded sections
- sheet and plate
- castings and mouldings
- forgings and pressings.

Availability varies with the material type and the form in which it is supplied. The more common plain-carbon steels and aluminium alloys are readily available as bar, sheet, tube, etc. in a range of standard sizes and from a number of alternative suppliers. The same applies to the more common raw polymer materials, which are readily available in the form of granules or resins. Newly developed materials may have desirable properties but may be available only from a single supplier in limited quantities.

Castings, forgings, mouldings and pressings require the production of patterns and dies before material can be supplied in these forms. This is an expensive operation and can take quite a long time. It can only be justified if large quantities will be required. Once the patterns and dies have been made, supplies can be scheduled to arrive at regular intervals.

As has been stated, there is a danger of becoming too reliant on a single supplier. An interruption in the supply of an essential material can lead to costly production stoppages and the cancellation of orders. If reliability is critical, it is good practice to have a back-up supplier. When choosing materials and suppliers, care should also be taken to ensure that supplies will be available to meet possible increases in production.

Materials and components for engineered products

Case study

You probably have an electric kettle in your kitchen. Electric kettles were first produced in the 1920s and the earliest ones were made from copper. They were almost the same as ordinary copper kettles except that, instead of being heated on a fire or gas ring, they were heated internally by means of a heating element.

Copper is a good conductor of heat and an ideal material for externally heated kettles. Electric kettles, however, should ideally be made from a poor conductor of heat. When heat is being lost through the kettle walls, electricity is being wasted and the water takes longer to boil. The material should also be a poor conductor of electricity to make the kettle safer to handle.

Later electric kettles were made from stainless steel and aluminium and, although cheaper than copper, these metals too are good electrical and thermal conductors. A ceramic or plastic material was required, but glass is fragile and the early plastics were unsuitable. Nylon absorbs water, Perspex softens in boiling water and the early thermosets, such as Bakelite, contain dangerous chemicals. It was not until the 1970s that plastics that were sufficiently non-absorbent, non-softening, non-toxic and economic to manufacture became available.

Polypropene (polypropylene) is the most widely used plastic material for making electric kettles. It does not absorb water, it is non-toxic and it stays rigid at temperatures up to 150°C. It also has a smooth surface when moulded to shape and can be produced in a range of light attractive colours. In 1979, the first jug kettles were introduced, and these are almost all made from plastics. More recently, the 'cordless' electric kettle was introduced, which becomes disconnected from the power supply when lifted off its base. These too are made mostly from polypropylene.

Progress check

1. How is the ultimate tensile strength (UTS) of a material determined?
2. What kind of test is commonly used to measure the surface hardness of a metal?
3. What is the property required in materials that are formed to shape by forging and rolling?
4. Define what is meant by toughness in materials.
5. What are the properties required for the material used to make electric kettles.
6. Which thermal property of a material describes the effect of temperature rise on its dimensions?
7. Under what different conditions do rust and millscale form on the surface of iron and steel?
8. List three metals used to give a protective coating to steel components.
9. Give three forms of supply in which metals can be purchased in standard sizes.
10. Why might it take longer to obtain the castings and forgings for new components than other material supplies?

29

Engineering materials and processes

Electrical and electronic components for electrical products

Electrical products used in the home and in industry are made up of circuits that contain a number of common components and materials. These include:
- connecting wire and cable
- insulators and insulating materials
- resistors
- capacitors
- inductors
- diodes
- transistors.

Connecting wire and cable

The most commonly used connecting wire is copper, insulated with PVC. This is available in a variety of conductor cross-sectional areas and insulation colours. The copper may be single core or multistranded. Multistranded wire is more flexible and less likely to break in applications in which movement or vibration is present. The copper may be coated with tin, and sometimes silver, to aid soldering. Silicon rubber and PTFE are used as alternative insulating materials. Silicon rubber is used for applications in which solvents, radiation and ozone may be present, and PTFE is used for higher temperature applications up to around 200°C.

Electrical cable is available in a wide variety of types, ranging from the twin-cored flexible cable used for domestic appliance connections to the mineral insulated and armoured cables used for heavy-duty industrial and commercial installations.

For specialist applications such as signal transmission, multicored cable is available containing up to 50 separate insulated and colour-coded conductors.

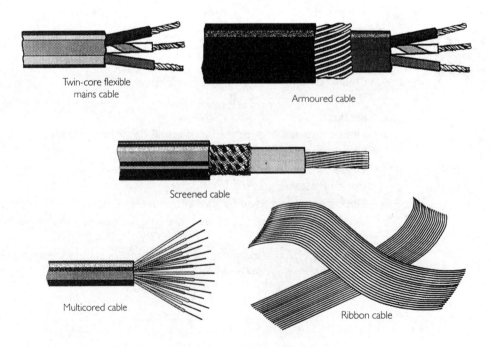

Figure 1.17 Cable types

The conductors may also be surrounded by metal foil or braided metal, underneath the outer insulating sheath, to prevent interference from stray magnetic fields. Ribbon cable is another specialist type, which is widely used in computer systems. It can contain up to 64 parallel conductors with colour-coded PVC insulation that are sandwiched between outer layers of PVC to form a wide ribbon.

Insulators and insulating materials

> **Insulators**
> Materials that are poor conductors of electricity

Insulators and insulating materials prevent unwanted contact between live conductors and components. They also protect the persons using electrical equipment from contact with the live parts. As has been stated, plastics, rubbers and ceramic materials are good electrical insulators. Thermoplastic materials, such as PVC and PTFE, and rubbers are widely used to insulate cables and connecting wire. Thermoplastic materials are also used for the moulded cases and enclosures for electrical equipment and appliances.

Thermosetting plastics are generally less flexible and not as easy to mould into intricate shapes. Nevertheless, formaldehydes, such as Bakelite and melamine, are widely used for insulating covers, control knobs and the insulated parts of switches, plugs and sockets. They are especially suited to applications in which temperature rise might cause thermoplastic materials to deform. Printed circuit board is also a thermosetting plastic composite material.

Ceramics such as glass and porcelain are used to make a variety of insulators and are particularly suitable for outdoor, high-voltage applications such as on overhead power lines. They are also used to mount the electrical heating elements in heaters and dryers and for a variety of insulating spacers and distance pieces.

Resistors

Resistors are used in a great many electric and electronic circuits to control the flow of current. They are made from different materials, the most common being carbon film and metal film. Wire-wound resistors are also used when particularly high precision is required and for high-power applications.

Carbon- and metal-film resistors are marked with a colour code, or a number and letter code, which gives their value in ohms (Ω), their power rating and degree of accuracy. The power rating can vary from 1/8 watt (W) upwards and their accuracy is expressed as a percentage tolerance, i.e. a tolerance of 5% indicates that the resistance is within 5% of the stated value.

Capacitors

> **Capacitance**
> A capacitor has a capacitance of 1 farad when a potential difference of 1 volt between its plates enables a charge of 1 coulomb of electricity to be stored

In their basic form, capacitors consist of two metal plates separated by an insulating material, which is known as a 'dielectric'. Capacitors are able to store electric charge and also have the property of blocking the flow of direct current while allowing an alternating current to pass. The unit of capacitance is the farad, although in electronic circuits the microfarad (µF) and the picofarad (pF) are more convenient units.

Capacitors are sometimes colour coded in the same way as resistors, the difference being that their value is measured in picofarads rather than ohms. In the smaller capacitors, the plates are usually strips of metal foil separated by a dielectric, which may be a ceramic or a plastic material. The larger values of capacitor are generally of the electrolytic type. Here the dielectric is formed by a

Engineering materials and processes

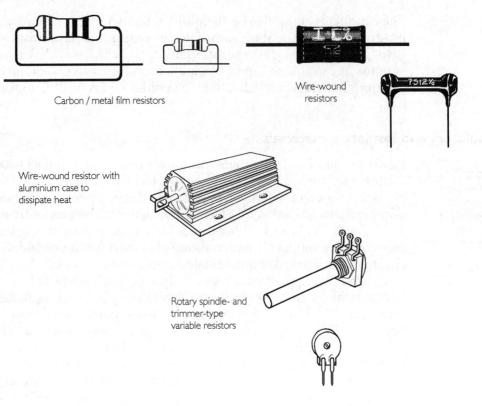

Figure 1.18 Resistors

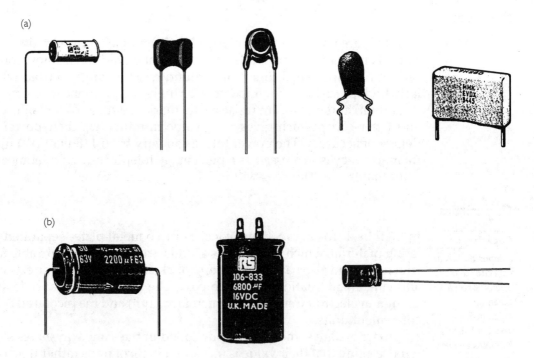

Figure 1.19 Capacitors

Materials and components for engineered products

non-conducting layer of metal oxide that covers one of the plates. The connections to the plates of electrolytic capacitors are marked positive and negative, and they must always be connected the correct way round to avoid damage.

Inductors

Inductance

An inductor has an inductance of 1 henry when a change of current of 1 ampere in 1 second produces a back-emf of 1 volt to oppose the change

In their basic form, inductors consist of a tightly wound coil of insulated wire. They may be wound on a core of insulating material or of a ferromagnetic material, such as soft laminated iron, compressed iron dust or ferrite, which is a moulded compound containing iron.

An inductor will allow the passage of a steady current but will oppose any change in the current flowing through it. If the current should rise, it will slow down its growth, and if it falls it will slow down its decay. It can completely block the flow of high-frequency alternating current, which is the exact opposite of the way in which a capacitor behaves. The unit of inductance is the henry, although for electronic applications the millihenry (mH) and the microhenry (µH) are more convenient units.

Figure 1.20 Inductors

Diodes

A diode is a component made from semiconductor materials that will allow electric current to pass through it in one direction only. The basic form is the junction diode, which contains pieces of n-type and p-type semiconductor material in contact.

Current may flow in a direction from the p-type to the n-type. When a voltage is applied in this direction, the diode is said to be 'forward biased'. When a voltage is applied in the opposite direction, no current will flow and the diode is said to be 'reverse biased'.

Photodiodes and light-emitting diodes (LEDs) are similar to the junction diode in construction but have special optical properties. Photodiodes have a transparent window at one end through which light can enter. No current flows when they are reverse biased in darkness, as with the junction diode. When light enters, however, its effect on the semiconductor materials is to allow current to flow in the reverse direction. Photodiodes can thus be used as light-sensitive switches.

Engineering materials and processes

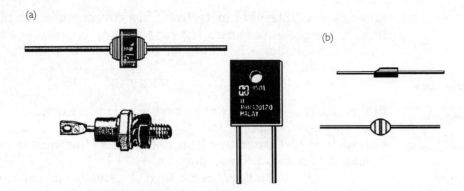

Figure 1.21 Diodes: (a) power diodes; (b) signal diodes

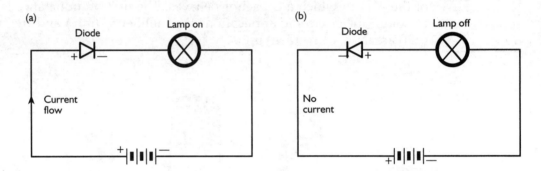

Figure 1.22 Diode circuits: (a) forward biased; (b) reverse biased

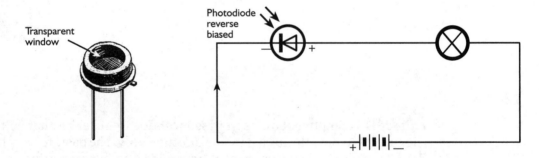

Figure 1.23 Photodiode circuit

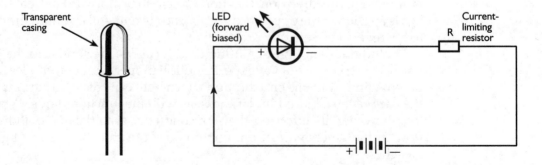

Figure 1.24 Light-emitting diode circuit

Materials and components for engineered products

LEDs also have a transparent lens or casing. When current passes in the forward direction, it causes light to be emitted. This can be red, green or yellow depending on the materials used in the construction. LEDs are used as a visual indicator that current is flowing in a circuit. They are also widely used to display the digits 0 to 9 on the seven-segment numerical displays used on electronic equipment and instrument systems.

Transistors

The basic type of transistor, known as the 'bipolar' transistor, is widely used in electronic circuits as a high-speed switch and as a current amplifier.

Figure 1.25 Transistors

Transistors are made from p-type and n-type semiconductor materials that are assembled together as a thin sandwich. They are broadly divided into n-p-n transistors and p-n-p transistors depending on the form of the sandwich. Transistors have three leads, or connections, known as the collector, the emitter and the base.

When the current in the circuit connecting the base and emitter reaches a particular value, it switches on the much larger current in the collector–emitter circuit, which contains the lamp (Figure 1.26). By this means, a weak current of, say, 1 milliampere (mA) from a sensor can control a much larger current of, say, 100 mA to light a warning lamp or perhaps operate a buzzer. When a transistor is used as a signal amplifier, small variations in a weak input signal to the base–emitter

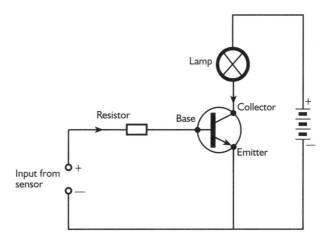

Figure 1.26 Transistor circuit

Engineering materials and processes

circuit are copied and magnified as larger changes of the much stronger current in the collector–emitter circuit.

Progress check

1. Name three different materials that are used to insulate the conductors in electrical cables.
2. Why are copper-wire conductors sometimes coated with tin?
3. For what kind of application is armoured cable used?
4. Why are the conductors in some electric cables surrounded by a layer of braided metal or metal foil?
5. What is the main function of a resistor in an electric or electronic circuit?
6. How do capacitors affect the flow of current in (a) a direct current circuit and (b) an alternating current circuit?
7. What precautions must be taken when connecting an electrolytic capacitor in a circuit?
8. How do inductors affect the flow of current in (a) a direct current circuit and (b) a high-frequency alternating current circuit?
9. What is the basic function of a junction diode?
10. State two basic uses of bipolar transistors in electronic circuits.

Activity 1.5

The BBC began regular radio broadcasts in 1923, and some of the very first radio receivers were called 'crystal sets'. The simplest ones contained only four components. These were an inductor, a variable capacitor, a diode (the crystal) and a headset for listening to the broadcasts. They did not need a battery or any other power supply to operate. The only other things required were a good outdoor aerial, which was usually a long length of wire, and a good earth connection, which was usually a metal cold water pipe.

Find out how the above components were connected together and how a crystal set operated. Your science or craft design and technology teacher will be able to help you, or you may be able to obtain the information from your library. Having obtained this information, you might then choose to make a crystal set yourself.

Selecting electrical and electronic components for an electrical product

Before suitable electrical and electronic components can be selected for a product it is necessary to know what their circuit requirements and operating conditions will be. These may include the maximum current or maximum voltage conditions under which they will be required to operate. Alternatively, the required power rating and operating temperature may be the factors that govern component selection. The component and material selection criteria used by circuit and product designers are as follows.

Connecting wire and cable

Connecting wire and cable is often listed in catalogues under the following headings, which cover the main areas of application:

Materials and components for engineered products

- electrical installation cable
- flexible mains cable
- data transmission cable
- signal and telephone cable
- audiovisual cable
- equipment connecting wire.

Each category is covered by one or more British Standard specifications, which specify the quality of the materials used and the applications for which it is suited. Having located the appropriate category, the engineer or designer can then make a selection on the basis of the required current, voltage, power and temperature rating. Suitability for a particular service environment should also be taken into account. This might include the presence of chemical solvents, gases and radiation, which may attack cable materials, and the possibility of mechanical damage from moving parts or loads.

Insulators and insulating materials

The range of insulators and insulating materials covered in suppliers' catalogues includes:

- high-voltage insulators
- panel and printed circuit board
- spacers and distance pieces
- caps, covers and control knobs
- insulated sleeving
- insulation tape.

A great many of the above components are produced to standard designs and in a range of sizes. They are generally listed with information on their suitability for different service conditions, which might include their working temperatures and resistance to environmental attack. Insulators and insulating materials for high-voltage applications are selected by their voltage rating. This is the voltage they can withstand before their resistance is broken down.

Resistors

Fixed-value and variable resistors are selected by their required value, measured in ohms, their tolerance and their power rating. Environmental factors may also need to be taken into account, such as operating temperature range and resistance to solvents and mechanical damage. The most common types listed in suppliers' catalogues are:

- metal-film resistors
- carbon-film resistors
- wire-wound resistors
- trimmers and potentiometers.

Metal-film and carbon-film resistors are colour coded as shown in Figure 1.27.

The British Standard number and letter code for small resistors, BS 1852, is gradually replacing the colour code. Examples are 2R7 = 2.7 Ω, 27R0 = 27 Ω, 2K7 = 2.7 kΩ, 27K0 = 27 kΩ, 2M7 = 2.7 MΩ, 27M0 = 27 MΩ, etc.

An extra letter is added to indicate tolerance, i.e. F = ± 1%, G = ± 2%, J = ± 5%, K = ± 10% and M = ± 20%. Thus, 2K7J is a 2.7-kΩ resistor with a tolerance of ± 5%.

The property ranges of the most common types of fixed resistor are shown in Table 1.9.

Engineering materials and processes

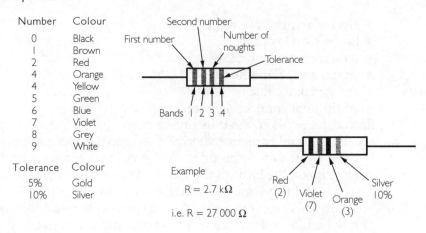

Figure 1.27 Resistor colour codes

Table 1.9 The most common types of property resistor

Type of resistor	Property		
	Maximum value	Minimum tolerance (%)	Power rating (W)
Metal film	1 MΩ	± 1	0.25–0.5
Carbon film	10 MΩ	± 5	0.125–2
Wire-wound	100 kΩ	± 5	2.5–300

Trimmer resistors and potentiometers are variable resistors that are available in a variety of designs and power ratings. Trimmers are small variable resistors with a power ratings usually less than 1 W. Their resistance is adjusted by means of a screwdriver. Potentiometers are larger variable resistors that are operated by rotating a central spindle or moving a slider along a linear track. They are used to adjust the voltage and current flow in electric and electronic circuits.

Capacitors

Fixed-value and variable capacitors are selected by their required value, tolerance, working voltage and leakage current. Their working temperature range might also need to be included in the selection criteria. Capacitors are listed in suppliers' catalogues by their dielectric material and according to whether they are of the non-polarised or electrolytic types. The main non-polarised types are:
- polypropylene
- polycarbonate
- polyester
- polystyrene
- mica
- ceramic materials.

Electrolytic types are listed according to the material whose oxide film is the dielectric. The main ones are:
- aluminium
- tantalum.

Materials and components for engineered products

As has been stated previously, some of the smaller value, non-polarised capacitors are colour coded in the same way as resistors, with their value being measured in picofarads (pF). More generally, however, capacitors have their nominal value, tolerance and working voltage stamped on the outer casing. This may be in the form of a four-digit number code, in which the third digit gives the number of zeros to be added on the end, e.g. 223 = 22 000 pF.

The leads to electrolytic capacitors are marked positive (+) and negative (−). They must be connected the correct way round. The larger values of capacitor for low-frequency applications are generally of the electrolytic type. Typical value and property ranges of the most common types of capacitor are shown in Table 1.10.

Table 1.10 Typical value and property ranges of the most common types of capacitor

	Values	Tolerances (%)	Uses
Non-polarised types of capacitor			
Polystyrene	10 pF to 0.01 µF	± 2 to ± 5	General
Polypropylene	0.001–0.022 µF	± 5 to ± 20	General
Polyester	0.01–10 µF	± 10 to ± 20	General
Polycarbonate	0.22–10 µF	± 5 to ± 20	General
Ceramic	10 pF to 1 µF	± 2 to ± 20	High frequency
Electrolytic types of capacitor			
Tantalum	0.1 to 100 µF	± 10 to ± 20	Low voltage
Aluminium	1 to 100 000 µF	± 10 to ± 20	Low frequency

The polystyrene, polypropylene, polyester and polycarbonate types are cheap and ideal for general use in electrical and electronics circuits. The ceramic types are used for high-frequency applications and can be made to handle high working voltages. Electrolytic capacitors are available in higher values of capacitance and used for low-frequency applications.

Trimmer capacitors are small-value compression-type variable capacitors that are adjusted by means of a screwdriver. These, and rotary variable capacitors, which are adjusted by turning a central spindle, are used in tuned circuits such as those used to select the required frequency in radio receivers.

Inductors

Inductors are selected by their required value, which is measured in microhenrys (µH), or millihenrys (mH), and their tolerance. Their direct current resistance and frequency ratings may also need to be considered during selection. Inductors may be classified according to the type of core on which the coil is wound. The three main types are:
- air cored
- iron-dust and ferrite cored
- iron cored.

Air-cored inductors have small inductance values of typically around 1 mH. They are widely used in combination with capacitors in the tuning circuits of radio and communications equipment. Air-cored inductors, known as high-frequency chokes, are also used as a filter that will allow low-frequency signals to pass while acting as a block to high-frequency signals.

Iron dust- and ferrite-cored inductors are also used for high-frequency applications but have the advantage of being more compact than air-cored types with the same inductance value. The cores can often be screwed in or out of the coils to vary their inductance. The AM (amplitude modulation) aerials inside radio sets usually consist of coils of wire that are wound on a ferrite rod.

Iron-cored inductors are used for low-frequency applications. The core is made up of soft iron laminae, which greatly increases the strength of the magnetic field around the coil. As a result, the inductance of iron-cored inductors can be very high. A typical value might be 10 H. Iron-cored inductors, known as smoothing chokes, are used in power-supply units. Their purpose is to smooth out the pulsations in the direct current, which is produced by passing an alternating current through a combination of diodes known as a rectifier.

Diodes

Diodes are selected according to their voltage and current ratings. Their operating temperature might also need to be considered during selection. Diodes may be classified as:
- signal diodes
- power diodes
- light-emitting diodes (LEDs)
- photodiodes.

Signal diodes are designed for low-power applications in electronic circuits. They are listed with operating voltages ranging from around 4 volts (V) to 150 V and current ratings from around 10 mA to 500 mA. Power diodes, as the name suggests, are designed to handle much higher values of voltage and current. A typical application is the rectification of alternating current to direct current in power-supply units. They are listed with voltage ratings up to around 1.5 kV and currents of over 100 A.

LEDs are selected for the required output light intensity and colour. Their luminous intensity is measured in candelas (cd) and millicandelas (mcd). It can range from about 2 mcd for close viewing to around 3 cd for a high-intensity warning signal that can be seen at some distance. Also listed in suppliers' catalogues are their voltage and current ratings, their power consumption and operating temperatures.

Photodiodes are essentially low-power devices, available with different lens sizes for different applications. They are listed in catalogues with details of their operating temperature, the speed at which they can react to a change in light intensity and the range of light wavelengths to which they are sensitive.

Transistors

The selection of transistors for particular applications requires specialised knowledge and experience. The main ratings, or parameters, which must be considered by the electronics engineer are the maximum collector current, the maximum collector–emitter voltage, the maximum emitter–base voltage, the maximum power rating, the direct current gain and the transistor frequency. There are, however, some general-purpose transistors suitable for use in amplifying

Materials and components for engineered products

and switching circuits. Five popular silicon transistors have the codes BC108, ZTX300, 2N3705, BFY51 and 2N3053.

Suppliers' catalogues contain a very wide variety of transistors. They range from low power to high power and from low frequency to high frequency and switching applications. Transistors are identified by their code, but unfortunately there are several coding systems. In the continental system the code begins with a letter that indicates the semiconductor material. This is followed by a letter that gives the use for which it is suitable, and a number, e.g. BC108. Here the B indicates that the semiconductor material is silicon and the C indicates that it is suitable for use as an audio frequency amplifier. Other listed information indicates that the BC108 is a low-current, high-gain device.

The American system is different: the 2N3705 and the 2N3053 listed above are American codes. In addition, some manufacturers and the armed services have their own codes. One transistor can often replace another, and suppliers provide charts that show the alternatives.

Progress check

1. Under which cable category in a supplier's catalogue would you expect to find the coaxial cable used to connect a TV and a video recorder?
2. What is the main factor to be considered when selecting an insulator or an insulating material?
3. A 47-kΩ resistor with a tolerance of ± 10% is required – what will be its colour code?
4. What are the main factors that govern the selection of a capacitor for a given application?
5. What kind of inductor would you select for use in a power supply unit to smooth out the pulsations in a direct current?
6. What kind of diode would you use as a sensor to detect changes in light intensity?
7. What do the first two letters of the code indicate in the continental system of transistor coding?

Assignment 1
Selecting suitable materials and components for an electromechanical product

This assignment provides evidence for:
Element 1.1: Select materials and components for engineered products
and the following key skills:
Communication 2.1: Take part in discussions
Communication 2.2: Produce written material
Communication 2.3: Use images
Communication 2.4: Read and respond to written materials

Engineering materials and processes

Having gained knowledge of some of the more common engineering materials and components, it should now be possible to make a reasoned selection for a mechanical and an electrical product. The choice should not be limited to the materials and components that have been described. Additional information can be obtained from manufacturers' and suppliers' catalogues, computer databases and other textbooks.

Your tasks

Figure 1.28 shows an exploded view of a bicycle pump, which is a familiar mechanical product.

1. Draw up a parts list for the numbered items. State the function of each part and the properties that it must have to fulfil its function.

2. Identify possible materials or types of standard component suitable for each part. Your selection criteria should include mechanical properties, resistance to chemical attack, ease of handling, cost and availability.

3. Write brief notes explaining why these materials and components were selected and why other possible materials and components were not selected.

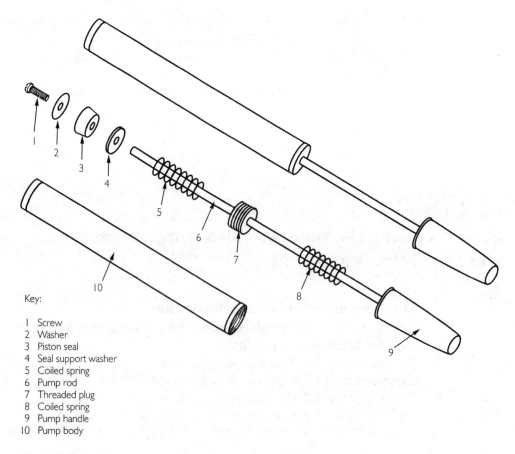

Key:

1. Screw
2. Washer
3. Piston seal
4. Seal support washer
5. Coiled spring
6. Pump rod
7. Threaded plug
8. Coiled spring
9. Pump handle
10. Pump body

Figure 1.28 A bicycle pump

Materials and components for engineered products

Figure 1.29 shows an exploded view of an electric soldering iron, which is a familiar electrical product – electric soldering irons contain other parts and components, but only the items which are numbered need to be considered.

1. Draw up a parts list for the numbered items. State the function of each part and the properties that it must have to fulfil its function.

2. Identify possible materials or types of standard component suitable for each part. Your selection criteria should include mechanical properties, electrical properties, thermal properties, resistance to chemical attack, ease of handling, cost and availability.

3. Write brief notes explaining why these materials and components were selected and why other possible materials and components were not selected.

Other portfolio suggestions for mechanical products are:
- a car jack
- a fishing reel
- a spring balance
- a skateboard.

Other portfolio suggestions for electrical products are:
- an electrician's screwdriver
- a continuity tester
- an extension lead
- a battery charger.

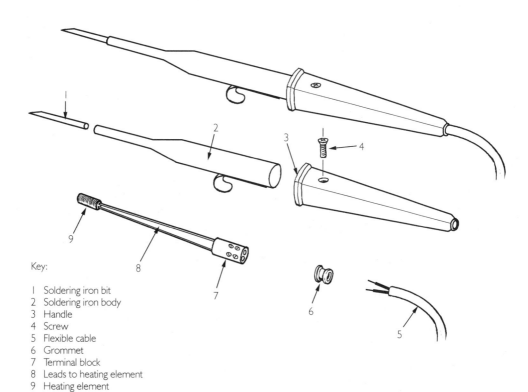

Key:
1 Soldering iron bit
2 Soldering iron body
3 Handle
4 Screw
5 Flexible cable
6 Grommet
7 Terminal block
8 Leads to heating element
9 Heating element

Figure 1.29 A soldering iron

Chapter 2: Engineering processes for electromechanical products

> **This chapter covers:**
> Element 1.2: Select processes needed to make electromechanical engineered products.
> **... and is divided into the following sections:**
> - Processes and specific techniques for making electromechanical products
> - Safety procedures and equipment for selected processes
> - Selecting processes and techniques for making a given electromechanical product.

Processes and specific techniques for making electromechanical products

The design and production planning of a product are activities that very often overlap. Design engineers discuss their ideas with production engineers to make sure that a product can be made economically while meeting the required quality standards. When the design has been finalised, production engineers must plan how the product is to be made. A number of production processes are often required. They may include:
- material removal
- joining and assembly
- heat treatment
- chemical treatment
- surface finishing.

Material removal

Some engineering components can be produced in a single forming operation by processes such as die casting and injection moulding. Others may have to be machined from barstock, castings and forgings because of the dimensional tolerances that are required and their material specifications. Three of the most common material removal processes are:
- turning
- drilling
- milling.

Turning

Turning is a process in which a single-point cutting tool is used to remove material from the surface of a rotating workpiece. It is carried out on a lathe and can be used to produce cylindrical surfaces by moving the tool along the workpiece, or flat surfaces by moving the tool across the end face of the workpiece. The

Engineering processes for electromechanical products

Turning

A process in which a single-point cutting tool is used to remove material from the surface of a rotating workpiece

simplest form of lathe is the centre lathe shown in Figure 2.1. Many of the lathes used for quantity production are complex computer-controlled machines, but they have all been developed from the centre lathe and function in much the same way.

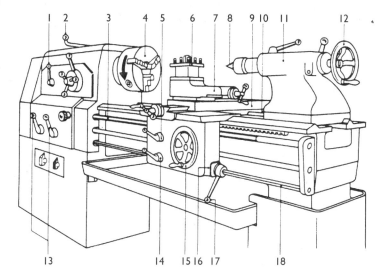

1. Reversing lever
2. Clutch lever
3. Lead screw
4. Self-centring chuck
5. Half-nut lever
6. Tool-post
7. Tool-slide or compound slide
8. Tailstock centre
9. Cross-slide
10. Bed
11. Tailstock
12. Tailstock hand-wheel
13. Feed change lever
14. Feed shaft lever
15. Saddle hand-wheel
16. Apron
17. Motor switch lever
18. Feed shaft

Figure 2.1 The centre lathe

The main parts of the centre lathe are the bed, the headstock, the tailstock and the saddle or carriage assembly. The lathe bed is made of cast iron with hardened V-shaped slideways along which the tailstock and saddle can be moved. The headstock contains the gearbox, for selecting different cutting speeds, and the spindle. The spindle can carry a chuck, faceplate or a catchplate and centre for holding and rotating the workpiece.

The tailstock, with its centre, can be clamped in any position along the lathe bed to support the free end of a workpiece. The centre can be removed from the tailstock spindle and replaced by a taper-shank drill or Jacobs-type chuck for drilling operations. A handwheel moves the tailstock spindle towards and away from the workpiece.

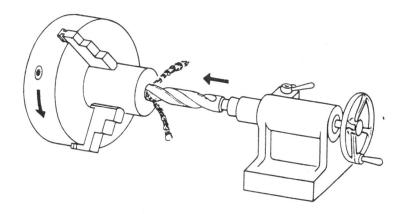

Figure 2.2 Drilling in a lathe

The saddle or carriage assembly rests on the slideways of the bed and contains the cross-slide, the compound slide and the toolpost. It may be moved along the bed manually, by means of a handwheel, or automatically, by engaging an automatic traverse mechanism. The cross-slide moves across the bed at right angles to the spindle axis, and is used for facing operations and for increasing the depth of cut when cylindrical turning.

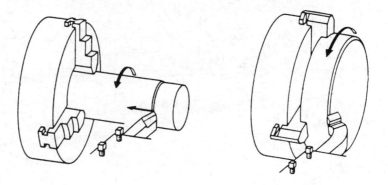

Figure 2.3 Turning operations: (a) parallel turning; (b) surfacing (transverse turning)

The tool slide or compound slide is located on top of the cross-slide and can be set at any angle for turning tapers and chamfers. The hand wheels of the cross-slide and compound slide contain scales calibrated in divisions of 0.01 mm that enable the depth of cut to be set accurately. The toolpost rests on top of the compound slide and, depending on the type, may hold up to four cutting tools.

The front of the carriage is called the apron. It contains the manual traverse handwheels and the levers that are used to engage automatic traverse for cylindrical turning, and automatic cross-traverse for facing. Automatic traverse is driven from the feedshaft, which runs along the bed, behind the apron. It can also be driven from the leadscrew, which runs alongside the feedshaft, but this should only be used for screwcutting operations.

A number of basic techniques and procedures are involved in the turning process. These include:
- workholding methods
- toolholding methods
- turning tool selection
- selection of cutting speeds and feeds
- use of coolant.

Workholding methods As stated above, the lathe spindle can hold a chuck, a faceplate or a catchplate and centre for holding and rotating the workpiece. Chucks may be of the three-jaw self-centring type or the four-jaw independent type.

The three-jaw self-centring chuck is used for gripping round and hexagonal workpieces. The jaws move inwards and outwards in unison, driven by a scroll plate that is turned by means of a chuck key. The three jaws ensure that the centre line of the workpiece lies on the spindle axis for turning external cylindrical surfaces, end facing, drilling and boring operations.

Before turning long, cylindrical workpieces, the free end should first be centre drilled. This is done using a special centre drill held in a Jacobs-type chuck in the tailstock. It produces a hole in the end of the workpiece that is part parallel and

Engineering processes for electromechanical products

Figure 2.4 Three-jaw self-centring chuck

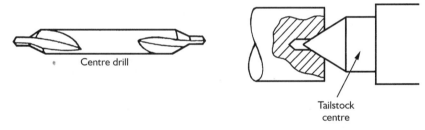

Figure 2.5 Use of centre drill

part tapered. The free end of the workpiece can then be supported on a centre that is held in the tailstock spindle. If a stationary centre is used, it must be well greased to prevent overheating. It is better to use a running centre if one is available. This rotates on ball bearings with the workpiece and generates very little heat.

The four-jaw independent chuck is used for square, rectangular and oddly shaped workpieces such as castings and forgings. Each jaw has its own chuck key socket to move it inwards and outwards. The position of the workpiece can thus be adjusted until its centre of rotation lies on the spindle axis. The four-jaw chuck has greater gripping power than the three-jaw type but takes longer to set up correctly.

The faceplate is used to hold irregular-shaped workpieces that may be too large or otherwise unsuitable for gripping in the four-jaw chuck. The workpiece may be clamped directly to the faceplate as shown or clamped to an angle plate. The procedure allows diameters and faces to be turned that are parallel or perpendicular to a premachined face. The premachined face locates against the

Engineering materials and processes

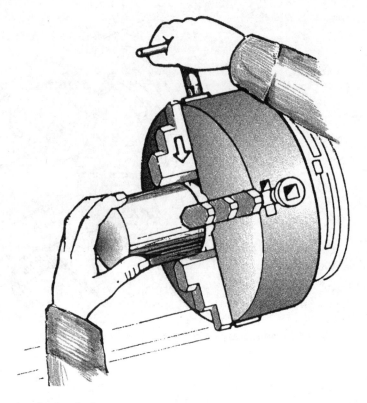

Figure 2.6 Independent four-jaw chuck

faceplate, or on the angle plate. For through-boring operations it is necessary to position parallel bars between the workpiece and the faceplate.

Irregularly shaped workpieces and the use of the angle plate can produce large out-of-balance forces. These cause vibration, which can damage the machine and result in a poor surface finish. Balance can be restored by fixing weights to the faceplate. They are positioned by trial and error on the opposite side to the out-of balance mass, until the faceplate can be placed in any angular position without rotating due to gravity.

The use of a catchplate, carrier and centres to support and rotate a workpiece is one of the oldest turning techniques. It is a very accurate method of turning long, cylindrical workpieces that have been centre drilled at each end. A carrier

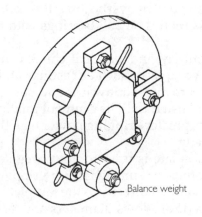

Figure 2.7 Use of faceplate

Engineering processes for electromechanical products

or 'dog' is clamped to one end of the workpiece, which is then held between the live centre in the spindle nose and the tailstock centre. The carrier and workpiece are rotated by the driving peg on the catchplate.

An advantage of turning between centres is that the workpiece can be taken out of the lathe for other machining operations and still run true when it is replaced. It can also be turned round and replaced between the centres without any loss of concentricity. A disadvantage of the method is that boring and full facing operations cannot be carried out.

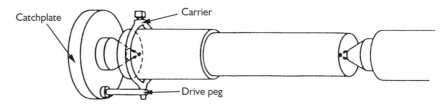

Figure 2.8 Turning between centres

Toolholding methods The single-point cutting tools used for cylindrical turning, facing and boring are clamped in the toolpost. A pillar-type toolpost is often used on small centre lathes, whereas the four-way and quick-release types are common on larger machines. The tools must be set with their cutting edges at the centre height of the lathe. The quick-release type usually has height-adjusting screws whilst packing pieces are placed beneath the cutting tools in the four-way toolpost.

The quickest way of setting the cutting tools to the machine centre height is to set the cutting edge level with the point of the tailstock centre or the centre in the spindle nose, if one is being used. Alternatively, a setting gauge or a steel rule can be used to measure the centre height above the lathe bed. A quick check on the centre height can be made by running the tool up to the stationary workpiece against an upright steel rule. The tool will be very close to centre height if the rule appears to stand vertical.

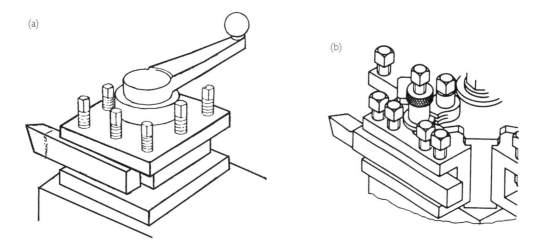

Figure 2.9 Toolposts: (a) four-way toolpost; (b) quick-release toolpost with height adjustment

49

The tailstock is used in drilling operations for holding and feeding the drill. Taper-shank twist drills or a Jacobs-type chuck are held in the internal morse taper in the tailstock spindle. Before drilling, the workpiece should first be centre drilled to start the hole if this has not already been done.

Turning tool selection Some of the more commonly used lathe cutting tools are shown in Figure 2.10. The roughing tool is used for making initial cuts where large amounts of material are to be removed. The knife tool is used for finishing cuts and producing right-angle corners. Facing cuts are taken using a cranked tool and finished components are cut off from bar using the thin parting-off tool. Boring tools have long shanks and are cranked at the end to prevent contact between the tool shank and the workpiece.

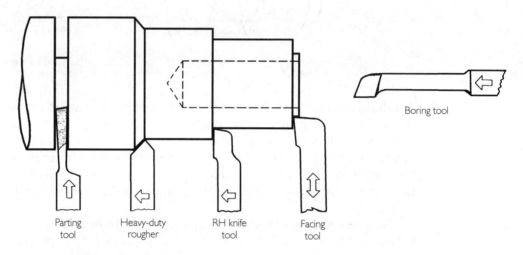

Figure 2.10 Lathe cutting tools

Selection of cutting speeds and feeds The cutting speed selected for a particular machining operation depends on the material being cut and the material from which the cutting tool is made. Table 2.1 can be used as a general guide for machining with high-speed steel cutting tools.

In the case of turning, the spindle speed, measured in revolutions per minute (r.p.m.), also depends on the diameter of the workpiece. It is given by the formula:

$$N = \frac{1000\,S}{\pi\,d}$$

where N = spindle speed in r.p.m., S = cutting speed in m/min and d = diameter of workpiece in mm.

Having selected the appropriate spindle speed, it is generally accepted that a deep cut and a fine feed rate is the best combination for fast removal of material, long tool life and a good surface finish. When taking roughing cuts, the depth may be as large as the machine can handle provided that the workpiece is securely held and the cutting tool is properly ground and set in position. Spindle speeds and automatic tool feed rates are obtained by setting selection levers or dials on the headstock to the appropriate position.

Engineering processes for electromechanical products

Table 2.1 Guide to cutting speeds for various materials

Material being cut	Cutting speed (m/min)
Aluminium	70–100
Brass	70–100
Phosphor-bronze	35–70
Mild steel	30–50
Grey cast iron	25–40
Thermosetting plastics	20–40

Use of coolant Most metal machining processes make use of a coolant. The purposes of a coolant are to:
- carry heat away from the workpiece and cutting tool
- lubricate the chip-cutting tool interface, thus reducing tool wear
- prevent chip particles from becoming welded to the tool face and forming a built-up edge
- improve the surface finish and wash the swarf away from the cutting edge
- prevent corrosion of the workpiece and the machine.

The coolant used on centre lathes is stored in a reservoir beneath the lathe bed. From here it is delivered by a pump to a control tap and an adjustable supply pipe on the carriage assembly. The most common coolant used for general machining is emulsified, or soluble, oil. This is a type of oil containing detergent and disinfectant that mixes with water. The mixture has a white, milky appearance and should be directed on to the cutting area in a steady stream. Cast iron can be machined without the use of a coolant. The graphite flakes in its structure make it self-lubricating and less heat is generated than when cutting other materials.

Progress check

1. What are the four main parts of a centre lathe?
2. What is the tailstock used for on a centre lathe?
3. What is the compound slide used for on a centre lathe?
4. What kinds of workpiece are suitable for being gripped in a three-jaw, self-centring chuck?
5. How is the free end of a cylindrical workpiece prepared so that it can be supported by the tailstock centre?
6. What kinds of workpiece are held on the faceplate in a centre lathe?
7. Give one advantage and one disadvantage of turning between centres.
8. What are the reasons for using a coolant in material removal processes?

Case study

A modern computer-controlled turning centre may not look very much like a centre lathe but it is a direct descendant. The last 30 years have seen rapid advances in machining technology as a result of the application of computer numerical control, more commonly referred to as CNC. Design and production can now be integrated in what are known as computer-aided manufacturing (CAM) systems. A designer is able to feed the dimensional details of a

51

component into the computer system, which then generates a parts program for the turning centre. The system enables components with complex shapes, such as that shown in Figure 2.11, to be produced quickly and accurately.

CNC machines are able to work to tolerances that could previously only be achieved by grinding operations. They are able to produce on a single machine components that otherwise might need to be transferred from one machine to another for different operations. They are also able to carry out their own quality monitoring and automatically compensate for tool wear. CAM systems are of course expensive and must earn their keep, but many engineering firms find that the degree of flexibility that they provide is well worth the investment.

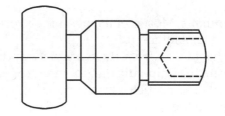

Figure 2.11 CNC-generated component

Drilling

Drilling
A process in which a rotating cutting tool removes material from a stationary workpiece to produce accurately positioned holes

In the drilling process, a rotating cutting tool, the twist drill, removes material from a stationary workpiece. The two most common types of drilling machine for small and medium-sized engineering work are the sensitive bench-mounted drill and the pillar drill.

The sensitive bench drill is used for light work, with hole diameters generally less than 10 mm, whereas the pillar drill can accommodate larger workpieces and hole sizes. The vertical spindle of the sensitive drill is belt driven and the speed is changed by altering the position of the drive belt on its stepped pulleys. The spindle of the larger pillar drills is driven through a gearbox equipped with speed-change levers to give a wider range of rotational speeds.

The drill spindle is fed downwards by hand through a rack and pinion mechanism, and automatic spindle feed is provided on some of the larger types of pillar drill. Both types of drilling machine have a worktable that can be adjusted for height and may also be set over at an angle on some machines. The worktable contains slots to which the workpiece or a machine vice may be bolted.

A number of basic techniques and procedures are involved in the drilling process. These include:
- workholding methods
- toolholding methods
- drill and reamer selection
- selection of cutting speeds
- hand-feeding procedure.

Workholding methods Large or irregular-shaped components, such as castings, may be secured directly to the machine worktable. Bolts and clamps are used with the bolts positioned as closely as possible to the workpiece. This enables them to exert the maximum clamping force. A single clamp may be sufficient for small workpieces but at least two should be used with the larger ones.

Engineering processes for electromechanical products

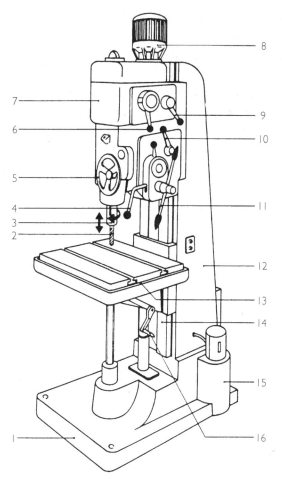

1 Base
2 Drill
3 Spindle
4 Automatic feed lever
5 Feed handwheel
6 Drill speed lever
7 Gearbox
8 Electric motor
9 Gear shift handle
10 Automatic feed adjusting lever
11 Fine feed lever
12 Column
13 Table
14 Vertical slide
15 Coolant pump
16 Height adjustment handle for table

Figure 2.12 Pillar drill

In some cases it may be necessary to clamp the workpiece to an angle plate which is itself bolted to the machine worktable. Securing cylindrical components can sometimes be a problem, and it is good practice to clamp them on V-blocks.

A machine vice, bolted to the machine worktable, is used to hold smaller components. Parallel bars are used to support the workpiece in the vice. These ensure that the drilled hole is perpendicular to the lower face of the workpiece.

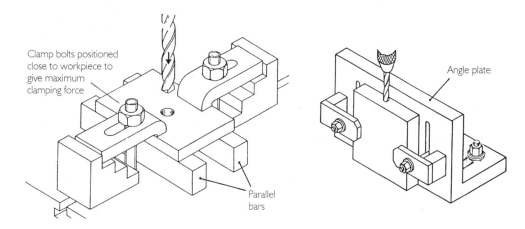

Figure 2.13 Workholding on the drilling machine

Engineering materials and processes

They also reduce the likelihood of damage from the drill striking the vice after breaking through the material.

Toolholding methods The smaller sizes of twist drill have parallel shanks and are gripped in a Jacobs-type chuck. This is tightened by means of a chuck key, which should then be removed and placed away from the workpiece. The chuck itself has a morse-tapered shank that locates in the drilling machine spindle. The larger diameter drills have morse-tapered shanks and locate directly in the spindle. It should be noted that the drive through a morse taper is by friction alone. The tang at the end of the tapered shank is used only when releasing the drill or chuck, using a tapered drift as shown in Figure 2.15.

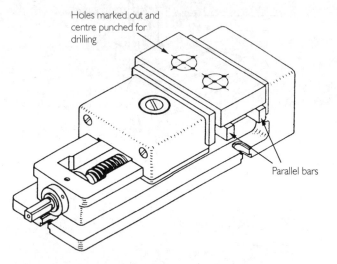

Figure 2.14 Machine vice for drilling

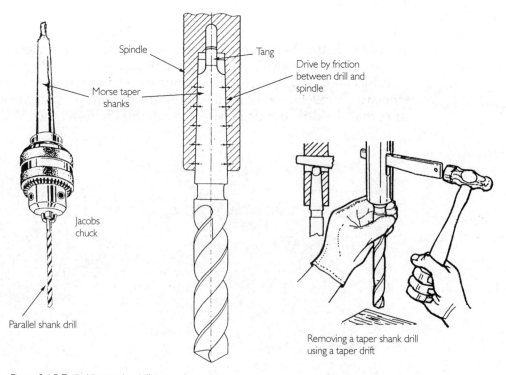

Figure 2.15 Toolholding in the drilling machine

Engineering processes for electromechanical products

Drill and reamer selection Twist drills have two cutting edges and two helical flutes that are cut along the body of the drill. Each flute has a raised 'land' running along its leading edge to reduce friction between the drill and the side of the hole. Twist dills for metal cutting are made from high-speed steel, whereas the cheaper types, suitable for DIY woodworking, may be made from a lower grade plain high-carbon steel. Twist dills are also made with different helix angles for drilling different materials. Those used for drilling metals have a faster helix than those used for drilling soft plastics, i.e. more twists.

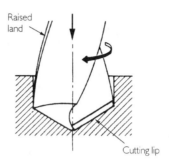

Figure 2.16 Twist drill

Twist drills should be thought of as roughing tools because the hole produced is often larger than the drill diameter and may also be slightly out of round. This may be due to damage or uneven wear on the cutting edges or it may be due to the drill being incorrectly ground. Hole positions should always be centre punched to help start the hole in the correct place. Where accurate positioning is required, or when drilling large-diameter holes, it is good practice to begin with a small pilot hole. If its position is satisfactory, this can then be opened out progressively using larger drills, up to the required diameter.

Holes that need to be finished very accurately to size and have a good surface finish should be drilled slightly under size and finished by reaming. Machine reamers are precision cutting tools that have more flutes than a twist drill. Furthermore, the helix is in the opposite direction to that on a twist drill. This has two purposes: it prevents the reamer from being drawn into the hole and stops chips from being drawn up the helix where they might spoil the surface finish.

There are two types of reamer. One type cuts only on the ends of the flutes and is best for use with mild steel. The other type, which is known as a rose-action reamer, cuts on both the ends and the edges of the flutes. It gives best results when used with brasses, bronzes, cast iron and plastics, which have a tendency to close on, or grip the reamer. Machine reamers should only be used with a low spindle speed.

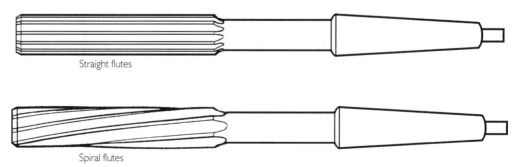

Figure 2.17 Machine reamers

Engineering materials and processes

Selection of cutting speeds In the case of drilling, the spindle speed, measured in r.p.m., depends on the workpiece material and the diameter of the drill that is being used. The cutting speed for a particular material can be obtained from Table 2.1. The spindle speed is then given by the formula:

$$N = \frac{1000\, S}{\pi d}$$

where N = spindle speed in r.p.m., S = cutting speed in m/min and d = diameter of workpiece in mm.

Hand-feeding procedure When feeding the drill by hand, care should be taken not to exert too much force. Small-diameter drills are easily broken and larger drills may overheat if they are fed too quickly. Care should also be taken to relieve the pressure on the drill as it is about to break through the material. At this point there is a tendency for the drill to 'grab'. This can cause smaller diameter drills to break and can also cause the workpiece to spin dangerously with the drill if it is not securely clamped. When deep holes are being drilled it is advisable to use a coolant and to remove the drill from the hole occasionally. This will lift out any swarf that has accumulated in the flutes of the drill.

Drilling machines are fitted with a depth gauge that can be used to measure the depth of penetration. The gauge can be set with a lock nut to limit the travel of the drill and it is useful when drilling 'blind' holes that do not pass all the way through the workpiece. It can also be used to limit the travel of the drill as it breaks through the material. This prevents possible damage to the drill, machine vice or worktable.

Activity 2.1

You are required to produce the component shown in Figure 2.18 on a centre lathe from 75-mm-diameter bright-drawn mild-steel bar. A high-speed steel turning tool is to be used to turn the 50-mm-diameter hole and a high-speed steel twist drill is to be used in the tailstock to produce the 10-mm-diameter hole.

Calculate (a) the spindle speed for turning down the 75-mm-outer diameter and (b) the spindle speed for drilling the 10-mm-diameter hole.

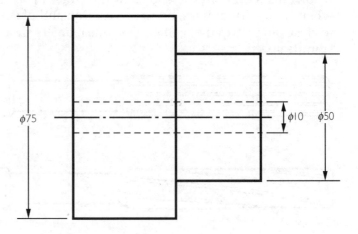

Figure 2.18 Turned component

Engineering processes for electromechanical products

Progress check

1. How should clamp bolts be positioned for the clamps to exert maximum force on the workpiece?
2. How can cylindrical components be securely held on the drilling machine worktable?
3. How can drilled holes be finished accurately to size?
4. Why is it good practice to relieve the pressure on a drill as it is about to break through the material?
5. How can 'blind' holes be drilled accurately to the required depth?
6. What procedure should be followed when drilling large-diameter holes to ensure that they are correctly positioned?

Milling

A process in which material is removed as the workpiece is fed past a rotating multitoothed cutter

Milling

In the milling process, material is removed as the workpiece is fed past a rotating multitoothed cutter. There are two basic types of milling machine: the horizontal mill and the vertical mill. With the horizontal mill, the cutter rotates about an axis in the horizontal plane and with the vertical mill the cutter axis is in the vertical plane. Both types of milling machine are used to machine flat surfaces, slots, channels and grooves.

On all except very small milling machines, in which the spindle may be belt driven, the drive is through a gearbox equipped with speed-change levers for selecting the desired cutting speed. With both types of milling machine, the worktable can be raised or lowered by means of a handwheel. It can also be traversed by hand in two perpendicular directions in the horizontal plane: from side to side and from front to rear. Automatic traverse with different rates of feed can also be engaged.

The basic techniques and procedures involved in the milling process include:
- workholding methods
- cutter selection
- cutter mounting
- selection of cutting speeds
- feeding the worktable.

Workholding methods The machine worktable on both horizontal and vertical mills is equipped with T-shaped slots, to which a large or awkwardly shaped workpiece may be clamped directly. Large cutting forces are present when milling and at least two clamps should be used. As when drilling, cylindrical workpieces should be supported on V-blocks.

Smaller workpieces may be held in a heavy-duty machine vice that is also bolted to the worktable. Vices with a swivel base that can be set at any angle in the horizontal plane are often used. Parallel bars should be used to support the workpiece in the machine vice. This ensures that the upper and lower machined surfaces of the workpiece are parallel.

Wear in the vice slideway often causes the workpiece to lift off the parallel bars. A blow from a hide hammer as the vice is being tightened usually ensures that the workpiece is sitting down tightly. If the vice is so worn that it is difficult to keep the workpiece in contact with the bars, it is then no longer suitable for precision work.

Cutter selection Four of the most commonly used cutters on the horizontal milling machine are slab cutters or slab mills, side and face cutters, slotting cutters and slitting saws.

Engineering materials and processes

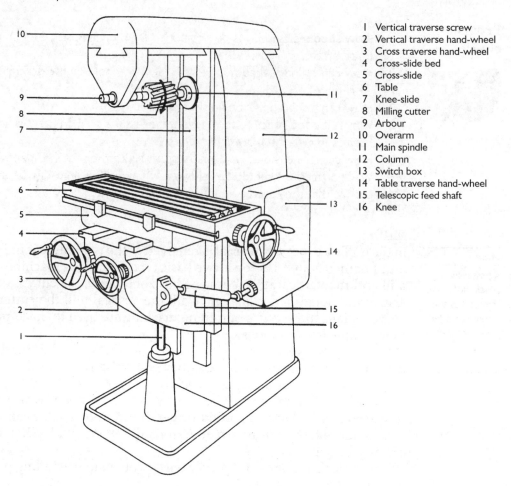

Figure 2.19 Horizontal milling machine

1. Vertical traverse screw
2. Vertical traverse hand-wheel
3. Cross traverse hand-wheel
4. Cross-slide bed
5. Cross-slide
6. Table
7. Knee-slide
8. Milling cutter
9. Arbour
10. Overarm
11. Main spindle
12. Column
13. Switch box
14. Table traverse hand-wheel
15. Telescopic feed shaft
16. Knee

Slab cutters are used to produce wide, flat surfaces. Side and face cutters have cutting edges on both the periphery and the sides of the teeth and are used for light facing operations and for cutting channels, slots and steps. Slotting cutters are thinner than side and face cutters and have teeth only on the periphery. They are used for cutting narrow slots and keyways. Slitting saws are very thin cutters used for cutting narrow slots and for parting off excess material.

Four of the most commonly used cutters on the vertical milling machine are face mills, shell-end mills, end mills and slot drills.

Face mills have cutting edges on both the end face and the sides. They are used to produce wide, flat surfaces, as are slab mills on the horizontal milling machine. They do, however, produce the surface more accurately because, unlike the slab mill, every part of each tooth on the cutter face passes over the whole surface. In this way, face mills are said to 'generate' a flat surface. Shell-end mills are smaller than, but similar to, face mills. They have the same cutting action and are used for generating smaller flat surfaces.

End mills are of smaller diameter than face mills but, like them, they have cutting edges on the end face and sides. They are used for light facing, profiling, recessing and milling slots. Slot drills are similar to end mills but have only two cutting edges. They are used for accurately milling slots and keyways.

Engineering processes for electromechanical products

1 Telescopic shaft for vertical knee traverse
2 Base
3 Knee
4 Table traverse hand-wheel
5 Switch box
6 Table
7 Column
8 Speed control lever
9 Graduated collar for head adjustment
10 Head
11 Collet bar
12 Shell end mill
13 Cross-slide

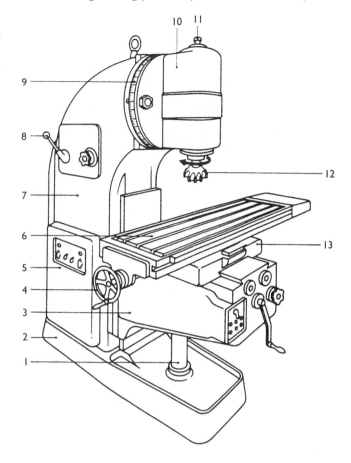

Figure 2.20 Vertical milling machine

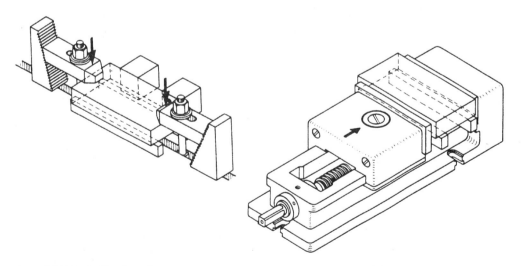

Figure 2.21 Workholding for milling

Engineering materials and processes

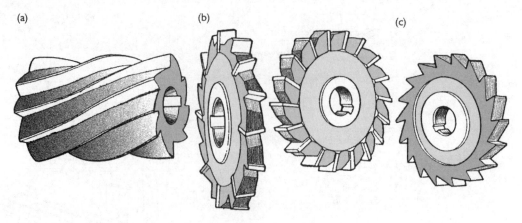

Figure 2.22 Cutters for the horizontal milling machine: (a) slab or roller mill; (b) side and face cutters; (c) slotting cutter

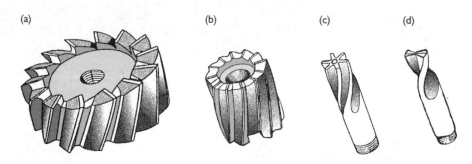

Figure 2.23 (a) Face mill; (b) shell end mill; (c) standard end mill; (d) two-lipped mill for slotting drill

Cutter mounting Horizontal milling machine cutters are mounted on a shaft known as an arbor. This has a taper at one end, which locates in the spindle nose of the milling machine, and a driving flange containing two slots. A draw bolt fastens the arbor to the spindle and two driving dogs locate in the slots. Unlike the morse-tapered drill, the drive to the arbor is not by friction on the taper but directly from the dogs to the driving flange.

The cutter is mounted with a key and spacing collars on the parallel part of the arbor, which has a keyway running along its length. The free end of the arbor passes through the overarm steady of the machine and a nut on the end secures the cutter and collars on the arbor. There is a tendency for the arbor to flex when cutting, and the cutter and overarm steady should be positioned so that there is as little overhang as possible on each side of the cutter.

The face mills and shell-end mills for vertical milling machines are mounted on a short 'stub arbor'. This has the same taper and driving flange as the long arbor described above and is fixed to the spindle nose of the vertical mill in the same way. The cutters locate on a spigot and are held in position by a locking screw.

End mills and slot drills that are used on the vertical mill have parallel shanks that are threaded at the end. They are held in an 'antilock' collet chuck fixed to the spindle nose. This is more complex than the Jacobs chuck used for drills. It is designed in such a way that, as the cutting forces increase, so does the grip on the cutter.

Engineering processes for electromechanical products

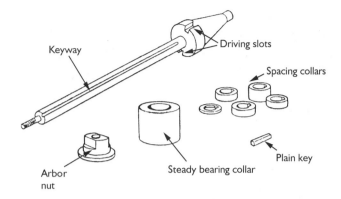

Figure 2.24 Long arbor for the horizontal mill

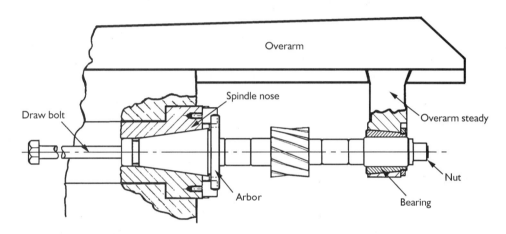

Figure 2.25 Cutter mounting for the horizontal mill

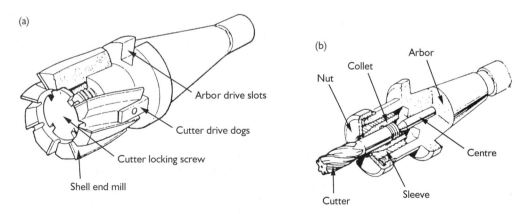

Figure 2.26 Cutter mounting for the vertical mill: (a) stub arbor; (b) antilock chuck

Selection of cutting speeds With both types of milling machine the spindle speed, measured in r.p.m., depends on the workpiece material and the diameter of the cutter that is being used. The cutting speed for a particular material can again be obtained from Table 2.1. The spindle speed is then given by the formula:

$$N = \frac{1000\, S}{\pi\, d}$$

where N = spindle speed in r.p.m., S = cutting speed in m/min and d = diameter of workpiece in mm.

Feeding the worktable There are two ways in which the workpiece can be fed under the cutter when using the horizontal milling machine. They are known as up-cut milling and down-cut milling.

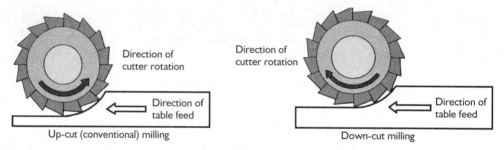

Figure 2.27 (a) Up-cut (conventional) milling; (b) down-cut milling

Up-cut milling should always be used unless the machine is specially designed for the other method. With down-cut milling there is a tendency for the cutter to climb on to the workpiece or for the workpiece to be dragged under the cutter. This is especially the case with machines that are not new and have some backlash in the worktable feed mechanism. At best, this can cause excessive vibration and a poor surface finish. At worst, the cutter can climb on to the workpiece, stall the machine and bend the arbor.

Up-cut milling also has some disadvantages. It tends to lift the workpiece from the table and the cutter teeth tend to rub on the workpiece before starting to cut. These are, however, far outweighed by the disadvantages of the other method. As with turning, it is advisable to use a coolant for all but the lightest milling operations. Soluble oil is again the most common coolant used, and for heavy-duty milling operations it is sometimes used in undiluted form.

Progress check

1. How should the workpiece be supported when held in the machine vice of a milling machine?
2. Why is the face mill on the vertical milling machine able to produce a truer flat surface than the slab mill on a horizontal milling machine?
3. Which cutter would be selected to produce narrow slots and keyways on the vertical milling machine?
4. What is the purpose of the overarm steady on the horizontal milling machine?
5. How are end mills and slot drills held in the vertical milling machine?
6. Why is up-cut milling the method that should be used on most horizontal milling machines?

Joining and assembly

Engineered products, such as a bicycle, are assembled from a number of fabricated parts, subassemblies of components and single-item components. The fabricated parts include the bicycle frame, which is made by welding or brazing the various tubular sections together. The subassemblies include the bicycle wheels made up of the hubs, spokes and rims. During final assembly, the fabricated parts and subassemblies are brought together and joined using single-item components such as the wheelnuts and other fastening devices.

The quality of an engineered product often depends on the care that has been taken during its assembly. The different joining and assembly processes have their own particular techniques and procedures. The most common methods are
- screwed fastening
- riveting
- soft soldering
- hard soldering
- welding
- adhesive bonding.

Screwed fastening

Before joining or assembling components by any of the above methods, care should be taken to see that the joint or locating faces are clean and undamaged. Any burred edges or surplus material left from machining or forming processes should be removed. The holes for screwed fastenings should also be inspected to ensure that they are not blocked with swarf and that internal screw threads are clean and undamaged. Components that are damaged in any way should be set aside for possible repair or reworking.

Fitted bolts are used in assemblies whose components must be positioned accurately. These have close tolerance diameters, and the holes in which they locate are drilled, and perhaps reamed to size, so that there is very little clearance.

For other applications, in which the positioning is less critical, black bolts are used. These are forged to shape with a larger tolerance on their diameters and the holes in which they fit are drilled so that there is ample clearance. This facilitates quick assembly and allows the position of the components to be adjusted slightly before the bolts are tightened.

The most common hand tools used to tighten hexagonal nuts, bolts and screws are open-ended spanners, ring spanners and socket spanners shown in Figures 2.28 and 2.29.

These should be of good quality so as not to damage the hexagonal faces of nuts and bolt heads. Spanners of imperial sizes, i.e. BSF (British Standard Fine thread) and UNF (Unified Fine thread), should not be used with metric fastenings and vice versa. Ring spanners and socket spanners are preferable to open-ended spanners as they give a firmer hold on the fastening.

Other hand tools include 'posidrive' or Phillips-type screwdrivers, cross-blade screwdrivers and the hexagonal keys for socket-headed screws. Again, it is important to use tools of an appropriate size and good quality.

If metric and non-metric screwed fasteners are in use, they should be clearly labelled and kept separate in the workplace. Only the type and size specified for assembly or servicing should be used.

Small-diameter bolts and setscrews can easily be sheared off and care should be taken not to overtighten them. Assembly drawings and servicing instructions often specify torque settings for screwed fasteners. Here a torque wrench that can be used with socket spanners is required.

Engineering materials and processes

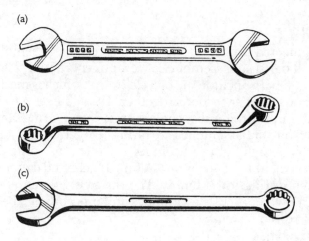

Figure 2.28 Spanners: (a) open-ended; (b) ring; (c) combination

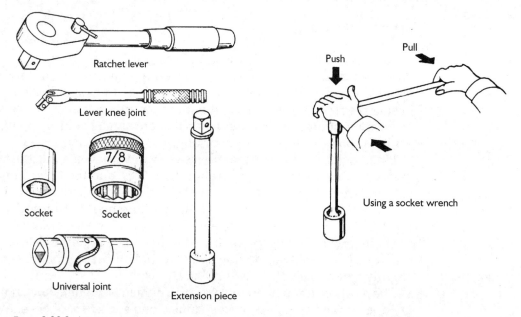

Figure 2.29 Socket spanners

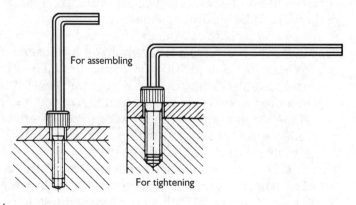

Figure 2.30 Hexagonal socket keys (Allen keys)

Engineering processes for electromechanical products

Torque

The product of the turning force on the wrench and the perpendicular distance between its line of action and the centre of rotation. The SI unit of torque is the newton metre (N m)

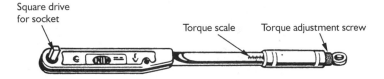

Figure 2.31 Torque wrench

Where components are joined by a number of screwed fasteners, the assembly instructions sometimes specify the order in which they should be tightened. This is to avoid distortion and it is particularly important when the joint must be gas tight or water tight. When no assembly instructions are given, and the fastenings lie on a pitch circle, it is good practice to follow a sequence of tightening those fastenings that are diametrically opposite. When the fastenings lie around a rectangle, it is good practice to start with those that are opposite and at the centre of the longer sides. Working alternately outwards, those in the corners should be tightened last.

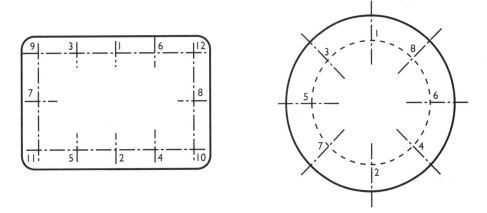

Figure 2.32 Order of tightening

Screwed fastenings are frequently used to make connections in electrical circuits. The regulations for electrical installations state that connections and joints must be mechanically and electrically sound. The termination of a conductor must be appropriate for its use. Typical light-duty terminations for use in domestic circuits and appliances are shown in Figure 2.33.

The simple loop termination is used in some plug tops in which the loop is secured between brass washers on the terminal screw. The Courtney washer encloses the loop and protects it during servicing and repair operations in which the conductor may need to be detached and reconnected. Crimped terminations are also used for these applications. They are usually of the loop or spade type, which are compressed on the end of a conductor using special pliers.

Heavy-duty screwed terminals are widely used with power distribution and supply cables. These may contain copper or aluminium conductors. With aluminium, a heavy-duty crimped lug is usually fitted. With copper the lug may be crimped or soldered on the end of the conductor.

65

Engineering materials and processes

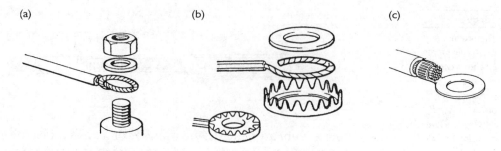

Figure 2.33 Electrical terminations: (a) loop termination; (b) Courtney washer; (c) Crimped loop

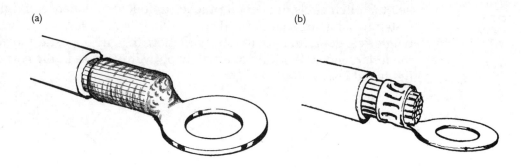

Figure 2.34 Heavy-duty terminations: (a) soldered lug; (b) crimped lug

Activity 2.2

The degree to which a screwed fastening must be tightened is specified as a torque setting. Look at a motor vehicle servicing manual. If you have not got one at home, your public library will have them for a range of modern cars. At the end of each section, such as that for the engine, you will find data that include the torque settings for the screwed fastenings.

Find out (a) how torque is defined, (b) the two different units in which torque is measured and (c) a way of converting from one set of units to the other.

Riveting

The joint surfaces that are to be riveted together should be inspected to ensure that they are clean and that the holes drilled for the rivets are clear and have no burred edges. The rivets used must be of the specified material, length, diameter and head shape. Rivets of different sizes and materials should be kept separate in the workplace and clearly labelled. This is especially important with aluminium alloy rivets, which should also be labelled with a 'use by' date. This is because they age-harden if they are not used within a given time and must then be resoftened by heat treatment. They are sometimes stored in a refrigerator to slow down the ageing process.

The parts to be joined by riveting may be positioned together and held by clamps. The rivets can then be inserted and correctly seated in their holes using a drawing-up tool.

The preformed head is rested on a support dolly, and a ball pein hammer can then be used to swell the shank and rough-form the head. A rivet snap of the appropriate shape can then be used to finish-form the head. Alternatively, a pneumatic riveting tool can be used to form the head in a single operation.

Engineering processes for electromechanical products

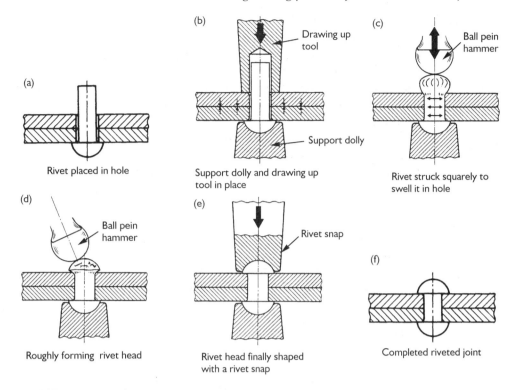

Figure 2.35 Riveting procedure

The same preparation of the joint surfaces is required when using pop rivets. Here again, the rivets used must be of the specified length and diameter. If a gas-tight or water-tight joint is required, care should be taken to use the sealed type. The rivet is formed using either a mechanical or pneumatic puller, which grips and pulls the central pin until it snaps off.

Soft soldering

Soft solder is an alloy of tin and lead to which a little antimony has also been added to improve its fluidity. It is used to join sheet-metal components made from mild steel, copper and brass, and to join the copper pipes used in plumbing. Mild steel is sometimes coated with tin and known as tinplate. The tin gives it corrosion protection and also makes it easier to join by soldering. Soft solder is also widely used to make joints in electrical and electronic circuits. It is made with different proportions of tin and lead for different applications. Some of the more common soft solders compositions are given in Table 2.2.

Table 2.2 Composition of common soft solders

British Standard type	Composition (%)			Melting temperature range (°C)
	Tin	Lead	Antimony	
A	65	34.4	0.6	183–185
K	60	39.5	0.5	183–188
F	50	49.5	0.5	183–212
G	40	59.6	0.4	183–234
J	30	69.7	0.3	183–255

Engineering materials and processes

Types K, F and G are used for sheet-metal work and type J is used for plumbing. They are supplied mainly in stick form. Type A is used for joints in electrical and electronic circuits. It is supplied mainly in wire form with a core of resin flux. The melting temperature range gives an indication of the time that it takes for them to solidify. Type A solidifies very quickly, whereas type J takes the longest. A special lead-free solder composed of tin and silver should be used to join pipes containing drinking water.

It is especially important to clean the joint surfaces before soft soldering or the solder may not form a good bond. They must be wiped free of any oil or grease, and metal oxides such as rust must be completely removed using emery cloth or wire wool. A flux, in the form of a liquid or paste, is then applied to the surfaces being joined.

A flux may be active or passive. A traditional active type is acidified zinc chloride known as Baker's fluid. When the surfaces are heated, this has the effect of cleaning away any remaining dirt or oxide and protecting the surfaces against further oxidation. Active fluxes can be corrosive, and all remaining traces should be washed off the joint immediately after it has been made. Resin, in the form of a paste or in the core of wire solder, is a passive type of flux that does not have a chemical cleaning action. It merely protects the cleaned surfaces against oxidation and is the type used for soldered electrical connections.

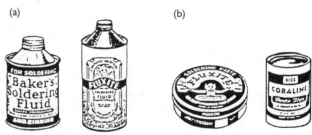

Figure 2.36 Commercial fluxes: (a) zinc chloride flux in trade packs; (b) resin-based flux in trade packs

After applying a flux, the next stage of joint preparation is to 'tin' the joint surfaces. This is done using a soldering iron that has been heated in a gas flame. The iron is usually hot enough when the flame turns a greenish colour around the copper bit. The bit is then quickly cleaned with a file, dipped in flux and loaded with molten solder. This is then applied in an even coating to the joint surfaces. The solder combines with the material surface to form what is known as an amalgam.

The tinned surfaces are then placed together and heated with a soldering iron while pressure is being applied. The soldering iron is drawn slowly along the joint until molten solder appears around the joint edges. This process is known as sweating the joint, in which the molten solder is drawn between the joint faces by capillary action. The soldering iron is then removed, but pressure is still applied until the solder has solidified. Any active flux that remains can then be washed off and, if the material being joined is bare mild steel, it is a good idea to apply a rust inhibitor.

Soft-soldered joints in copper pipes are generally made using standard fittings that contain an insert of solder. They include elbows, T-pieces, nipples and reducers. The pipes are cleaned with wire wool, coated with passive resin flux and the joint assembled. Heat is then applied from a blowpipe fed with natural gas or propane. As soon as solder appears at the edges of the joint, the heat is removed and the solder allowed to solidify.

Engineering processes for electromechanical products

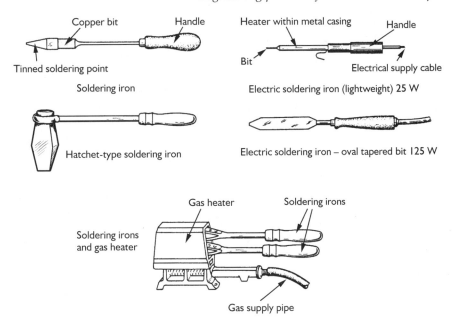

Figure 2.37 Soldering irons

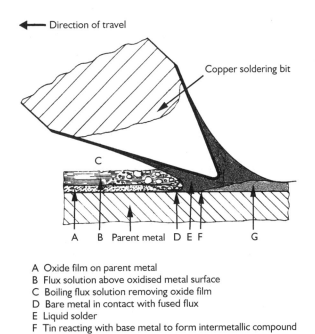

A Oxide film on parent metal
B Flux solution above oxidised metal surface
C Boiling flux solution removing oxide film
D Bare metal in contact with fused flux
E Liquid solder
F Tin reacting with base metal to form intermetallic compound
G Solidifying solder

Figure 2.38 The tinning process

For fittings that do not contain a solder insert, solder is applied to the edge of the joint when it is judged to be hot enough. The solder melts and is drawn into the joint by capillary action. The same process is carried out when soldering electrical terminations to copper cable. Here, care must be taken not to overheat the joint as excess heat may be conducted back along the conductor and damage the insulation.

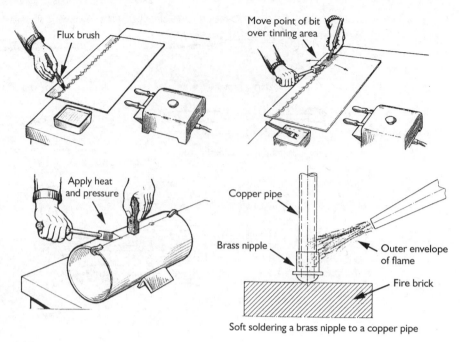

Figure 2.39 Soldering processes

The electronic components used with printed circuit boards (PCBs) usually have leads that are already coated with tin. After placing them in position and cutting the leads down to size, heat is applied from a soldering iron bit loaded with just enough resin-cored solder to make the joint. This should be done as quickly as possible to avoid damaging sensitive components. The leads on the opposite side of the PCB to the joint can sometimes be gripped with pliers to conduct away excess heat.

Progress check

1. Why is it sometimes important to tighten screwed fasteners in a given order?
2. How are crimped terminations fitted to an electrical conductor?
3. What is the tool used to ensure that rivets are seated properly in their holes?
4. What is a 'rivet snap'?
5. Which grade of solder is the quickest to solidify and what is it used for?
6. What is the difference between an active and a passive flux?
7. Describe the procedure for tinning a joint surface.
8. Why should all traces of an active flux be washed off a completed joint?
9. What precautions should be taken when soldering electronic components to a printed circuit board?

Hard soldering

Hard solder can be used to join mild steel, cast iron, copper and brass. It includes brazing, in which brass, known as brazing 'spelter', is the bonding metal and silver soldering, in which the solder is an alloy of silver, copper zinc and cadmium. Hard solder has a higher melting point than soft solder and produces a stronger joint. Like soft solder, it combines with the material at the joint surface to form an amalgam. Silver solder has a lower melting point than brazing spelter but does not produce as strong a joint. It is used for joining brasses and bronzes

Engineering processes for electromechanical products

whose melting point is very close to that of brazing spelter. Two typical hard solders are shown in Table 2.3.

Table 2.3 Two typical hard solders

British Standard	Category	Composition (%) Silver	Copper	Zinc	Cadmium	Melting temperature range (°C)
3	Silver solder	50	15	15	20	620–640
10	Brazing spelter	60	40	–		885–890

As with soft soldering, surface preparation is important. The joint surfaces should be cleaned free of oil or grease and surface oxide removed using a wire brush, emery cloth or wire wool. Hard-soldering fluxes are of the active type and mostly contain borax. This is a chemical compound of sodium and boron which is supplied in the form of a white powder. It is mixed with water to form a stiff paste, which is then applied to the joint surfaces.

The fluxed components are assembled on the brazing hearth surrounded by fire bricks. This is to reflect and contain the heat, which is supplied from a blowtorch. As the temperature of the components rises the flux will be seen to be spreading and cleaning the joint surfaces. The hard-solder rod or wire can then be touched against the joint and if the temperature is correct it will be seen to melt and be drawn into the joint by capillary action.

After the joint has cooled, the remaining flux will be seen to have set in a glass-like film. This can usually be removed by emery cloth or wire brushing. Caustic soda and sulphuric acid solutions are also used, followed by a thorough washing in water.

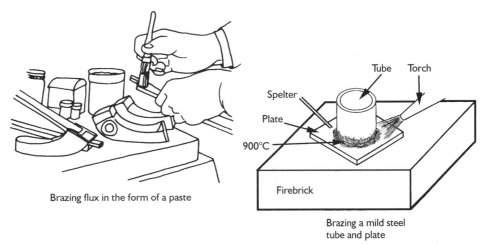

Brazing flux in the form of a paste

Brazing a mild steel tube and plate

Figure 2.40 Hard soldering

Welding
Fusion welding is a process in which the materials to be joined are fused together in their molten state. Additional material may be added to the joint from a filler rod that is of the same material composition. If the process is carried out properly, the strength of the joint should be very close to that of the parent materials.

Engineering materials and processes

Welding technology covers a wide range of procedures and techniques for joining different materials of various thicknesses. These are continually being added to as new methods of welding and improving weld quality are developed. There are two basic manual welding processes. They are:

- oxy-acetylene welding
- manual metal arc welding.

Oxy-acetylene welding With this process, acetylene and oxygen gases are mixed together in a blowpipe and burn to supply the heat for welding. The gases are stored in steel cylinders fitted with pressure regulators and flashback arrestors. Each pressure regulator has two pressure gauges. The right-hand gauge measures the pressure of the stored gas inside the cylinder, and the left-hand gauge measures the regulated welding pressure in the line to the blowpipe.

The flashback arrestors prevent the gas flame from travelling back along the connecting hoses to the cylinders. The blowpipe is fitted with control valves to adjust the flow of the gases. Temperatures in the order of 3150°C are generated in the welding flame, which is well above the melting point of most engineering metals. Approved goggles are essential when welding, together with other items of safety wear that will be discussed later.

As with other joining processes, the edges of the material to be joined should be free from rust and scale. When welding all but the thinnest plate and sheet metal, the joint edges are ground to an angle to assist weld penetration.

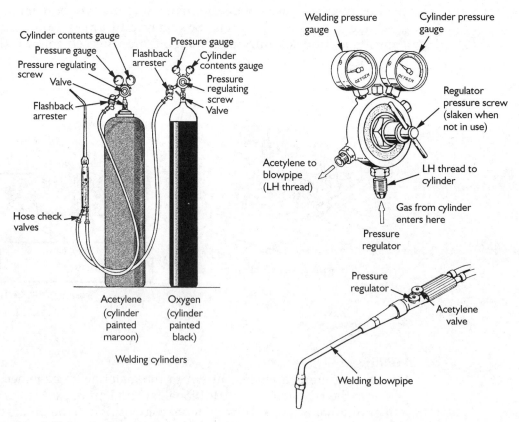

Figure 2.41 Oxy-acetylene welding equipment

Engineering processes for electromechanical products

Joint	Type of edge preparation	Weld symbol BS 499
 (flanged edge joint diagram)	Flanged edge (up to 3 mm plate thickness)	 (symbol)
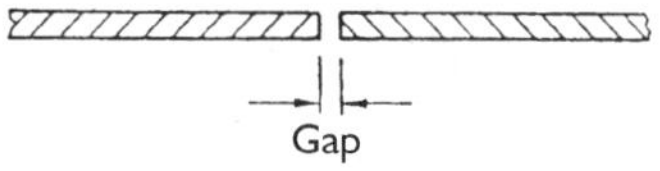 (square butt joint diagram with Gap)	Square butt (up to 3 mm plate thickness)	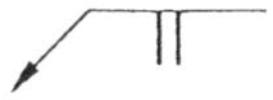 (symbol)
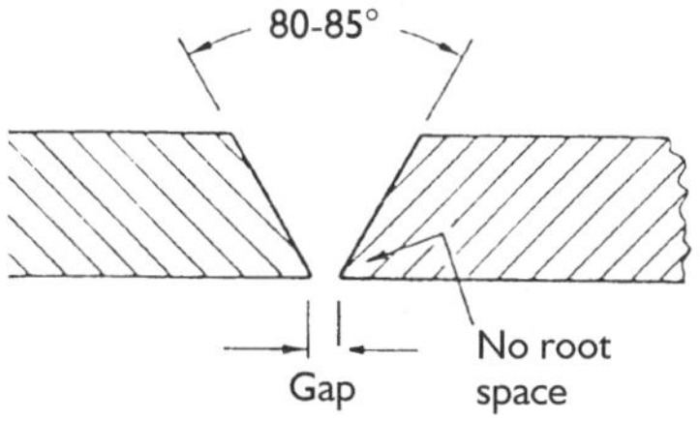 (single vee joint diagram, 80–85°, Gap, No root space)	Single vee oxy-acetylene (above 3 mm plate thickness)	 (symbol)

Figure 2.42 Edge preparation for oxy-acetylene welding

The makers of oxy-acetylene welding equipment supply charts that show the blowpipe-nozzle sizes and the acetylene and oxygen welding pressures that should be used for welding different thicknesses of material. Having set the line pressures separately, the acetylene control valve is opened and the gas ignited using a spark lighter. The yellow flame is adjusted until it ceases producing smoke and the oxygen control valve is then slowly opened. The colour of the flame now changes to blue with two lighter coloured inner cones. For welding mild steel, the oxygen supply is adjusted until the outer cone just disappears to leave a sharply defined inner cone. This gives what is known as a neutral flame in which acetylene and oxygen are being supplied in equal volumes.

Manual welding is a highly skilled operation, and a great deal of practice is required to produce welds of a reliable quality and tidy appearance. The recommended techniques for welding with oxy-acetylene equipment are shown in Figure 2.44. Leftward welding is used for plate thicknesses up to around 6 mm and the rightward method for greater thicknesses. Judgement of the correct speed and movement of the blowpipe and filler rod is critical. If it is too fast, the weld penetration will be incomplete, and if it is too slow, the flame may burn through the metal.

Engineering materials and processes

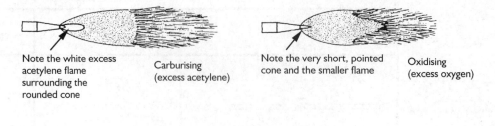

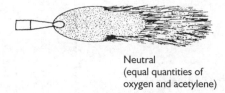

Figure 2.43 Types of welding flame

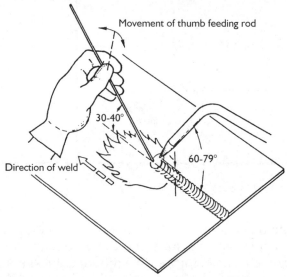

Figure 2.44 Leftward welding technique

Manual metal arc welding With this method, an electric arc is struck between the filler rod, which acts as an electrode, and the joint material. This supplies the heat for welding and the temperature generated in the arc can approach 6000°C. At this temperature, oxidation can affect weld quality and, to overcome this, the filler rod is coated with a flux. This burns to give a gas that shields the weld against the oxygen in the atmosphere.

Arc-welding machines may be powered from the AC mains supply or from a portable generator. They contain a transformer, which gives an output of about 100 V before the arc is struck, falling to about 25 V during welding. The operator is able to control the current for welding and the equipment suppliers often provide information on suitable current settings and recommended types of filler rod for different applications.

Ancillary equipment includes an insulated electrode holder and its lead cable, and a return lead cable and clamp, which is attached to the workpiece or the metal table on which it rests. A chipping hammer is required for removing solidified flux from the finished weld. An approved safety visor, through which

Engineering processes for electromechanical products

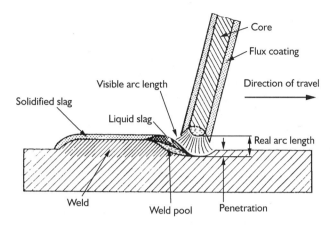

Figure 2.45 The electric arc

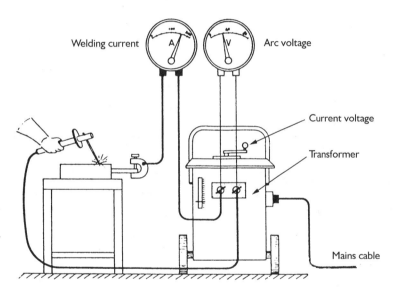

Figure 2.46 Circuit for arc welding

the workpiece must be viewed, is essential, as are other items of safety wear that will be discussed later.

After cleaning and preparing the joint edges, the components are assembled in position on a metal worktable. The return cable clamp is fixed to the table or directly to the workpiece and a filler rod of the correct size is inserted in the electrode holder. Having adjusted the current control to a setting suitable for the size of weld and filler rod, welding can commence. The recommended angle at which the filler rod electrode should be held is shown in Figure 2.47, and for thick plate the weld is built up in layers.

The view of the weld through the safety visor is more restricted than for oxyacetylene welding. Manual metal arc welding is thus a highly skilled occupation. Striking and maintaining the arc and moving it at the correct speed takes a great deal of practice before welds of industrial quality can be made.

Engineering materials and processes

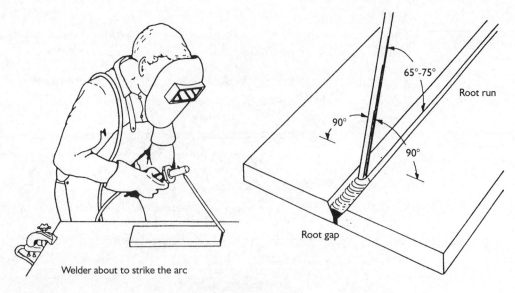

Figure 2.47 Manual metal arc welding technique

Joint	Type of edge preparation	Weld symbol BS 499
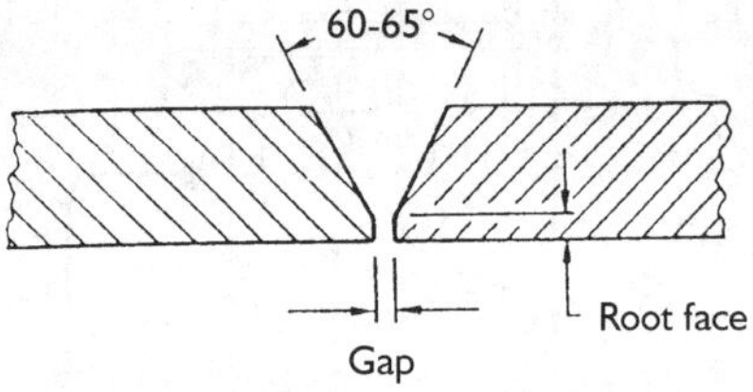 60–65°, Gap, Root face	Single vee manual metal arc	
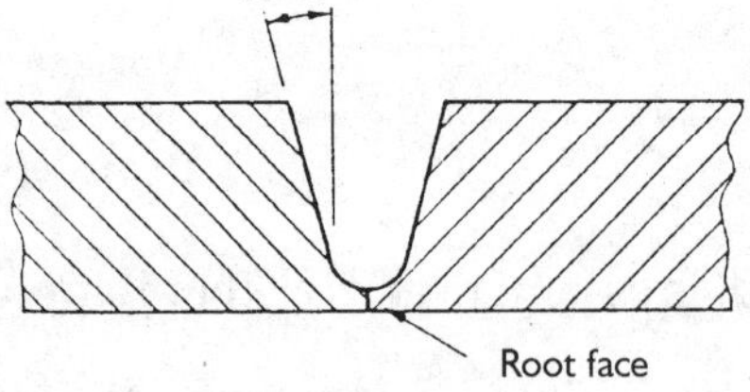 10–15°, Root face	Single 'U'	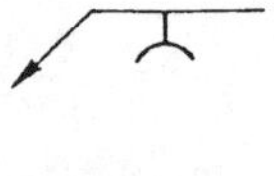
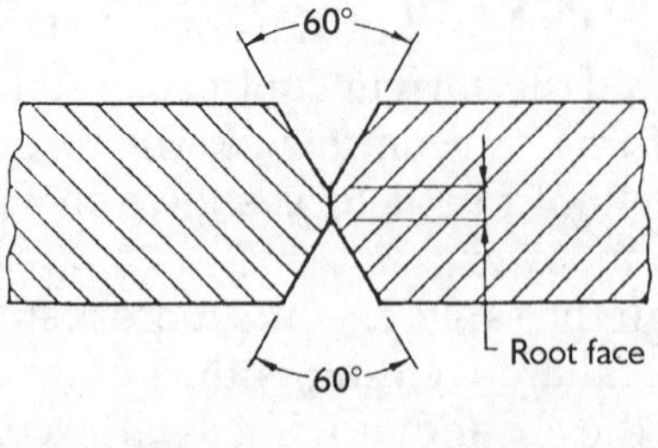 60°, 60°, Root face	Double vee	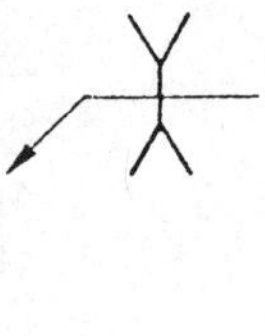

Figure 2.48 Edge preparation for manual metal arc welding

Engineering processes for electromechanical products

Adhesive bonding

Adhesion

The ability of an adhesive to form bonds with the joint materials; in contrast, cohesion is a measure of the strength of an adhesive itself

The use of animal and vegetable glues and pastes dates back many centuries. Crafts such as furniture making and book binding have always made use of them. Their use has increased in recent years as a result of the development of new synthetic adhesives by the plastics and rubber industries. Adhesive bonding is now used for many applications that previously used screwed fasteners, rivets, soldering and welding.

When choosing an adhesive, consideration should be given to the loading conditions and service environment that it will encounter. Where a mechanical fixing is to be replaced by adhesive bonding, the joint may need to be redesigned. Adhesives are strong when loaded in tension and shear but tend to be weak when subjected to peeling and cleavage forces.

All surfaces to be joined by adhesives must be completely free of oil, grease and loose oxides such as rust. They should be finally cleaned with a detergent or

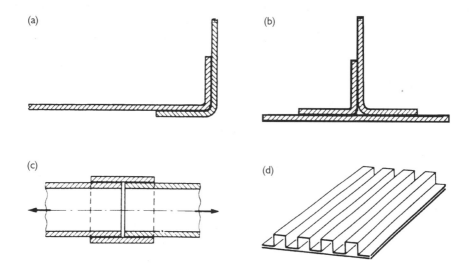

Figure 2.49 Typical adhesive joints: (a) bonded corner joint; (b) bonded flange joint; (c) pipe joint; (d) corrugated sheet bonded to a flat sheet

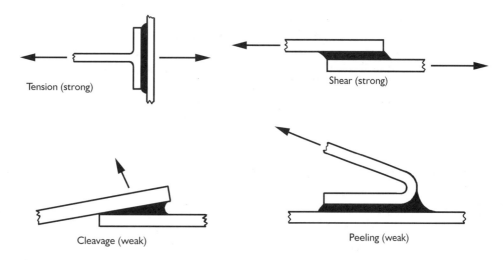

Figure 2.50 Loading of bonded joints

77

a chemical solvent and may benefit from being roughened to provide a 'key' for the adhesive. The manufacturer's instructions for applying the adhesive should be followed carefully, particularly health and safety instructions. The safety procedures and equipment associated with adhesive bonding will be discussed later.

The most common adhesives can be categorised as:
- thermoplastic resins
- thermosetting resins
- elastomers
- cyanoacrylate adhesives.

Thermoplastic resins These may be supplied in liquid or solid form. The solid form must be melted by heating it before use. It is then applied to the joint surfaces, which are held together under pressure until the adhesive has cooled and set. The liquid type is applied in the same way but sets as a result of the evaporation of a solvent.

Solvent-based adhesives should not be used on joints that have a large surface area, unless the materials are porous, as the solvent must be able to evaporate. As a general rule, thermoplastic adhesives should not be used where the joint will be subjected to heat as this may cause softening and lead to failure.

Thermoplastic resins, of which Bostik is a familiar brand name, are used to bond glass, Perspex and wood. Their bond strength is less than that of thermosetting adhesives but they are less brittle and will allow a joint to flex slightly.

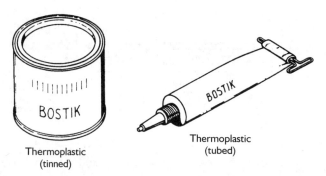

Figure 2.51 Commercial thermoplastic adhesives

Thermosetting resins These are usually two-part adhesives, consisting of the resin and a chemical hardener. These are mixed together in specified amounts and applied to the joint faces. The joint is then assembled and held together while the adhesive hardens. Hardening is caused by a chemical reaction taking place in the adhesive. This generates heat, which causes cross-links to form between the polymers as the resin solidifies.

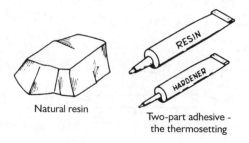

Figure 2.52 Commercial thermosetting plastic adhesive

Engineering processes for electromechanical products

Thermosetting resins give a strong hard bond that is unaffected by temperature change and is resistant to most chemicals. They are used to join metals, glass, ceramics and plastics. Two-pack Araldite adhesive is a familiar brand name.

Elastomers These have a synthetic rubber base and contain a solvent. They are called contact or impact adhesives. The adhesive is applied to both joint surfaces and left for a short time until it becomes 'tacky'. The surfaces are then pressed together to exclude any air and a bond is immediately formed. It is important that they are correctly positioned as the joint cannot be adjusted once contact has been made.

Elastomers are used to join rubber plastics and laminates and, because the solvent evaporates before the joint is made, they are particularly suited to joints with large surface areas. Evostik is a popular brand name.

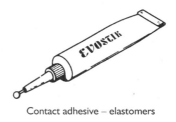

Contact adhesive – elastomers

Figure 2.53 Commercial contact adhesive

Cyanoacrylate adhesives These are more commonly known as 'super glues'. They are supplied in thin liquid form and as a gel, and cure in the presence of moisture. A thin coating is applied to one of the joint surfaces only. The surfaces are then brought into contact and light pressure is applied. A bond is formed almost immediately and, as with impact adhesives, it is important that the joint surfaces are correctly positioned when contact is made.

Cyanoacrylate adhesives will join wood, metals, ceramics, and most plastics and rubbers. They are used extensively in the assembly of small electrical and electronic components.

Their use requires particular care as they will readily form a bond with human skin that is difficult to separate.

Progress check

1. What are the constituents of silver solder?
2. What are the temperatures required for (a) silver soldering and (b) brazing?
3. Why are the components for hard soldering surrounded by fire bricks?
4. The pressure regulators of gas welding equipment have two pressure gauges. What pressures do they indicate?
5. How is a neutral flame obtained with oxy-acetylene welding equipment?
6. Why is the filler rod coated with a flux for manual metal arc welding?
7. Why do surfaces that are to be joined with adhesives often benefit from being roughened?
8. Why should solvent-based adhesives not be used for joints with large, non-porous surfaces?
9. What do the two-part packs of thermosetting resin adhesive consist of?
10. What precautions should be taken when using cyanoacrylate (superglue) adhesives?

Engineering materials and processes

Heat treatment

Heat treatment

A process by which the properties of a material can be changed or modified by heating and cooling techniques

Temperature rise affects materials in different ways. Metals oxidise on the surface and sometimes undergo a change in structure before finally melting. Thermoplastics become softer and melt before eventually starting to burn. Ceramics and thermosetting plastics are less affected by temperature rise, but eventually they too will char and degrade if subjected to high enough temperatures.

By heating and cooling metals and alloys in different ways, their properties can be changed. Some heat-treatment processes are designed to make the material more malleable and ductile, whereas others seek to increase its hardness or toughness. The temperature to which the material is raised and the rate at which it is cooled down can have a critical effect on its final properties. The most common heat-treatment processes are:

- annealing
- normalising
- quench hardening
- tempering
- case hardening.

Annealing

When metals are cold worked by pressing, rolling, extruding or drawing, the crystals or 'grains' of which they are composed become distorted. This makes the material harder and it is said to have become work hardened. Sometimes this is a good thing, particularly in products such as pressings, which can be formed in a single cold-working operation. If it is necessary to deform the material further, however, it may need to be softened to avoid breaking it. The process used to restore the malleability and ductility to a material is annealing.

Annealing involves heating the material in a furnace to what is known as its recrystallisation temperature, and holding it there for a period of time. At this temperature, new, undeformed crystals or grains start to form from the points where the old ones are most distorted. These grow until the old structure has been completely replaced by new, undeformed grains. If the period of 'soaking' is prolonged, the new grains feed off each other and grain growth is said to occur. The bigger the grains, the softer the material will be, and if the process is carried on for too long the material may become weakened.

Materials experts are able to specify the time for which a material should be held in the annealing furnace and the way in which it should be cooled down when the new grains are of the correct size. Plain-carbon steel, aluminium, copper and brass can all be softened by annealing. Figure 2.55 shows the tempera-

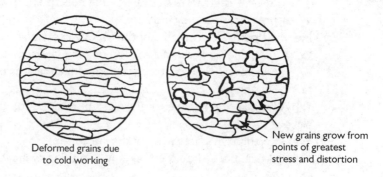

Figure 2.54 Recrystallisation

Engineering processes for electromechanical products

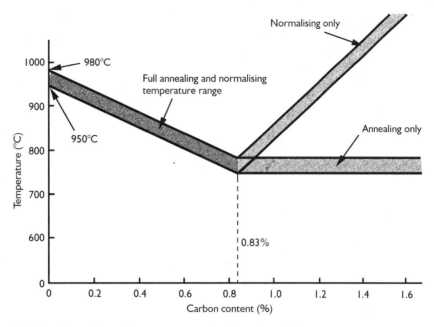

Figure 2.55 Annealing and normalising temperatures for plain-carbon steels

ture band in which recrystallisation occurs for different carbon compositions of plain-carbon steels.

When the steel has been soaked for a sufficient time, the furnace is turned off and the material is allowed to cool down slowly in the 'dying' furnace. A general guide to the recrystallisation temperatures for other materials is given in Table 2.4.

Unlike steels, these materials do not need to be cooled down slowly. They may be quenched after a suitable period of soaking.

Normalising
Steel components that have been formed to shape by hot-working processes, such as forging and hot pressing, often contain internal stresses and grains of unequal size. This is mainly due to uneven cooling. Normalisation is a process, closely related to annealing, which relieves the stresses and refines the grain structure. It involves heating the components to within the normalising temperature band shown in Figure 2.55. After a suitable soaking period, during which recrystallisation takes place and the stresses are relieved, the components are removed from the furnace and allowed to cool down in still air. Because the cooling rate is faster than for annealing, the normalised grains are smaller and the material is stronger than if annealed.

It will be noted from Figure 2.55 that normalising and annealing are carried out at the same temperatures for steels with a carbon content up to 0.83%. The

Table 2.4 Some recrystallisation temperatures

Material	Recrystallisation temperature (°C)
Pure aluminium	500–550
Cold-working brass	600–650
Copper	650–750

high-carbon steels, whose content is above this value, are normalised at a higher temperature than for annealing.

Quench hardening

Plain-carbon steel with a carbon content above 0.3%, i.e. medium- and high-carbon steels, can be hardened by heating them to within the same temperature band as for annealing and then quenching them in oil or water. When steel is heated to these high temperatures, a structural change takes place in the way that the iron and carbon atoms are positioned and combined together. If the steel is allowed to cool slowly, the structure changes back again but quenching does not allow time for this to happen. As a result, the steel becomes very hard and brittle.

Water quenching gives the fastest cooling rate and maximum hardness. Oil quenching is slower and leaves the steel a little less hard and brittle. High-carbon steels, with a carbon content above about 0.9%, should only be quenched in oil, as water quenching is too violent, and can cause cracking. Mild steels, with a carbon content below 0.3%, have insufficient carbon atoms in their structure to be hardened in this way. They can however be 'case hardened' as described below.

Tempering

Steel components that have been quench hardened are often too hard and brittle for use. Tempering is a process that removes some of the hardness and makes the steel tougher. It involves reheating the components to between 200 and 600°C and then quenching them again in oil or water. The tempering temperature chosen for a particular component depends on its final use.

Special furnaces, in which the temperature can be accurately controlled, are used for tempering large batches of components, but small single items can be tempered in the workshop using a gas blowtorch. Before heating, they are polished and, when placed in the flame, the coloured oxide films that spread along the surface are used as a temperature indicator. The tempering temperature for different components and the corresponding oxide colour are given in Table 2.5.

Table 2.5 Tempering temperatures for various components

Components	Tempering temperature (°C)	Oxide colour
Trimming knives and other edge tools	220	Pale straw
Turning tools	230	Medium straw
Twist drills	240	Dark straw
Screwcutting taps and dies	250	Brown
Press tools	260	Brownish-purple
Cold chisels	280	Purple
Springs	300	Blue

Other components such as crankshafts, gears, transmission shafts, etc., which have to be both hard wearing and tough, are tempered at various temperatures up to 600°C.

Case hardening

As stated above, mild steel cannot be quench hardened. Case hardening is a method of increasing the surface hardness of the steel while leaving a soft and tough core. This is an ideal combination of properties for components that must be both wear and impact resistant. If it is not necessary to case harden the complete component, it is first copper plated. The surfaces to be case hardened are then machined and it is only these that are affected by the process.

The first part of the process is carburising. This involves soaking the components in a carbon-bearing material at the temperature used for annealing. Over a period of time the carbon soaks into the surface layer of steel and increases its carbon content. The traditional method is known as 'pack hardening' in which the components are packed in iron boxes with a carbon-bearing powder. This often contains ground up charcoal, bones and leather scrap, and can be purchased with different trade names. After soaking and removal from the carbon, the components are reheated to refine the grain size and the outer case is hardened and tempered.

Activity 2.3

Take a sample of mild-steel bar and a similar sample of high- or medium-carbon steel bar. (About 10 mm diameter and 30 mm long is ideal.) Normalise each specimen by heating to a bright cherry red and cooling in still air, and then place each one in a vice. Using the same force on the hammer, make a centre punch mark on each specimen and then give each one a few strokes with a file. Note the depth of the centre punch marks and how it feels to file each specimen. Heat the specimens to a bright cherry red and quench in water. Once again, test each one with the file and centre punch. Note any differences in the size of the centre punch marks and how it feels to file them.

Reheat the mild-steel specimen to a cherry red and place it in case-hardening compound. Allow it to soak up the carbon as it cools and, if time permits, repeat the process. Finally, reheat the mild steel and quench it in water. Test once more with a file and centre punch and note the effects.

Progress check

1. What happens to a metal that has been work hardened when it is heated to its recrystallisation temperature?
2. How are steel components cooled down in (a) the annealing process and (b) the normalisation process?
3. What are the grades of plain-carbon steel that can be quench hardened?
4. Why should high-carbon steels be quench hardened using oil rather than water?
5. How can mild steel be given a hard outer case while retaining a soft, tough core?

Engineering materials and processes

Chemical treatment

Materials are given chemical and electrochemical treatment to prepare them for other processes, to remove material, to add material and as a finishing process. Treatments include:
- chemical cleaning
- pickling
- etching
- electroplating.

Chemical cleaning

Material and components are often coated with oil or grease from their forming processes. Sometimes they are also sprayed with a film of oil or wax to protect them while being stored. It may be possible to remove the coating by washing before use in white spirits or paraffin. In some cases, this may not leave the surfaces clean and dry enough and a more effective process is needed.

Special degreasing equipment that uses more powerful solvents such as trichlorethylene is widely used in industry. Very often, however, the solvents give off toxic fumes and may also be flammable. Such equipment should be installed in a well-ventilated area and used only by operators who are properly trained and protected.

Pickling

Metals that have been hot formed are often covered with a layer of oxide. In particular, steel that has been forged or hot rolled is covered with a black flaky coat of iron oxide, known as millscale. Pickling is used to remove this layer. It is a chemical process in which the components are immersed in sulphuric acid, phosphoric acid or a mixture of the two. Phosphoric acid has an advantage when used with steel. In addition to removing the scale, it combines with the metal to form a protective surface layer of iron phosphate.

Etching

Chemical etching is used both for surface preparation and as a material removal process. When used for surface preparation, the etching solution removes only a small amount of surface material, mainly from around the grain boundaries. In this way it cleans the surface and roughens it slightly, ready for painting or plating with another material. Aluminium components in particular need to be etched in this way before painting.

Etching is used to remove larger amounts of material in the production of printed circuit boards and in chemical milling. It enables thin sheets or thin layers of material to be formed into very complex shapes. The areas that are to be left untouched are 'masked' off with a material not attacked by the etchant and applied by a photographic process.

Different etching solutions are used for different materials. Ferric chloride is commonly used in PCB production when the material removed is copper. The chemical substances used in etching processes can be hazardous to health. They can cause irritation to the skin and eyes and give off toxic fumes. Their use requires the provision of good ventilation, safety wear and training. These will be discussed later.

Electroplating

This is an 'electrochemical' process commonly used for plating components with copper, cadmium, zinc, nickel and silver. These are mainly intended to provide

Electrolyte

A solution that is able to conduct electricity; the chemical changes that occur as a result are called electrolysis

corrosion protection but can also improve the wear resistance and appearance of a product.

Electroplating requires a direct current supply and a suitable chemical solution known as an electrolyte. The component to be plated is connected to the negative terminal of the supply and is known as the cathode. A slab or bar of the plating metal is connected to the positive terminal and is known as the anode. They are then immersed in the electrolyte and the current is switched on. Over a period of time metal from the anode dissolves, passes through the electrolyte and is deposited on the cathode. The surface layer is evenly distributed and has a uniform thickness that can be accurately controlled.

The process for chromium plating is a little more complex. With steel, the material is first plated with nickel and then copper to fill any small pinholes in the nickel. A further layer of nickel is then applied, followed by a thin coating of chromium. The chromium does not come from the anode but from the electrolyte itself, which contains chromium in chemical form.

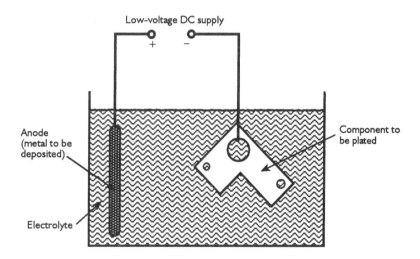

Figure 2.56 Electroplating process

Case study

Very thin metal components that contain a lot of fine detail, such as small holes and engraved patterns, can be made by a process called 'electroforming'. Thin nickel and silver components are sometimes produced in this way. An example is the thin metal foil used on electric shavers. The process is very similar to electroplating except that the metal is plated on to a special cathode known as a 'mandrel'.

The surface of the mandrel contains all the details required on the finished component. They are made from stainless steel, copper, brass or nickel, depending on the metal that is being plated. It is important that the finished product can be easily separated, or 'lifted off' the mandrel when a sufficient thickness of metal has been deposited. The main disadvantage of the process, as with electroplating, is the length of time that it takes.

Surface finishing

Finishing processes are chosen to meet the specified dimensional tolerances, service conditions and aesthetic requirements of a product. For finish machined surfaces, the tolerance, surface roughness and the machining process to be used are often specified on engineering drawings. The service conditions and aesthetic requirements might require a product to be finished by painting or plating with another material. Some of the more commonly used finishing processes are:
- grinding
- polishing
- painting
- plating
- phosphating
- oxidising.

Grinding

This is a material removal process in which components can be accurately machine finished to close dimensional tolerances. The two main grinding processes are surface grinding and cylindrical grinding. In each case the material is removed by a rotating abrasive grinding wheel, as shown in Figure 2.57.

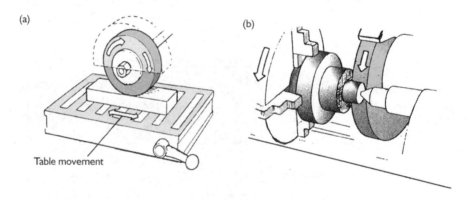

Figure 2.57 Grinding processes: (a) surface grinding; (b) cylindrical grinding

Steel workpieces are held on a magnetic worktable for surface grinding. The process is rather like horizontal milling, but with the grinding wheel replacing the milling cutter. Cylindrical grinding can be used to finish external and internal cylindrical surfaces. The process is rather like turning, but with the grinding wheel replacing the cutting tool or boring bar.

Polishing

Polishing and the related processes, lapping and honing, use an abrasive substance in the form of a paste or a liquid containing abrasive particles. Surfaces that are to be finished by these processes should already be quite smooth as they only remove a small amount of material.

With polishing, the abrasive substance is loaded on to a pad, which is also known as a buff. This rotates on the spindle of the polisher, which may be of a hand-held or stationary-machine type. With the stationary polishing machine, components are held against the rotating buff. With the portable hand-held type, the polisher is held against and moved over the workpiece. It can be used to polish the surface metal or to add shine and lustre to surface coatings such as cellulose paint, which is used on motor vehicles.

Painting

Paint is made up of fine solid particles suspended in a liquid. The particles are called pigments, and it is these which give the paint its colour. When paint is applied to a surface, the liquid dries and sets owing to the evaporation of a solvent or by chemical action. Engineered products may be painted purely for decoration, but in the main it is to protect them against corrosion and heavy use. They are also sometimes painted to identify them by means of a colour code.

All traces of dirt, grease, rust and scale must be removed before painting. This can be done using one or other of the chemical processes that have already been described, such as degreasing, pickling and etching. Alternatively, a mechanical process such as wire brushing or sand blasting may be used. Some of the more commonly used paints are:
- oil paints and varnishes
- lacquers
- stoving paints and enamels
- catalytically drying paints.

Oil paints and varnishes These have their pigments suspended in natural oils such as linseed oil, which also contains a solvent such as turpentine or its substitutes. They generally dry slowly as the solvent evaporates. Oil paints and varnishes have been greatly improved in recent years and may contain resins, synthetic rubber, bitumen and asphalt to give added protection and strength in hostile service environments. They are used both indoors and outdoors for the general protection of wood and metal.

Lacquers These include enamels and cellulose paints whose chemical solvents make them much quicker drying than oil-based paints. They may be applied by brush but are more commonly applied by a spray gun. This is much quicker and gives a more even distribution of the paint. Lacquers give a thin, hard-wearing finish which, in the case of cellulose paints, can be further enhanced by polishing with a fine abrasive.

Lacquers are widely used on furniture and motor vehicle bodies when a high-quality hard-wearing finish is required. Shellac varnish, which is a clear, quick-drying lacquer, belongs to this category of paint. It is used by the manufacturers of electric motors and transformers to insulate the coils and core laminations.

Stoving paints and enamels Stoving paints and enamels consist of pigments and thermosetting resins mixed with other materials called plasticisers. These are natural oils or synthetic materials that help the paint to adhere to a surface and also make it less brittle when set. When heat is applied from an infrared source, cross-links are formed between the polymers in the resin and the paint hardens and dries in minutes. A conveyer system is often used to transport mass production components through the stages of paint spraying and drying.

Stoved paints and enamels give a very smooth and hard-wearing finish that is resistant to temperature and solvents. They are used for domestic equipment such as washing machines, cookers and refrigerators, and also for some motor vehicle bodies and components.

Catalytically drying paints Like stoving paints, these are thermosetting resin-based and also contain a chemical hardener and a solvent that keeps the paint in liquid form while in its container. When the paint is applied to a surface, the solvent evaporates and a chemical reaction takes place between the resin and the hardener. This causes the resin to harden quickly as cross-links are formed be-

Engineering materials and processes

tween its polymers. The paint film is hard and tough and gives better protection to wood and metal than oil-based paint.

Plating

In addition to electroplating, which has already been described, sheet steel and steel components can be plated with another metal by:

- hot dipping
- flame spraying
- powder bonding.

Hot dipping Both tin and zinc are plated on steel using this process. Steel strip that is plated with tin is called tinplate. The tin gives corrosion protection and is not attacked by the natural chemicals found in foodstuffs. It is for these reasons that tinplate is widely used for food canning.

When zinc is plated on steel the process is called hot-dip galvanising. Zinc gives sacrificial protection to the steel. That is to say, if the zinc coating is damaged and the steel is exposed, it is the zinc that will corrode in the presence of moisture and not the steel. Because zinc corrodes at a very slow rate, the protection will usually last for the working life of a steel component. Tinplate does not behave in this way and even a thin scratch through the tin coating will allow corrosion to start.

Tinplating proceeds by passing a continuous steel strip through a flux and then through a bath of molten tin. The tin forms an amalgam with the steel surface and the strip then passes through rollers that give the coating a uniform thickness. Hot-dip galvanising is carried out on both steel sheet and finished components. The process begins with degreasing and fluxing, after which the steel is dipped in a bath of molten zinc and quenched in water.

Flame spraying In flame spraying or 'hot spraying', the coating metal is fed continuously into a blowpipe, where it is melted and emerges as a molten spray. Aluminium, cadmium and zinc can be flame sprayed in this way. As with hot dipping, the receiving surface must be properly cleaned and fluxed. The advantage of the process is that a smooth, uniform coating of metal can be applied to components with irregular shapes, rather like paint spraying.

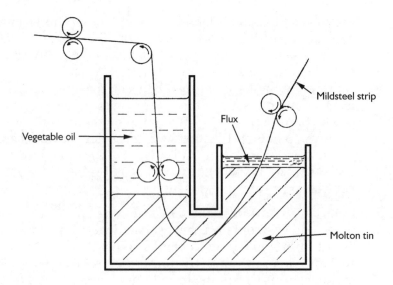

Figure 2.58 Tinplate production

Powder bonding By this method, zinc can be plated on steel components by a process known as Sherardising, and aluminium can be plated by a process known as Calorising. The components are first cleaned by sand blasting. They are then placed in a revolving cylinder that contains zinc and zinc oxide dust or aluminium and aluminium oxide dust. As the temperature of the components is increased, a thin film of zinc or aluminium forms on the surface.

Phosphating

This is a chemical finishing process in which steel components are immersed in a bath containing acid phosphates of zinc or manganese. A protective phosphate film forms on the metal surface, which, when washed and dried, gives a good base for paint. Alternatively, the components can be washed, dried and finally dipped in hot oil to seal the coating. This process is known as Parkerizing.

Oxidising

When steel components with a clean polished surface are heated at around 300°C, a blue oxide film is formed on the surface. This takes on a dark-blue colour as it thickens and, after a sufficient period of time, the component is oil dipped to seal the film. The process, which is known as bluing, gives good corrosion protection and a decorative appearance. High-quality sporting firearms have traditionally been given a 'blued' finish.

Progress check

1. Name a solvent that is widely used in industrial degreasing equipment.
2. Why do engineering components sometimes need to pass through a pickling process?
3. Which chemical process is used in the production of printed circuit boards?
4. Describe how the equipment used for electroplating is arranged and connected to its power supply.
5. What are the two basic types of grinding process that are used to produce a smooth surface finish?
6. What forms of abrasive substance are used in polishing processes?
7. What are the main constituents of oil-based paints?
8. How are stoving paints and enamels hardened?
9. What are the main constituents of catalytically drying paints?
10. Describe how a steel component can be given a 'blued' finish.

Safety procedures and equipment for selected processes

Notifiable accidents

Accidents that, because of their severity, must be reported to the inspectorate by employers are called 'notifiable' accidents

The law relating to the welfare of people at work is contained in The Health and Safety at Work etc. Act of 1974. In addition to this, there are many other acts of parliament, such as the Offices, Shops and Railway Premises Act, parts of which are concerned with health and safety. The Health and Safety at Work Act provided for the creation of the Health and Safety Commission, whose duties are to oversee the implementation and functioning of the Act. The Commission, which is made up of members appointed from industry, the trade unions, local authorities and the general public, also oversees the working of the Health and Safety Executive.

The Health and Safety Executive is the body that is empowered to carry out workplace inspections. Its function is to ensure that all the laws and regulations are being complied with and to investigate the causes of accidents. The inspectors can prosecute employers and employees who have seriously contravened the law and safety regulations.

Where there is a fault in the workplace that can cause serious injury, the inspectors can issue a prohibition notice. This closes the workplace until the fault is corrected. For lesser faults, the inspectors can issue an improvement notice. This allows work to continue but specifies that the fault must be rectified within a given period of time.

The Health and Safety Act lays down the employer's and the employee's responsibilities. The employer must provide a safe and healthy working environment and appropriate safety equipment. Plant and equipment must be properly maintained and guarded, and all process regulations must be followed. Installations such as lifting gear, electrical equipment and pressure vessels must be regularly inspected and up-to-date inspection and maintenance records kept. The employer is also required to use safe working procedures and keep a record of all accidents and dangerous incidents that occur in the workplace.

Certain types of accident that require more than first aid treatment or cause an employee to be absent from work for more than a given number of days must be reported to the inspectors. The inspectors have a right to see the register at any time and investigate the cause of any accident or incident.

The employer is also required under the Act to provide supervision, information, instruction and training, particularly for new or young and inexperienced employees. Employees must be made fully aware of hazardous substances and processes and be trained in their use. They must also know how to act in an emergency situation and be properly trained in the use of emergency equipment.

Under the Health and Safety at Work Act, the employee can also be prosecuted for breaking the law and failing to observe safety regulations. It is the legal responsibility of the employee to cooperate with the employer in fulfilling the requirements of the Act. Employees must follow safe working procedures and make full use of the safety wear and safety equipment provided by the employer. They must not endanger other workers or members of the public and can be prosecuted for interfering with or misusing health and safety equipment. They thus have a responsibility to take reasonable care of their own health and act in a responsible manner towards others.

The health and safety procedures relevant to the engineering workplace may be divided into:
- general health and safety procedures
- use of health and safety equipment
- specific process safety procedures.

General health and safety procedures

General health and safety procedures, in which all employees should be instructed, and which are relevant to all processing, assembly and storage areas are:
- personal hygiene
- personal conduct
- reporting accidents
- responding to an alarm.

Personal hygiene

Personal cleanliness contributes to the health of an employee. In the engineering workplace it is good practice to rub a barrier cream on the hands and forearms before starting work. Barrier creams prevent dirt from entering the pores and becoming ingrained. They also have antiseptic properties that protect the skin from infection and, being water soluble, they are easily washed off. Finger nails should be kept short and any cuts or abrasions to the skin should be covered with a suitable dressing.

Chemical solvents should not be used to clean the hands as they may cause skin irritation and may be toxic. Hands should always be washed before meal breaks and after using the toilet. Barrier cream should then be rubbed on again before restarting work. Overalls should be removed after work and regularly washed. They should not be reworn if they become soaked in oil or glazed with grease as this can penetrate the clothing and cause skin irritation. It can also be a fire risk.

Personal conduct

Employees should at all times behave responsibly. Practical jokes and horseplay can cause accidents and damage. They have no place in an industrial environment. Compressed air lines, pressure hoses, power tools and chemical substances can be lethal. Equipment and materials should never be used for anything other than their intended purpose.

Food and drink should not be consumed at the workstation and employees should obey 'no smoking' instructions. On no account should employees be under the influence of alcohol or illicit drugs while in the workplace.

Reporting accidents

All accidents and dangerous incidents should be reported to the area supervisor. In cases of personal injury, first-aid treatment should be sought and given. The nature of the accident, however minor, must be recorded, together with the treatment given.

Responding to an alarm

Whenever an emergency situation, such as an outbreak of fire or the escape of toxic substances, is discovered, the alarm signal should be triggered. Employees should be made aware of the location of alarm points and the way in which the alarm is sounded. It generally requires the breaking of a glass cover with a striker to release a spring-loaded button. This triggers the alarm, which is usually in the form of a siren.

On hearing the alarm, all process equipment should be turned off and isolated. The building should then be evacuated in an orderly manner without running or panic. The evacuation route should be marked with green arrows in passageways and on stairs. Employees should be made familiar with this, and the outside assembly point, during regular evacuation drills. A roll call should be taken at the assembly point to ensure that all personnel are present.

Activity 2.4

Draw a floor plan of your workshop or classroom block that contains the following items of information:
1. the emergency evacuation routes
2. the positions of emergency alarms
3. the positions of emergency stop buttons

Engineering materials and processes

    4.   the positions of electrical isolation switches.

Also, write down the location of your assembly point outside the building and the name of the person responsible for carrying out a roll call to make sure that everyone is present.

Use of health and safety equipment

Under the Health and Safety at Work Act, an employer is required to make the workplace as safe as is reasonably possible. The various items of safety equipment needed to achieve this depend on the type or work being carried out and its associated hazards. Health and safety equipment can be categorised into:
- protective equipment
- safety wear
- first-aid equipment
- emergency equipment.

Protective equipment

Employers must provide guards and barriers to protect employees from the moving parts of machinery, hot surfaces and hazardous substances. It is an offence to sell or use machinery that is not adequately guarded. All gear trains, belt drives, chain drives and transmission shafts must be fitted with guards. Guards must also be fitted to the exposed moving parts of material-removal machinery. The workpiece and cutter guards fitted to lathes, drills and milling machines will be discussed later.

    It is an offence to remove or tamper with machine guards, and if they are found to be faulty or damaged the matter should be reported to the supervisor. Operators and technicians should check to see that the guards are functioning correctly before starting work and then use them in the proper manner. When guards need to be removed for maintenance work on a machine, the machine should be stopped and isolated from its power supply. To ensure that the machine cannot be started accidentally, the key should be removed from the electrical isolating switch. If no key is provided, the fuses should be removed.

    Guard rails are often positioned around furnaces, chemical processes and automated equipment such as robots. They are coloured with diagonal black and yellow lines to warn of a potential hazard. Portable barriers that are colour coded in the same way are used to prevent employees from entering areas where maintenance work is in progress or some other temporary hazard is present. As with machine guards, it is an offence to remove or tamper with guard rails and safety barriers.

Safety wear

Safety wear is designed to give protection when working with hazardous materials and processes. Many larger employers provide all necessary items free of charge. Others may require employees to provide items such as overalls and safety footwear themselves or provide them at reduced cost. It is the employee's responsibility to dress in a manner appropriate to the workplace and to make full use of the safety wear provided. This might include:
- protective clothing
- protective footwear
- hand and arm protection
- head and ear protection
- eye protection

Engineering processes for electromechanical products

- face masks and respirators.

Protective clothing The most appropriate item of outer clothing for general engineering work is a one-piece boiler suit. This should be close fitting but not too tight. The pockets should not be loaded down unnecessarily with tools or components, although it is reasonable to carry a small steel rule, notebook and pen or pencil. The cuffs should be fastened or safely turned back and the front should be buttoned or zipped. There must be no loose flapping edges that can become caught up in machinery.

Long hair is potentially as dangerous as loose, flapping clothing. It should be held back in a band or, better still, contained under a cap. When working away from process areas such as in an inspection department or stores, or when working in a seated position, it may be appropriate to wear a good-fitting warehouse coat. Certain processes require specialist safety clothing, particularly where there are sparks or splashes from chemicals, and these will be described later.

Figure 2.59 Suitable and unsuitable dress

Protective footwear Protective footwear is available in a variety of boot and shoe styles. It may be provided free of charge or at a reduced cost through the firm. Safety boots and shoes are made with metal-reinforced, non-slip soles and metal toecaps. The soles protect the feet when treading on sharp objects such as metal swarf or up-turned nails. The toecaps protect the feet when heavy objects are being handled such as lathe chucks and heavy components for machining or assembly. Rubber 'wellington' boots can also be obtained with metal reinforced soles and toecaps.

Engineering materials and processes

Figure 2.60 Sectioned safety boot

Hand and arm protection A wide variety of protective gloves and gauntlets are available to protect the hands and forearms. Some are designed to give protection from sparks or when handling hot objects. Others are designed to protect the hands from cuts when handling sharp-edged materials such as sheet metal and glass. Others still are designed to give protection from corrosive substances and irritant oils, greases and paints. Safety gloves and gauntlets are mostly made from tough leather, rubber and plastic materials.

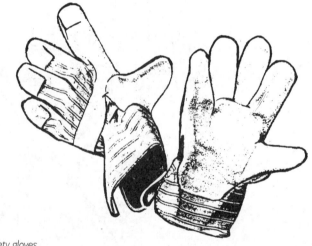

Figure 2.61 Safety gloves

Head and ear protection This is most important when working in an area where overhead cranes are in use or where overhead maintenance or construction work is in progress. In these cases a safety helmet should be worn, and very often entry to such areas is forbidden without one. Safety helmets are made from lightweight plastic materials that are non-flammable and have a high resistance to impact. They also protect the head when working in confined spaces and against electric shock.

In years gone by, deafness caused by working in a noisy environment was looked upon as an unfortunate occupational hazard and no compensation was given or expected. Nowadays, it is realised that ear protection is necessary for noise levels above a certain intensity. Headsets are provided with lightweight plastic ear protectors containing a sound-absorbing material.

Engineering processes for electromechanical products

Figure 2.62 Safety helmet

Eye protection A wide range of safety goggles, spectacles and visors have been designed to suit different processes. Eye protection is essential for material-removal processes such as turning, drilling, milling, grinding and polishing. For each of these, clear safety goggles or safety spectacles with side shields should be worn. The lenses are made from a clear, toughened, plastic material that will not shatter and is not easily penetrated by flying objects.

Full face protection is required for some operations such as when quenching hot metal, working with volatile chemicals and using portable grinding equipment. Here a clear visor may be worn that can be tilted back when not in use. Welding operations require the provision of heavily tinted goggles and visors to protect the eyes not only from sparks but also against the harmful radiation that is emitted from arc-welding processes.

Employees not directly involved in welding operations but who work in the vicinity of arc- and spot-welding processes are advised to wear clear safety spectacles whose lenses are designed to filter out harmful radiation.

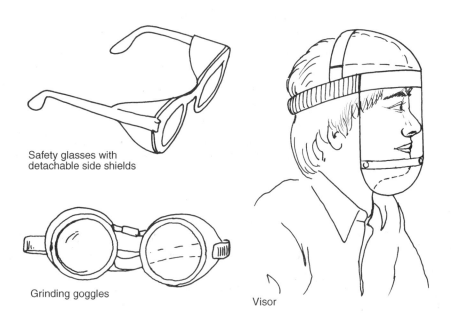

Figure 2.63 Eye and face protection

95

Engineering materials and processes

Face masks and respirators These are designed to protect the lungs from dust and fumes. The simplest is made from permeable moulded plastic which can be discarded after use. It is the type which is also worn by doctors and dentists. Another type is the Martindale mask. This consists of a curved metal outer plate that is pierced where it fits across the mouth and also extends up each side of the nose. A throw-away filter fits behind the plate, covering the mouth and nostrils.

Although these are suitable for filtering out dust particles, more specialised respirators may be required where there are fumes. These have cartridge-type filters and may also incorporate goggles to give full-face protection.

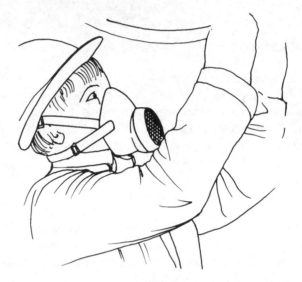

Figure 2.64 Use of a respirator

Case study

Bux v Slough Metals Ltd (1974)
In this case, a metal worker had worked for many years without wearing eye protection. The company manager then supplied goggles but after a while the worker went back to his old practice of working without them. One day he was splashed with molten metal and as a result he lost his sight.

The Court of Appeal decided that, although the goggles would not have prevented the worker from becoming splashed, they would have reduced the seriousness of his injuries. The worker was said to have contributed to his injuries by being negligent and, as a result, the compensation payment that he received was reduced. The Court also found that the employers had been negligent because, although they had provided goggles, they had failed to instruct and persuade the worker how important it was to wear them.

Progress check

1. Which act of parliament relates to the welfare of people at work?
2. What kind of training is an employer legally required to provide?
3. Why is it advisable to apply a barrier cream to the hands and forearms before starting work?
4. What are the legal responsibilities of an employee under health and safety legislation?
5. How should an employee respond to an alarm signal?

> 6. What is the procedure that must be followed after an accident or dangerous incident has occurred?
> 7. For what kind of offences can an employee be prosecuted under health and safety legislation?
> 8. What are the most appropriate items of outer clothing for the engineering workshop?
> 9. Why are employees who work in the vicinity of arc- and spot-welding processes advised to wear eye protection?
> 10. How do safety boots protect the feet?

First-aid equipment

The number of first-aid stations and the number of employees trained to give first aid varies with the size of the company. Large engineering firms have well-equipped medical centres supervised by trained nurses. They provide first aid and also monitor the health of the employees by periodically carrying out medical examinations.

Small engineering firms are not as well equipped, but as an absolute minimum they should have a trained first aider and a first-aid station equipped with dressings and solutions for treating cuts and burns. They should also have slings, splints, neck braces and a stretcher for dealing with more serious accidents. In areas where toxic and caustic chemicals are used there should be a first-aid post equipped with neutralising solutions and a trained person who knows how to use them. Affected parts of the body, particularly the eyes, can then be irrigated quickly before calling for outside medical help.

Emergency equipment

An emergency can arise because of fire or the escape of toxic fumes and high-energy radiation. It can also arise as a result of natural causes such as flooding and storm damage. In such cases, the workplace would be evacuated, but emergency teams might then have to enter the area to rectify or assess the situation and possibly to rescue stranded personnel. Emergency equipment in the form of respirators, protective clothing and monitoring equipment might then be required.

Outbreak of fire is the emergency most likely to arise in engineering workshops and offices. Employers have a legal requirement to carry out periodic evacuation drills and also to provide appropriate fire-fighting equipment. If a fire is seen to be well alight and spreading, the best course of action is to sound the alarm, evacuate the workplace and call the emergency services. If, however, the fire is small, consisting perhaps of smouldering material caused by overheated equipment, it may be judged possible to deal with it using a suitable fire extinguisher. Employees should be trained to recognise the different types of fire extinguisher and know where they are stationed and how to operate them.

It is well known that a fire needs three things to continue burning – a fuel, an oxygen supply and a sustained high temperature. If any one of these is removed, the fire will die. Fire extinguishers aim to achieve this, the most common types being:
- fire blankets
- hose reels and pressurised water extinguishers
- foam extinguishers
- carbon dioxide extinguishers
- vaporising liquid extinguishers

Engineering materials and processes

- dry-powder extinguishers.

Fire blankets These put out the fire by smothering it, i.e. by cutting off the oxygen supply. They are made of fire-retardant synthetic fabrics and are very useful for wrapping around a person whose clothes are on fire. They are usually kept rolled up in a cylindrical container and can be released quickly by pulling on a tape.

Hose reel and pressurised water extinguishers Water lowers the temperature of a fire and, when it has been turned into steam, it has the effect of cutting off the oxygen supply. Hose reels and water extinguishers are coloured red. They should be used only for fires in which the fuel is a solid material such as wood, paper or cloth. Water should not be used on burning liquids as this tends to spread the flames and it should not be used where there is live electrical equipment.

Foam extinguishers These contain chemicals that produce a jet of foam. They operate by cutting off the oxygen supply and are coloured cream. Foam extinguishers are the type most suitable for use on burning liquids such as petrol, oil and chemical solvents. They can also be used effectively on solid materials.

Carbon dioxide extinguishers These produce a jet of carbon dioxide gas, which cuts off the oxygen supply and does not leave a residue to be cleaned up after the fire has been extinguished. Carbon dioxide extinguishers should be used for extinguishing fires in electrical equipment where there may be live conductors. They may also be used effectively with burning liquids and solid materials.

Vaporising liquid extinguishers These produce a jet of vapour that cuts off the oxygen supply in the same way as the carbon dioxide extinguisher. They are coloured green and used for fires in electrical equipment, burning liquids and solids. The vapour is, however, toxic and they are best used outdoors.

Figure 2.65 Fire extinguisher

Engineering processes for electromechanical products

Dry-powder extinguishers These cut off the oxygen supply to a fire in two ways. First, the powder has a smothering effect and, when heated, it produces carbon dioxide gas, which also cuts off the oxygen supply. They are coloured blue and are useful for fighting flammable liquid fires, particularly in kitchens and food stores. The powder is non-toxic and it can be swept up or removed with a suction hose afterwards.

Activity 2.5

Find out and make a note of the following items of safety information.
1. Where are the fire points in your workshop or classroom block?
2. What types of fire fighting appliance are they equipped with and how are these operated?
3. What does the emergency alarm sound like?
4. On hearing the alarm, what action should you take before leaving your workstation or study area?

Specific process safety procedures

The general safety procedures and items of safety equipment used in engineering have been discussed above. In addition, there are certain other procedures and items that are specific to the different production processes. The processes are:
- material removal
- joining and assembly
- heat treatment
- chemical treatment
- surface finishing.

Some reference has already been made to their safety aspects, and these can now be discussed in more detail.

Material removal

The material-removal processes that have been described are turning, drilling and milling. Before using any of these process machines it is essential that operators fully understand the controls. They should be able to identify the isolator switch and the motor start and stop switches. The isolator may be in the form of a key switch or a lever switch and the stop and start switches are usually of the push-button type, coloured green and red respectively. In the case of sensitive bench drills and very small centre lathes, where there is a direct drive from the motor to the spindle, these switches also start and stop the spindle. It is thus important to know their position so that the machine can be stopped quickly in an emergency. Larger machines are fitted with a clutch mechanism, operated by a lever. Again, it is important to know exactly where the lever is positioned so that the machine can be stopped quickly.

In addition to being able to stop an individual machine in an emergency, it is important to know the position of the emergency cut-out switch for the work area. This is in case one sees a fellow worker in difficulties who might be some distance away. Emergency switches are usually of the push-button type and are situated at strategic positions around the work area.

Mention has already been made of the chucks that are used on both lathes and drilling machines. Chuck keys should not be left in the chucks. Injury to the

Engineering materials and processes

operator or damage to the machine can result from starting the spindle with the chuck key still in position. Chuck keys and other hand tools should not be allowed to accumulate and clutter up the machine or its work table. They should be kept in the machine tool cabinet and returned there after use.

The danger of wearing loosely hanging clothes and long hair when operating machines cannot be stressed too greatly. Some of the most horrific industrial accidents have resulted from operators becoming caught up in machines. The importance of wearing protective spectacles and goggles has also been mentioned. This applies especially where very small chips are thrown out by the cutting process.

Lathes and drills are fitted with different designs of chuck guard and milling machines are fitted with cutter guards. It is essential that the guards are properly

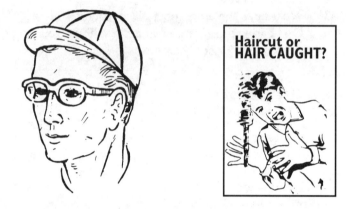

Figure 2.66 Protective glasses and cap

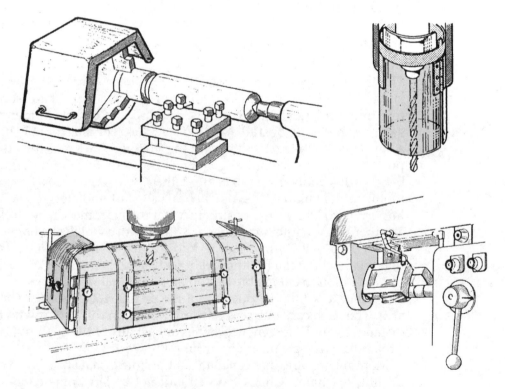

Figure 2.67 Machine guards

Engineering processes for electromechanical products

used and in position before starting the machine and that they are left in position until the spindle is again stationary.

Care should be taken when handling swarf. This can be both hot and sharp and should only be handled with protective gloves. Furthermore, swarf should not be allowed to accumulate on a machine and it should be removed only when the machine is stopped. It is also advisable to wear protective gloves when handling milling cutters and large twist drills.

Joining and assembly

Joining and assembly operations can be hazardous where lifting gear such as cranes, hoists or fork-lifts are required to raise and position heavy components. Here it is important not to exceed the maximum safe working load (SWL) for which the equipment is designed. It is a requirement that this should be prominently displayed on the lifting device. Chain, wire and rope slings should also have their SWL displayed on a tag.

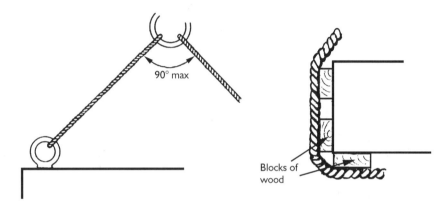

Figure 2.68 Safe use of lifting slings

When the included angle between the two sides of a sling is 120°, the load in each side is equal to the load being lifted. At greater angles the load in the sling is greater than the weight being lifted and may exceed the safe working load. As a safe working practice, it is recommended that the angle should not be more than 90°.

When heavy objects with sharp corners are being lifted, the sling should be protected by placing pieces of wood around them. Damaged slings should not be used and reported immediately to a supervisor. Slings may be attached to the load by positive means such as hooks or shackles that locate in eye-bolts. Alternatively, they may pass around the object and rely on friction for their lifting grip. This is potentially more dangerous and should only be done by employees who have received training in safe methods of handling.

Safe working practices and the correct use of safety equipment are also essential when joining materials by soft soldering, hard soldering and welding. Heat is involved in each of these processes and heat-resistant gloves should be made available for handling the hot materials. Welding requires the provision of additional items of safety wear because of the hot sparks and splashes of molten metal that are produced. These include leather gaiters to protect the legs and feet, a leather apron or jacket to protect the front of the body and gauntlets to protect the arms.

Depending on the type of welding, the above items need not all be worn, but it is advisable to wear gloves and a protective apron for all welding operations.

Engineering materials and processes

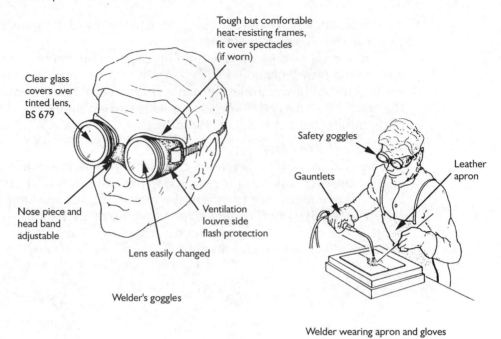

Figure 2.69 *Safety equipment for oxy-acetylene welding*

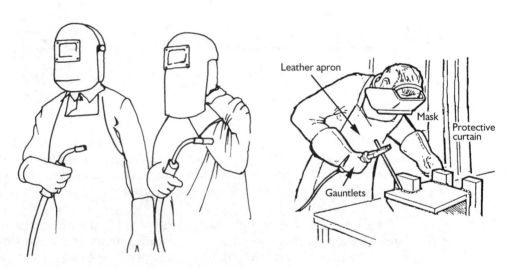

Figure 2.70 *Safety equipment for arc welding*

Heavy arc-welding processes, particularly overhead welding, require full-body protection. Here, all of these items are worn together with a combined helmet and safety visor to protect the head. Fumes are produced during arc welding and indoor welding areas should be equipped with fume-extraction equipment. A respirator might be needed when welding in confined spaces with poor ventilation.

Soft soldering, hard soldering and welding require the provision of eye protection. The acidified zinc chloride, used as a flux when soft soldering, can spit when heated, and clear spectacles or goggles should be worn for this process. They should also be worn when hard soldering to protect the eyes from sparks.

Engineering processes for electromechanical products

As has been stated, welding requires the wearing of tinted goggles and visors. These protect the eyes not only from heat and sparks but also from the harmful radiation that is produced when arc welding. Arc-welding operations should be carried out in a screened booth or behind a portable screen to protect fellow workers from the intense flash and radiation. Where this is not possible, workers in the vicinity should be warned to turn away as the arc is struck. In areas where unscreened arc welding is an on-going process, protective spectacles that filter out harmful radiation should be provided for all the employees.

Heat treatment

Particular care is required in heat-treatment areas when furnaces are being loaded or unloaded. This sometimes requires the expert use of lifting and positioning equipment in the presence of intense heat. Heat-treatment personnel engaged in these operations should be fully equipped with heat-resistant clothing and visors. Other employees should be kept clear of the area.

Care is also required during quenching operations where hot oil or hot water and steam are likely to splash or spit out from the quenching tank. Again, protective clothing and eye protection are required. Components that have been normalised are required to cool down in still air. The area reserved for this should be fenced off and have warning signs to indicate the presence of hot material. All heat-treatment areas should be well ventilated.

Chemical processes

The chemical substances used in processes such as pickling, etching, electroplating and phosphating must be properly stored and labelled. In addition to the trade name, the labels should give details of the chemical composition of the substance together with one of the signs shown in Figure 2.71.

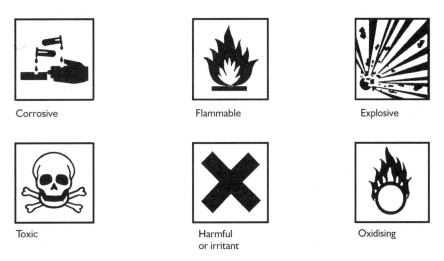

Figure 2.71 Hazardous substances

These indicate at a glance the hazardous nature of a substance. The Care of Substances Hazardous to Health (COSHH) regulations specify the ways in which hazardous substances should be transported and stored.

As has already been stated, first-aid posts in chemical process areas should be equipped with the neutralising solutions for the chemicals that are in use and personnel who know how to apply them should be available. Process operators should be provided with appropriate safety wear and eye protection and there should be good ventilation. Personal cleanliness is most important where corrosive or toxic chemicals are in use. The hands should be thoroughly washed before meal breaks and food should not be consumed in the process area.

Finishing processes

Of the finishing processes described previously, grinding and polishing are essentially material removal processes. The safe working procedures described above for turning, drilling and milling generally apply, particularly in the use of guards and the wearing of eye protection. These processes often produce sparks and dust. Stationary machines such as surface and cylindrical grinders should be fitted with dust extraction equipment. When using portable grinders and polishers, a face mask with a throw-away filter should be worn

Grinding wheels are supplied to suit different materials. As a general rule, a soft grinding wheel is used with hard materials and a hard wheel is used with soft materials. Grinding should always be carried out using the periphery of the wheel. Using the side face can cause the wheel to shatter. The remounting and truing of grinding wheels is a skilled operation and should only be carried out by a qualified person.

Paint spraying by hand requires the provision of a face mask, eye protection and protective overalls. These often incorporate a hood to keep the paint mist out of the hair. Cotton gloves are also worn to protect the hands from the paint and solvent. The process should always be carried out in a well-ventilated area.

Hot-dipping, hot spraying, powder bonding and oxidising are finishing processes that make use of molten metal or the application of heat. They have much in common with heat treatment and welding processes and share the same general safety procedures and items of safety wear.

More detailed information on the items of safety equipment and safety procedures described in this chapter can be obtained from British Standard Specifications BS 3456, BS 4163, BS 5304, BS 5378 and PD 7304.

Activity 2.6

Find out and make a note of the following information.
1. Who is the safety officer at your school, college or place of work and where can he/she be contacted?
2. Who is responsible for giving first-aid treatment at your school, college or place of work and where do you report to for first-aid treatment?
3. What is the name, address and telephone number of your Health and Safety Inspector?

Progress check

1. With what essential items should the first-aid post in an area where there are toxic and caustic chemicals be equipped?
2. What is the course of action which should be followed on finding a fire which is well alight?
3. What are the three things which a fire needs to continue burning?
4. To what kind of fires should hose reels and pressurised water extinguishers be confined?
5. What kind of fire extinguisher should be used with fires in electrical equipment and what is its colour code?
6. What precautions should be taken when removing swarf from a machine?

7. What is the maximum recommended angle between the two sides of a lifting sling?
8. Why is it advisable to wear eye protection when soft and hard soldering?
9. What kind of protection is required during heat-treatment quenching operations?
10. What is the name of the regulations which govern the handling, storage and transportation of hazardous chemical substances?

Selecting processes and techniques for making a given electromechanical product

It should now be possible to make a reasoned selection of the processes and specific techniques for the manufacture of given electromechanical products. This selection will be governed by the following criteria:
- the properties of the raw materials
- the final material properties required
- the desired quality
- the dimensional tolerances
- the likely quantity required
- the cost.

It should also now be possible to identify the hazards associated with a range of production processes, together with the appropriate safety procedures that should be followed and the safety equipment that should be provided.

Assignment 2
Selecting processes and techniques for making a given electromechanical product

This assignment provides evidence for:
Element 1.2: Select processes needed to make electromechanical engineered products
and the following key skills:
Communication 2.1: Take part in discussions
Communication 2.2: Produce written material
Communication 2.3: Use images
Communication 2.4: Read and respond to written materials

The following 10 products are made using one or more of the processes that have been described.

Engineering materials and processes

1. A centre punch made from medium carbon-steel bar.

2. A V-block made from mild steel.

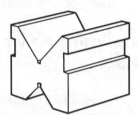

3. An extension lead made up from components and cable that are bought complete.

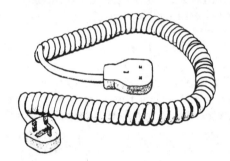

4. A partitioned tray for holding different sizes of fasteners and made from tinplate.

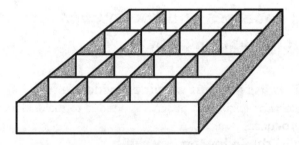

5. A helical spring made from a high-carbon steel rod.

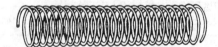

Engineering processes for electromechanical products

6. A tailstock centre for a lathe made from mild-steel bar.

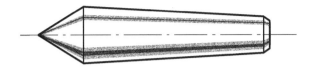

7. A printed circuit board for an electronic product.

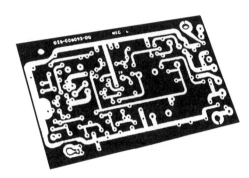

8. Mild-steel hexagonal nuts and bolts that have been produced but now require corrosion protection.

9. A mild-steel rubbish bin that will receive heavy outdoor use and also requires corrosion protection.

Engineering materials and processes

10. A bicycle frame that has been fabricated and requires an aesthetic and protective finish.

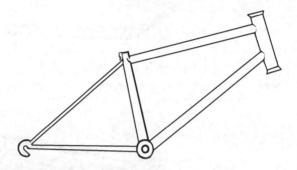

Your tasks

Carry out the following tasks and write a report.

1. Identify the likely processes used to produce the given products considering the above selection criteria.

2. Identify the specific techniques appropriate to each process.

3. Identify any safety procedures and equipment associated with these processes.

Your report should contain general notes explaining why certain processes and techniques were selected whereas other possible alternatives were not.

Chapter 3: Producing electromechanical products to specification

> **This chapter covers:**
> Element 1.3: Produce an electromechanical engineered product to specification.
> **...and is divided into the following sections:**
> - Selecting the sequence of processes and process techniques to make a product to specification
> - Maintaining tools, equipment and the working area in good order
> - Selecting equipment, tools, materials and components to make a product to specification.

Selecting the sequence of processes and process techniques to make a product to specification

The design and production planning of a new product is often carried out with the design and production engineers working in close cooperation. As a design develops, the production engineers might suggest modifications that will make it easier and cheaper to make. When the design is finalised, the production engineers must make a detailed production plan that contains all the information required to make the product. Production planning involves:
- selecting processes to meet the product specifications
- selecting the sequence of processes
- selecting the sequence of process techniques.

Selecting processes to meet a product specification

When selecting processes, production engineers must take careful note of the product specifications. These are to be found on the engineering drawings and memoranda issued by the design team. They will include:
- material specifications
- dimensions and tolerances
- surface finish and texture
- required test readings
- production quantities.

Material specifications
Engineering drawings are issued for each component part that is to be manufactured. The drawings specify the material that is to be used, and may also specify its composition in the form of a British Standards specification number, e.g.

Mild steel BS 970: 040A10 or

Engineering materials and processes

Brass BS 2780/: CZ106.

The drawings also specify the type of heat treatment that might be required, e.g.

Normalise at 900°C or

Oil quench from 850°C and temper at 300°C.

Dimensions and tolerances

Tolerance

The acceptable range of a dimension between an upper and lower limit

The dimensions on component drawings are usually given from one of the surfaces of the components. This is known as the 'datum surface'. On rough castings and forgings, this surface is the first to be machined and all other measurements are then taken from it. Working in this way, it is possible to avoid 'multiple' or 'cumulative' error, which can occur if measurements are made progressively from point to point. All of the dimensions may not be displayed on a drawing, but it should be possible to calculate them from the major ones that are given.

A dimension that is given with a 'tolerance' states the allowable limits within which it must be. The method of stating the tolerance can differ slightly. Two typical examples are shown in Figure 3.1.

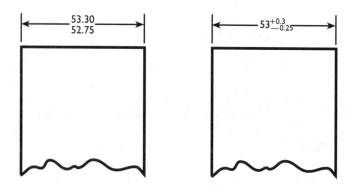

Figure 3.1 Tolerances

Dimensions that are not given with a tolerance are often called 'open dimensions'. The title block on the drawing usually gives guidance as to the degree of accuracy that they require, e.g. 'All dimensions to be ± 0.25 mm unless otherwise stated'.

The closeness of the tolerance sometimes determines the choice of machining process, although with modern computer-aided turning and milling systems it is possible to achieve tolerances that could previously be obtained only by grinding. When milling, turning and grinding on conventional small to medium-sized machines, the limits shown in Table 3.1 should be within the capability of skilled operators.

Surface finish and texture

The engineering drawing often specifies the finishing process that must be carried out on a component or product. In the case of painting, the type of paint and the number of coats would be specified. In the case of plating with another material, the material and the thickness of the plate would be given. Surfaces that are to be machine finished, plated or chemically treated to a particular degree of

Producing electromechanical products to specification

Table 3.1　Possible tolerances

Milling parallel surfaces	± 0.1 mm
Turning cylindrical surfaces	± 0.03 mm
Surface and cylindrical grinding	± 0.008 mm

surface roughness are indicated on drawings with the symbol shown in Figure 3.2.

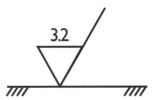

Figure 3.2 Surface finish symbol

The number on the symbol indicates the required degree of surface roughness. Its units are micrometres (μm), which are thousandths of a millimetre and also called microns. Special-purpose inspection equipment is required to measure these small surface variations. The symbol may also specify the finishing process that is to be used and the limits of surface roughness, as in Figure 3.3.

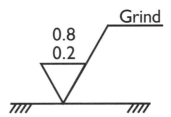

Figure 3.3 Surface finish specifications

Even when the finishing process is not specified, the required surface roughness value often dictates the process that must be used, e.g. for a machined component a surface roughness between 0.2 μm and 0.8 μm, as specified in Figure 3.3, would be difficult to achieve by a process other than grinding. Components that have to be processed all over to the same surface finish need only have the information given once in the title block of the drawing. An example is given in Figure 3.4.

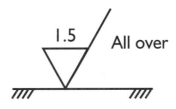

Figure 3.4 Finish all-over specification

Test readings

The specifications for a component or product might require tests to be carried out at key stages of processing. These often involve the use of specialist equipment to check material quality, the quality of welded joints or the functioning of electrical and electronic circuits, etc.

The required test readings or results may be indicated on engineering and circuit drawings or listed in memoranda from the design engineers. A typical example is the required surface hardness of a material following heat treatment. This might be given on a drawing as follows:

Surface hardness VPN 850

where VPN stands for Vickers pyramid hardness number. The specification indicates that testing of all or a sample number of components is required, using a Vickers pyramid hardness tester to confirm their quality. The specified quality indicators can affect the choice of process and the choice of process technique. Quality testing and checking must be included in the production plan together with any specialised equipment that is needed.

Production quantities

For any new product, the production engineers need to know the quantity required and, in the case of an on-going product, the likely rate at which it will be required. These determine the scale of production, which can range from single-item or 'jobbing' production, through batch production to continuous production.

Small-scale production requires flexible production systems that can be adapted quickly to make different products. They often consist of general-purpose machines and equipment, but may also include computer-controlled machining centres that can be quickly reprogrammed.

If the requirements are sufficient to support continuous or large repetitive batch production, it may warrant the setting up of a dedicated production system. These systems are often made up of specialist automated machinery and equipment that is confined to producing one product or a range of similar products.

Selecting the sequence of processes

Having decided on the processes needed to make a product, they must be arranged to take place in a logical sequence. It is useful to display them on a block diagram or flow chart, which is sometimes called a process layout as shown in Figure 3.5.

As can be seen, the flow chart can be made to show not only the sequence of processes and techniques, but also how the 'made-in' and bought-out components must be routed to meet for subassembly, final assembly and finishing.

The process layout flow chart might also show the estimated time required for each process. This information is useful to the purchasing staff and the production schedulers. It is their responsibility to make sure that the material and components are available when required and that production is scheduled to meet agreed delivery dates. Further information on block diagrams and flow charts can be obtained from British Standard specification BS 4048.

Selecting the sequence of process techniques

Having selected a sequence of processes, the next step is to produce a detailed processing plan. In some firms these are also called process specifications. They

Producing electromechanical products to specification

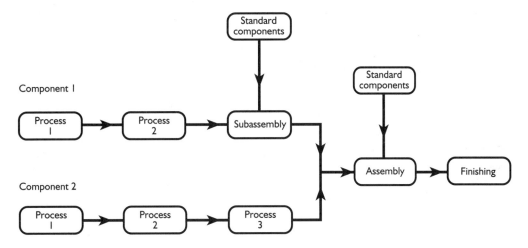

Figure 3.5 Process flow chart

are drawn up on a document known as a process planning or process specification sheet. Each manufactured component has its own process plan, and separate ones are drawn up for the assembly and finishing stages.

The format varies from firm to firm, but the plans usually contain the sequence of processing operations, the tools and equipment required, quality specifications and estimated operation times. The process plans might also make reference to safe working procedures that are to be followed and the safety equipment that is to be used. A typical format is shown in Figure 3.6.

Part no.	Description			Material	
Operation number	Description of operation	Machine/ equipment	Tools/ guages	Quality indicator	Estimated time
10					
20					
30					
40					

Figure 3.6 Typical process specification sheet

It is usual to number the process operations 10, 20, 30, etc. so that if additional operations are necessary they can be inserted as 15 or 25, etc. The 'Description' column should state exactly what is required such as 'Drill 8 mm hole' or 'Heat to 850°C and quench harden'. The 'Machine/equipment' column should contain a brief description of items, such as 'Pillar drill, Centre lathe, Plating bath', etc.

The 'Tools/gauges' column should list items such as special jigs and fixtures. These are devices designed to hold or position a component during processing. Hand tools and the gauges needed for checking critical dimensions should also be listed in this column. The 'Quality indicator' column should contain the readings required from test equipment or the results of a visual inspection. The 'Estimated time' column should give the processing time per piece or, in the case of processes such as heat treatment, the batch processing time.

Engineering materials and processes

In small engineering firms, hard copies of the process plans and engineering drawings might be kept in a master file, but in larger firms they are more likely to be held on a computer database. With either system, the drawings and process plans must be readily available on screen or as hard copies for the process technicians and operators to access.

Case study

A possible sequence of operations for producing the pin shown in Figure 3.7 is given in Figure 3.8.

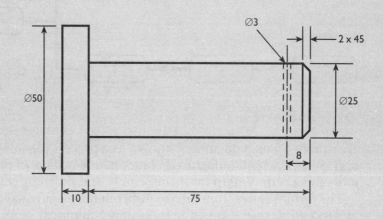

Figure 3.7 Hardened steel pin. Material: 0.5% carbon-steel bar, 50-mm-diameter bright bar; heat treatment, quench harden in oil at 900°C and temper at 300°C; finish, zinc phosphate; quantity, one off.

Part no. 123	Description: Pin			Material: 0.5% carbon steel	
Operation number	Description of operation	Machine/ equipment	Tools/ gauges	Quality indicator	Estimated time
10	Turn 25 mm diameter over 75 mm length	Centre lathe	Micrometer	$\varnothing 25^{+0.05}_{-0.05}$	12 min
20	Turn chamfer	Centre lathe	–	–	2 min
30	Part off to length	Centre lathe	–	–	3 min
40	Drill 3-mm-diameter pin hole	Sensitive drill	3-mm drill	–	5 min
50	Quench harden in oil from 900°C	Furnace, oil bath	–	–	30 min
60	Temper at 300°C and quench in oil	Furnace, oil bath	–	VPN 650	30 min
70	Pickle to remove scale	Acid bath	–	–	15 min
80	Phosphate	Phosphate acid bath	–	–	20 min

Figure 3.8 Sequence of operations for producing a pin

Producing electromechanical products to specification

Maintaining tools, equipment and the working area in good order

Maintenance activities should be part of the workplace routine in order to keep tools, machines and other items of equipment in good, and safe, working order. The work area itself should be maintained in a safe and tidy condition. Poor maintenance can cause accidents and production stoppages. It can also affect the quality of a product. Maintenance activities can be divided into:
- routine maintenance
- scheduled maintenance.

Routine maintenance

This should be an on-going workplace activity. If the work area has been left in an untidy state, it should be cleared of unwanted tools, unused material and components or waste material before starting work. The floor area should be clear of any obstructions, as should adjacent passageways and gangways.

The work area should be kept in a safe and tidy condition throughout the working day. Tools and materials should be returned to their proper place when not in use, waste material placed in the containers provided and any spillages should be cleaned up immediately. The following are general guidelines for the routine maintenance of:
- machine tools
- power tools
- hand tools.

Machine tools
Lathes, drilling machines and milling machines should be given a thorough visual inspection before use. In particular, the guards should be in safe working order, the lighting on the machine should be working and there should be no worn or damaged electrical connections.

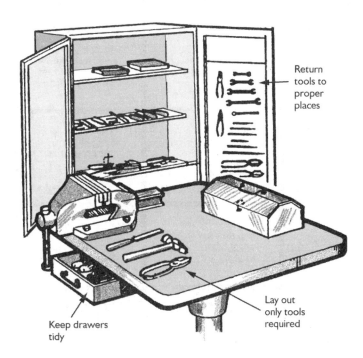

Figure 3.9 A tidy workstation

If a coolant delivery system is provided, it should be in working order and if necessary its reservoir should be topped up. Similarly, the level of lubricating oil in a machine gearbox should be checked and replenished if required. From time to time, lubricating oil should be applied to the machine slideways, leadscrews and bearings. Swarf and tools should not be allowed to accumulate on a machine and any faults that occur during the working day should be reported immediately.

Power tools

As with stationary machine tools, portable power tools such as drills, grinders and polishers should be inspected visually before use. The condition of guards and electrical connections should be checked, together with that of the outer casing and grips. While being held in a safe position, the tool should be switched on and off to check that it is running smoothly and that the automatic cut-off switch is working. Figure 3.10 lists other safety and maintenance hints.

During the working day, power tools should be kept in a clean condition and stored in a dry place when not in use. Any signs of damage or malfunction should be reported immediately.

Safety hints for portable electric tools

Make sure that the equipment is correctly and securely earthed before using it	Examine the casing for damage. If it is damaged don't use the tool	See that cable and plug are sound, and that the plug is properly connected.
The power cable must be long enough to reach the working place without straining it.	Use tools only on the correct power-supply as indicated on makers' label.	Keep cable off the floor to prevent damage and to avoid tripping people up.
Never stand on a damp or wet surface when using equipment and keep the tool clean and dry	Never use worn or damaged bits or other accessories	Connect the tool to a power point– NOT a lighting socket
Disconnect the tool from the power supply when not in use.	Store tools carefully after use, with cables coiled and cutting tools removed from chucks	Tools must be regularly inspected tested and properly maintained

NOTE: Double-insulated and all-insulated tools do not have an earth lead

Figure 3.10 Safety poster

Hand tools

Hand cutting tools such as files, hacksaws, punches and cold chisels should be inspected before use. Damaged or badly fitting handles should be replaced, as should hacksaw blades if they are cracked or have broken teeth.

File teeth often become clogged or 'pinned' with waste material, and they should be cleaned periodically with a wire brush or file card. File teeth are hard and brittle. They are easily damaged and can also damage other tools with which they come into contact. To prevent this, they should be stored separately, preferably on a rack.

The heads and cutting edges of punches and chisels should be checked regularly. The heads may show signs of mushrooming and this should be ground off. The cutting edges should be inspected for sharpness and also reground if required. Chisels and punches that have been allowed to get in a very bad state can usually be reconditioned. First, they are annealed to remove any hardness. They can then be filed or machined to the correct shape and rehardened and tempered. Finally, the cutting edge can be ground to the required sharpness.

As with files, the handles of hand scrapers should be replaced if they are loose or damaged. Hand scrapers must have a very sharp and smooth cutting edge. They need to be sharpened quite often during use but only on an oilstone.

Measuring and marking-out tools require special care. Micrometers, vernier callipers and dial test indicators are precision instruments that should be handled carefully, kept in a clean condition and returned to their protective cases when not in use. Engineers' rules, squares, callipers, scribers, centre punches and dividers should also be treated with care. They should be kept lightly oiled and never used for anything other than their intended purpose. The points of scribers, centre punches and dividers should be touched against a grinding wheel from time to time to keep them sharp.

Scheduled maintenance

Scheduled or 'preventative' maintenance activities are programmed to be carried out at specified intervals in order to prevent, or reduce, the occurrence of production stoppages and accidents. Scheduled maintenance can also lead to improved product quality and reduced operating costs. The activities are planned to be done outside production periods, such as at weekends or during night shifts. They may be the duty of a specialist maintenance team or be carried out by the production workers themselves.

Scheduled maintenance activities include the cleaning, lubricating and adjusting of machinery and replacing worn components. They also include replenishing lubricating oils and carrying out safety checks on guards and other items of emergency equipment. Finally, they might include the running and testing of equipment in readiness for production to restart.

Maintenance plans are drawn up for particular process areas or items of machinery and may be presented on a maintenance planning sheet. As with the process plan, these may be filed as hard copies or stored on a computer database. A typical format is shown in Figure 3.11.

The frequency column states how often an activity should be carried out. It is usual to list the operations in descending order of frequency, i.e. that which is to be done the most often is operation 10. A separate record is usually kept that must be signed and dated as evidence that the maintenance activities have been carried out.

Engineering materials and processes

Process		Location			
Maintenance operation	Description of operation	Equipment	Tools	Materials	Frequency of operation
10					
20					
30					

Figure 3.11 A typical maintenance specification sheet

Progress check

1. What are the maximum and minimum allowable sizes for a component whose dimension is specified as $75^{+0.05}_{-0.05}$?
2. What does the following symbol indicate on an engineering drawing?

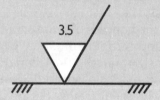

3. What does the abbreviation VPN stand for?
4. What kind of machines would you expect to find in a flexible production system for producing small quantities of different components?
5. Why are the process operations often numbered 10, 20, 30, etc. on a process plan?
6. Describe the maintenance activities that may be necessary in the workplace before starting work.
7. What typical maintenance activities need to be carried out from time to time on machine tools?
8. How should files be maintained in good working order?
9. What special precautions should be taken to maintain measuring and marking-out tools in good condition?
10. What advantages can result from a programme of scheduled maintenance?

Assignment 3
Selecting equipment, tools, materials and components to make a product to specification

This assignment provides evidence for:
Element 1.3 Produce an electromechanical engineered product to specification
and the following key skills:
Communication 2.1: Take part in discussions
Communication 2.2: Produce written materials

Producing electromechanical products to specification

Communication 2.3: Use images
Communication 2.4: Read and respond to written material
Application of number 2.2: Tackle problems
Application of number 2.3: Interpret and present data

You should now be able to select the materials, components, processes and specific process techniques required to produce an engineered product. Having drawn up a production plan, the product can then be realised using the correct procedures and making full use of safety equipment. The following drawings and diagrams are for two engineered products that should be within your capabilities.

Your tasks

To plan and produce these, or some other electromechanical product of your choice, you will need to carry out the following tasks.

1. Select appropriate materials and components for the product.

2. Identify a sequence of processes to produce the product to specification.

3. Select the specific process techniques and their sequence to produce the product to the specification.

4. Select appropriate tools and equipment, including safety equipment, to process the product to specification.

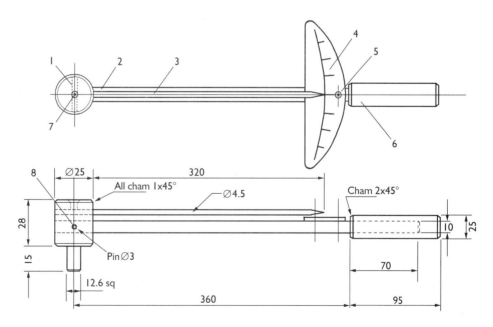

Figure 3.12 Product 1: torque wrench that is required to measure right- and left-hand applied torque up to a value of 100 N m. 1, Square socket drive; 2, torque arm; 3, pointer; 4, torque scale; 5, scale fixing screws (2 off); 6, handle; 7, pointer fixing screw; 8, pin

Engineering materials and processes

5. Carry out the processes in sequence to realise the product making correct use of relevant safety procedures and equipment.

6. Carry out any required maintenance activities on tools, equipment and the working area during and after processing.

As you perform each task, keep a log of your planning and production activities. In the case of the torque wrench, your log might contain details of how you calibrated the device. In the case of the continuity tester, it might contain an evaluation of the device when tested on a range of lamps and resistors.

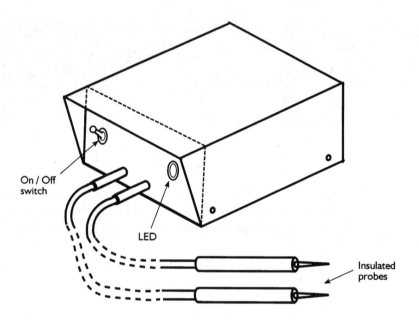

Figure 3.13 Product 2: continuity tester that can be used to test diodes, lamp filaments and parts of a circuit for continuity. The tester will have an audio means of identifying continuity and give a visual indication of battery condition

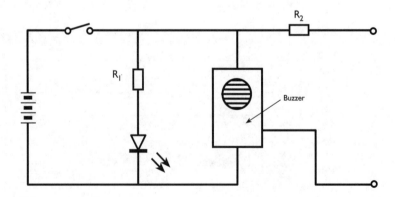

Figure 3.14 Circuit diagram. Circuit components: 9-V battery and battery holder, switch, buzzer, light-emitting diode (LED); resistors to suit above components, terminal sockets, test probes, leads and plugs. Circuit to be assembled on stripboard or suitable alternative that is attached to the enclosure but insulated from it

Producing electromechanical products to specification

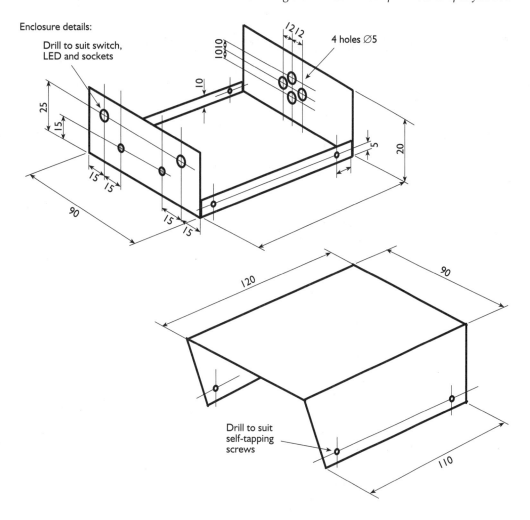

Figure 3.15 Enclosure details (note that the dimensions suggested may be modified to suit particular components, material and production methods). Enclosure parts: upper and lower parts of the enclosure to be made from sheet metal of an appropriate gauge, joined together by self-tapping screws. The material should have, or be given, an aesthetically pleasing surface finish

Sample unit test for Unit 1

The answers are given on page 460.

1. The carbon content of mild steel is between:
   a 0.1% and 0.15%
   b 0.15% and 0.3%
   c 0.3% and 0.8%
   d 0.8% and 1.4%.

2. Grey cast iron is:
   a Strong in compression
   b Difficult to machine
   c Strong in tension
   d Malleable and ductile.

3. The chief ingredients of brass are:
   a Copper and aluminium
   b Tin and zinc
   c Copper and zinc
   d Aluminium and tin.

4. 'Silver' coinage is nowadays made from an alloy whose main ingredients are:
   a Copper and zinc
   b Iron and carbon
   c Tin and lead
   d Copper and nickel.

5. Polychloroethene is the chemical name for:
   a PTFE
   b Perspex
   c PVC
   d Terylene.

6. Polyamides are thermoplastics that include:
   a Perspex
   b Nylon
   c Terylene
   d PTFE.

Sample unit test for Unit 1

7 Boat hulls and vehicle body panels are sometimes made from a composite of glass fibre and:
   a Perspex
   b Formica
   c Epoxy resin
   d Polystyrene.

8 Amorphous ceramics are a group of materials that include the different types of:
   a Glass
   b Abrasive
   c Cement
   d Semiconductor.

9 Ceramic materials are generally:
   a Strong in tension
   b Good conductors of heat
   c Malleable and ductile
   d Good electrical insulators.

10 Toughness is a measure of a material's:
   a Wear resistance
   b Strength in tension
   c Resistance to impact
   d Corrosion resistance.

11 Inspection covers that have to be removed regularly are often secured by:
   a Studs and nuts
   b Pop rivets
   c High-tensile bolts
   d Grub screws.

12 The component that allows electric current to flow through it in one direction only is:
   a A capacitor
   b An inductor
   c A diode
   d A resistor.

13 A 270-kΩ resistor with a tolerance of ± 5% is required. Its colour code will be:
   a Red, violet, yellow, gold
   b Brown, black, yellow, silver
   c Orange, white, red, gold
   d Yellow, red, blue, silver.

123

Engineering materials and processes

14 Machined holes may be finished off to an accurate size using a:
   a Machine reamer
   b Twist drill
   c Radiusing tool
   d End mill.

15 Internal diameters may be machined on the lathe using a:
   a Knife tool
   b Parting tool
   c Roughing tool
   d Boring bar.

16 Wide, flat surfaces may be machined on a horizontal milling machine using a:
   a Slab cutter
   b Side and face cutter
   c Slotting cutter
   d Slitting saw.

17 Soft solder is an alloy of:
   a Lead and tin
   b Tin and copper
   c Copper and zinc
   d Zinc and lead.

18 The term 'superglue' is a commonly used name for:
   a Thermoplastic adhesives
   b Cyanoacrylate adhesives
   c Impact adhesives
   d Thermosetting adhesives.

19 Etching, pickling and anodising are all:
   a Heat-treatment processes
   b Machining processes
   c Chemical-treatment processes
   d Joining processes.

20 In the drilling process, parallel-shank twist drills are are held in:
   a A Jacobs chuck
   b The machine spindle
   c A four-jaw independent chuck
   d A machine vice.

21 Large, awkwardly shaped components may be turned in the centre lathe while held:
   a Between centres
   b In a three-jaw self-centring chuck
   c On a face plate
   d On the cross-slide.

Sample unit test for Unit 1

22 The cutters for horizontal and milling machines are held:
  a On an arbor
  b In a toolpost
  c On the overarm
  d In a Jacobs chuck.

23 Before the component parts of a soldered joint are placed together they should be:
  a Sweated
  b Heated
  c Tinned
  d Quenched.

24 In the normalising process, steel components are cooled down in:
  a The dying furnace
  b Water
  c Still air
  d Oil.

25 Pack hardening is a carburisation process carried out on components made from:
  a Aluminium
  b Mild steel
  c Cast iron
  d Brass.

26 In the hot-dip galvanising process, steel is coated with molten:
  a Tin
  b Aluminium
  c Lead
  d Zinc.

27 It is an offence for an operator to:
  a Remove the machine guards provided
  b Stop and isolate a machine
  c Inspect the condition of the machine guards
  d Carry out routine maintenance on a machine.

28 Fire extinguishers suitable for fighting fires in flammable liquids are painted cream and contain:
  a Water
  b Carbon dioxide
  c Dry powder
  d Foam.

Engineering materials and processes

29 Terms such as BHN 250 or VPN 800 on an engineering drawing of a component specify its:
   a Tensile strength
   b Surface texture
   c Toughness
   d Surface hardness.

30 The tolerances on an engineering drawing specify:
   a The production process to be used
   b The allowable limits of size
   c The quantity required
   d The material composition.

31 It is useful to display the sequence of processes to make a product on:
   a A flow chart
   b A maintenance plan
   c An assembly drawing
   d A product specification.

32 Scheduled maintenance activities are carried out:
   a When time permits
   b Throughout the working day
   c When a fault is reported
   d At specified intervals.

PART TWO: GRAPHICAL COMMUNICATION IN ENGINEERING

Chapter 4: Graphical methods for communicating engineering information
Chapter 5: Scale and schematic drawings in engineering
Chapter 6: Interpreting engineering drawings

Sample unit test for Unit 2
Engineering is fundamentally about ideas: ideas for new products, new systems and new services. But it is one thing to have new ideas and quite another to communicate these ideas to other people. Most engineering work is precise and detailed, so it is very important that we have methods of communication that enable us all to have a common understanding.

This part is all about graphical communication. It will show you the different methods for communicating graphical information and also how to interpret the material produced by other engineers.

Chapter 4: Graphical methods for communicating engineering information

This chapter covers:
Element 2.1: Select graphical methods for communicating engineering information.
... and is divided into the following sections:
- Graphical methods used in engineering
- How graphical methods relate to types of information
- Choosing graphical methods for different purposes
- Selection criteria for graphical methods
- Identification of graphical methods.

Engineering involves the design, manufacture, installation and maintenance of a wide variety of technically advanced products and services. It is a complex activity involving many people, often from different countries and speaking different languages. Precise information and instructions have to be communicated from one person to another. Written or spoken words, even if they are in a language the listener or reader understands, can often be misinterpreted or misunderstood. It also requires many words to convey the same information as can be contained in a picture or a diagram. An ancient Chinese proverb says, 'one picture is worth a thousand words', and there is much truth in this.

Figure 4.1 Say it with a picture!

Graphical communication in engineering

Graphics are, in effect, models that represent the real world. These models can be:

- Iconic – an iconic model is a drawing that looks like the thing it is representing. You are probably familiar with the icons used in computer software to represent different functions such as print (picture of a printer) and save (picture of a disk). The numbers on a digital clock are exactly like real numbers and so are iconic models. Examples of iconic models are drawings of real objects such as cars, engines and computers.
- Analogue – an analogue model tries to represent a quantity or property by using another quantity or property. The hands on a clock move through a distance to represent time and so are an analogue model of the time. Examples of analogue models are circuit diagrams, flow diagrams, contour lines, etc.

Some general everyday examples of the use of graphical methods are:
- photographs in newspapers, magazines and books
- maps showing contours, vegetation, buildings, roads, railways, etc.
- service manuals for cars, motor cycles, washing machines, etc.
- signs, e.g. road signs, signs on doors, safety signs in factories, signs on controls for electronic equipment and in vehicles, pub and shop signs
- advertising and corporate logos, such as the Peugeot 'lion' or Apple Macintosh's 'apple'.

Figure 4.2 shows a sign showing that ear protectors must be worn. Notice how the ear protectors are clearly shown to indicate visually what is meant. It is clearly an iconic model. No words are necessary, although 'ear protectors must be worn' is usually written also.

Activity 4.1

Using the examples given above as a general guide, find and sketch six examples of signs in your immediate environment. Explain the meaning that the sign is trying to convey and how well it conveys that meaning. Say whether the signs are iconic or analogue.

Figure 4.2 A sign showing that ear protectors must be worn

Graphical methods used in engineering

Integrated circuit

A solid state electronic component that is designed to function as a complete circuit on a single 'chip'

Engineering covers an enormous range of activities from building very large items such as bridges, ships and aircraft down to the very small scale of micro-electronics. The sort of engineering information that has to be communicated includes:

- electrical circuits and electrical devices such as motors, generators and transformers
- electronic circuits and electronic devices such as integrated circuits, diodes and resistors
- mechanical items such as engines, gearboxes, vehicles, machine tools and cranes
- fluid power (hydraulic and pneumatic circuits) and devices such as cylinders, motors and valves.

Programmable logic controller (PLC)

A control device that accepts signals in from sensors of values such as temperature, position etc., and outputs signals to control devices such as motors and valves

Graphical methods are used in all the activities that an engineering company is involved in. These activities include:

- research and development
- design
- manufacturing
- assembly
- testing
- costing and estimating
- marketing and publicity.

The various types of graphical methods will be fully covered in this chapter. Methods of producing engineering drawings are explained in Chapter 5. To help in the interpretation of drawings reference should be made to:

- BS 308 Engineering drawing practice
- PP 7308 Abridged version of BS 308 for schools and colleges.

Case study

A company designs, manufactures and markets a range of drilling heads for use on automated production lines. The drill is powered by an electric motor and linear actuation is by pneumatic cylinder. Control is from a programmable logic controller (PLC) via a solid-state switch for the motor and a solenoid valve for the pneumatic cylinder. Figure 4.3 is a freehand schematic drawing of this arrangement.

Sketches

Freehand sketches are the graphical equivalent of shorthand and are most widely used by designers when working on initial concepts. Figure 4.3 from the case study shows the use of a schematic sketch. Figure 4.4 shows a designer's rough sketch of a cement mixer that could be given to a detail designer to refine into a finished design.

In a more sophisticated version, rough sketches can be worked up into fully fledged paintings and artists' impressions for use in design presentations and for publicity brochures, etc. Figure 4.5 shows an artist's impression of a new car for a brochure.

Sketches and artists' impressions are usually done freehand with pencil, crayons or paint. However, they can also be produced using computer software packages such as MS Paint, with the output being from plotter or printer. The best-quality output will be from plotter or laser printer. Ink-jet or bubble-jet printers produce good-quality work, with dot-matrix printers being slower and of lower quality. All output devices can work in colour or black and white. If sizes larger than A4 are required, a plotter will have to be used. Plotters are available from size A4 up to extremely large sizes for producing full-size drawings for the shipbuilding and aircraft industries.

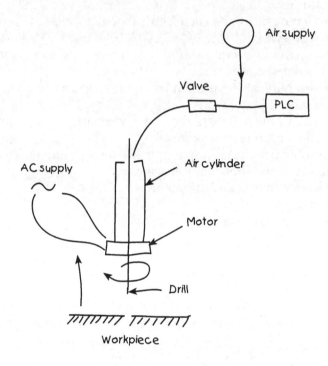

Figure 4.3 Schematic drawing of a drilling system controlled by a PLC (programmable logic controller)

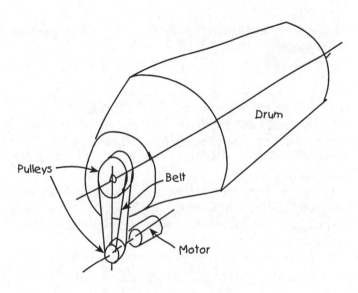

Figure 4.4 Designer's rough sketch of a cement mixer

Graphical methods for communicating in engineering

Figure 4.5 Artist's impression of a new car for a brochure

Activity 4.2

Find five engineering examples of sketches or artists' impressions. Describe the purpose of all your examples – are they for publicity, for design instructions, etc.? Comment on their effectiveness – are they eye catching, highly descriptive, very clear? If you can, find out by which method they were produced and comment on the quality of production.

Detail engineering drawings

All the individual parts in an assembly have to be made, measured and tested. They have to be described exactly so that they can be made in any part of the world. The most effective way of doing this is by representing the solid three-dimensional part on two-dimensional paper or computer screen. The two main

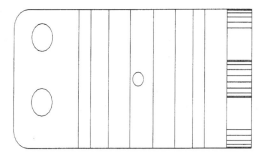
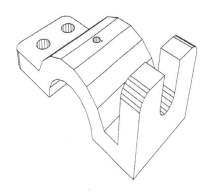
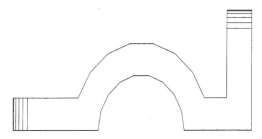

Figure 4.6 A bracket drawn with three orthographic views and one pictorial view (Leonardo Computer Systems, Reading)

133

Graphical communication in engineering

Tolerances

The total specified allowable amounts of variation on dimensions or other values

types of graphical representation are orthographic views and pictorial views. Orthographic views show what an observer would see looking at the elevation (front or back), the end view (sides) and plan (top or bottom) of an object. Pictorial views try to represent how an object appears to the eye when it is positioned so that the three dimensions of length, width and height can be seen simultaneously. Figure 4.6 shows a bracket drawn with three orthographic views and one pictorial view (in this case it is in isometric projection).

Notice how the use of all four views makes it much easier to visualise the bracket.

In the bracket drawing there are no dimensions so the bracket could not be made. The majority of drawings will have dimensions so that the part can be made. Figure 4.7 shows an isometric view of a casing with dimensions. Note that it is more usual to put manufacturing information on orthographic views, as shown in Figure 4.8 in the following case study.

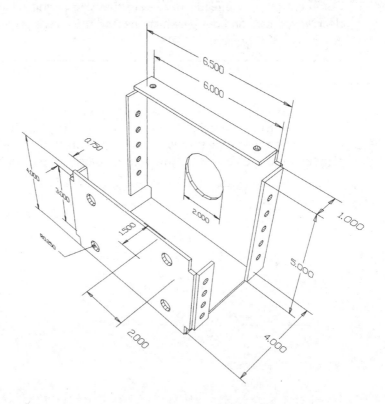

Figure 4.7 *Pictorial view of a casing with dimensions (Leonardo Computer Systems, Reading)*

Case study

The drilling head described in the previous case study will have to have its individual parts designed and manufactured. Both detail and assembly drawings will be required. Figure 4.8 shows the detail drawings for the pneumatic cylinder parts with all necessary information for manufacture, including dimensions, tolerances and materials.

Graphical methods for communicating in engineering

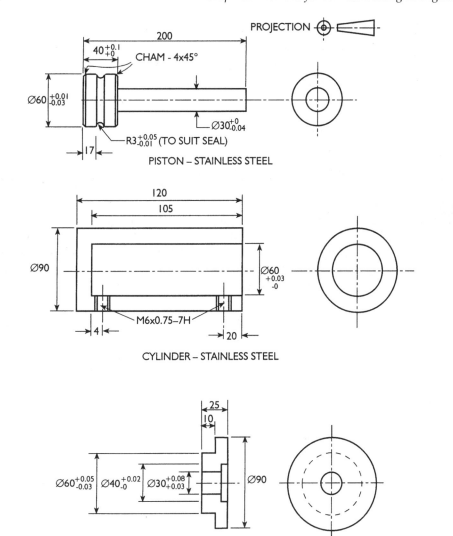

Figure 4.8 Detail drawing with manufacturing information

Activity 4.3

Find examples of engineering detail drawings that show:
- a mechanical part, e.g. pin, bearing, shaft, bracket
- an electrical part, e.g. insulator, motor brushes, contacts
- an electronic part, e.g. integrated circuit, printed circuit board, capacitor, resistor
- a fluid power (pneumatic/hydraulic) part, e.g. cylinder, piston, valve body.

Describe the specific information contained in each drawing, e.g. sizes, shapes, finishes, material.

Graphical communication in engineering

Assembly and subassembly drawings

Engineered products have to be assembled correctly. They will normally be made up of many individual parts and subassemblies. A subassembly is a self-contained assembly within an assembled product.

Case study

This case study presents two examples of assembly drawings of (a) a vice and (b) a carburettor.
(a) An assembly drawing has to show how many of each part are required to make the assembly and to show the different parts in their correct relationship to each other. Figure 4.9 shows an assembly drawing for a vice.
(b) Another type of assembly drawing is the 'exploded' diagram. This shows all the individual parts in the assembly separated for ease of identification. The main use for this is service manuals to show the individual parts and part numbers and their inter-relationships in the assembly. Figure 4.10 shows an exploded diagram of a carburettor with all parts numbered and listed.

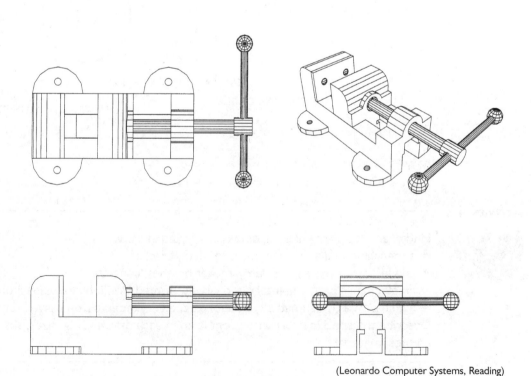

(Leonardo Computer Systems, Reading)

Figure 4.9 An assembly drawing for a vice

136

Graphical methods for communicating in engineering

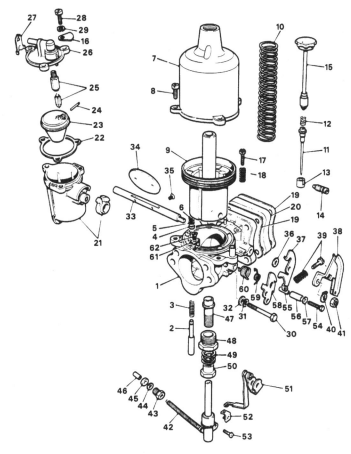

KEY TO THE CARBURETTOR COMPONENTS

No. Description	No. Description	No. Description
1. Body	21. Float-chamber and spacer	42. Jet assembly
2. Piston lifting pin	22. Joint washer-chamber	43. Sleeve nut-jet flexible pipe
3. Spring for pin	23. Float	44. Washer
4. Scaling washer	24. Hinge pin-float	45. Gland
5. Plain washer	25. Needle and seat	46. Ferrule
6. Circlip	26. Lid-float-chamber	47. Sealing washer
7. Piston chamber	27. Baffle plate	48. Jet locating nut
8. Screw-piston chamber	28. Screw-float-chamber lid	49. Spring
9. Piston	29. Spring washer	50. Jet adjusting nut
10. Spring	30. Bolt-securing float-chamber	51. Pick-up lever
11. Needle	31. Spring washer	52. Link-pick-up lever
12. Spring-needle	32. Plain washer	53. Screw-securing to jet
13. Support guide-needle	33. Throttle spindle	54. Pivot bolt
14. Locking screw-needle support guide	34. Throttle disc	55. Pivot bolt tube-inner
15. Piston damper	35. Screw-securing disc assembly	56. Pivot bolt tube-outer
16. Identification tag	36. Washer-throttle spindle	57. Distance washer
17. Throttle adjusting screw	37. Throttle return lever	58. Cam lever
18. Spring for screw	38. Progressive throttle (small cam)	59. Spring-cam lever
19. Joint washers	39. Fast idle screw and spring	60. Spring-pick-up lever
20. Insulator block	40. Lock washer-throttle spindle nut	61. Guide-section chamber piston
	41. Nut-throttle spindle	62. Screw-securing guide

Figure 4.10 An exploded view of a carburettor

Case study

The pneumatic cylinder is assembled from the different parts as shown in Figure 4.8. These individual parts will be put together to make the complete cylinder. The assembly drawing shows how the parts are put together. Figure 4.11a shows the assembled pneumatic cylinder using the parts detailed in Figure 4.8 and Figure 4.11b shows an exploded diagram of the assembly.

Figures 4.8 and 4.11 give all the information that is required for manufacture and assembly.

Graphical communication in engineering

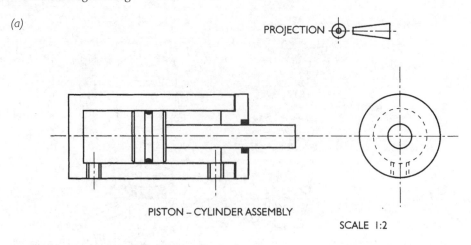

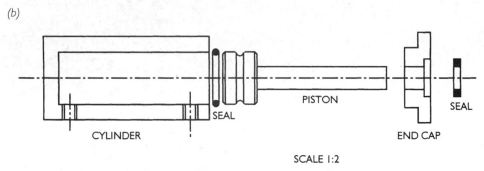

Figure 4.11 (a) The assembled pneumatic cylinder; (b) exploded diagram of the assembly

Activity 4.4

Actuator

A device for producing motion, such as a motor or a cylinder

Find examples of engineering assembly drawings that show:
- a mechanical assembly, e.g. engine, gearbox, clamp, lock mechanism
- an electronic assembly, e.g. circuit board with components, computer
- an electrical assembly, e.g. motor, switch, transformer
- a fluid power (hydraulic/pneumatic) assembly, e.g. pump, valve, actuator.

It may be possible to find a single drawing with examples of mechanical, electronic, electrical and fluid power assemblies. Describe the specific information contained in the drawing, e.g. part numbers, assembly and testing instructions, arrangement of parts.

Circuit diagrams

These are drawings using symbols to represent components and the way they are connected together. Their main use is to describe systems such as:
- electronic and electrical
- fluid power (hydraulic and pneumatic)
- process plant
- heating and ventilating plant.

These systems are made up of individual components connected together. The exact way in which the components are connected together is represented by a circuit diagram. The circuit diagram does not describe the actual positions of the parts in the assembled circuit. An assembly or layout drawing will still be required for this. This assembly drawing is, however, sometimes also called a cir-

Graphical methods for communicating in engineering

cuit diagram. The symbols used in circuit diagrams are standardised by the British Standards Institution (BSI) working in conjunction with the International Organization for Standardization (ISO). Reference should be made to the BSI publication PP7307 Graphical symbols for schools and colleges. Chapter 6 of this book deals with interpretation of circuit diagrams and should also be referred to as required.

A complete list of British Standards covering graphical symbols is given below:
- BS 3939 Electrical and electronic graphical symbols
- BS 2917 Symbols for fluid power systems
- BS 1553 Graphical symbols for general engineering (this covers all process plant as well as heating and ventilating)

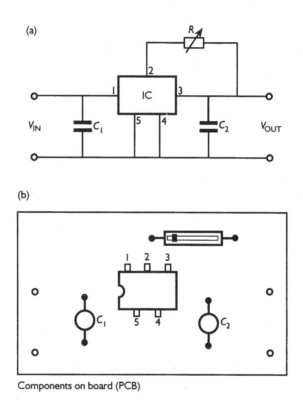

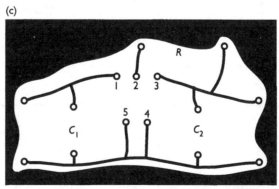

Figure 4.12 (a) Circuit diagram for voltage regulator; (b) possible arrangement of the components on the circuit board; (c) reverse side of board

Graphical communication in engineering

- BS 5070 Drawing practice for engineering diagrams (this covers the general principles to be followed in all types of engineering circuit and installation drawings)
- BS 4058 Specification for data processing, flow chart symbols, rules and conventions.

Activity 4.5

1. Consult the British Standards listed above. Find, and sketch, the symbols for a:
   - resistor
   - p-n diode
   - double-acting piston
   - hydraulic-pressure source
   - fan
   - flanged and bolted pipe joint.

   Comment on how well the graphical symbol is an analogue representation of the actual item.

2. Figure 4.12a shows a circuit diagram for an electronic circuit. It is a representation of how the system works. Figure 4.12b shows a possible arrangement of these components on the circuit board.

   Using the BSI publications PP7307 and/or BS 3939, identify all the electronic components used in Figure 4.12.

3. Figure 4.13a shows a circuit diagram for a hydraulic circuit. Like the electronic circuit, it represents how the system works. Figure 4.13b shows a drawing of a possible arrangement of the various components and pipes.

   Using BSI publications PP7307 and/or BS 2917, identify all the hydraulic components used in Figure 4.13.

4. Figure 4.14a shows a simple process plant circuit, with Figure 4.14b being a drawing of the actual layout.

   Using BSI publications PP7037 and/or BS 1553, identify all the items in Figure 4.14.

Case study

The electrical and pneumatic circuits for the drilling head are shown in Figure 4.15a and b. These circuits show how the electrical and pneumatic systems will operate and are vital information. Unless these circuits are correct, the drilling head will not function properly.

Activity 4.6

For Figure 4.15 use PP7307 and/or British Standards 3939 and 2917 to identify all the items used in the circuits.

Block diagrams

As the name implies, block diagrams use individual blocks or boxes. In fact, an alternative name for the blocks are 'black boxes'. The block is a simplified representation of a system in which the rectangular block can be a very complicated device or circuit, etc. This system can be part of a larger system, in which case it is called a subsystem. Within each block, the system uses whatever is put into it

Graphical methods for communicating in engineering

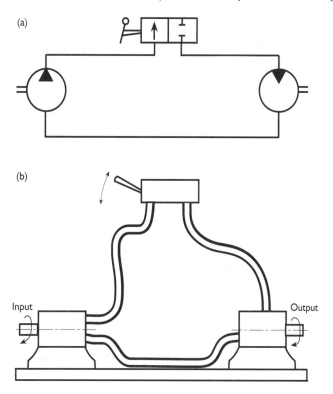

Figure 4.13 (a) Circuit diagram for a hydraulic circuit; (b) possible arrangement of components and pipes

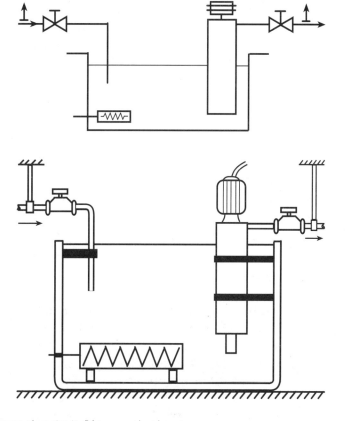

Figure 4.14 (a) Process plant circuit; (b) process plant layout

141

Graphical communication in engineering

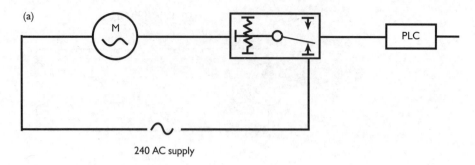

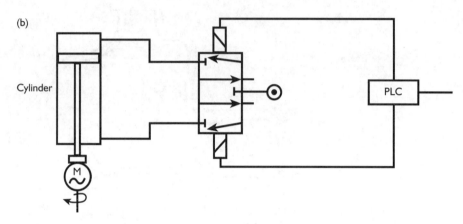

Figure 4.15 (a) Electrical circuit for drilling head; (b) pneumatic circuit for drilling head

to produce an output. Figure 4.16 shows a generalised block diagram with inputs, system and outputs. Because the system is operating in a larger environment, it has other, uncontrolled, inputs, which are called disturbances.

Note that the diagram implies that there is a relationship between the input and the output.

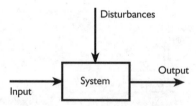

Figure 4.16 Generalised block diagram with inputs, system and outputs

Case study

This case study shows three block diagrams of (a) a telephone handset, (b) a lathe and (c) the organisational structure of an engineering company.
(a) Figure 4.17 shows the block diagram of a telephone handset system. Here the inputs of electronic signals are converted into outputs of sound by the system. The energy to do this conversion is provided by the input of electrical energy. Disturbances such as variations in current, electrical interference and so on may affect the output.

Graphical methods for communicating in engineering

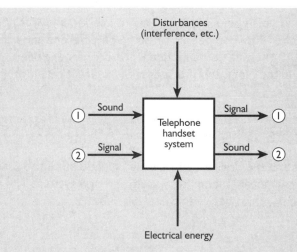

Figure 4.17 Block diagram of a telephone handset system

(b) Figure 4.18 shows the block diagram for an electrically powered lathe for producing cylindrical parts (the process of producing cylindrical parts is called turning).

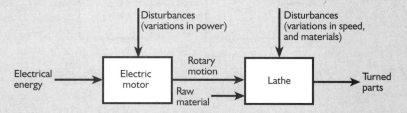

Figure 4.18 Block diagram of an electrically powered lathe

This time there are two subsystems that make up the complete system. In the first subsystem electrical energy is input into a motor to produce an output of mechanical energy that rotates the chuck. This, together with raw material, is fed as input to the turning system. This gives an output of turned parts.

(c)

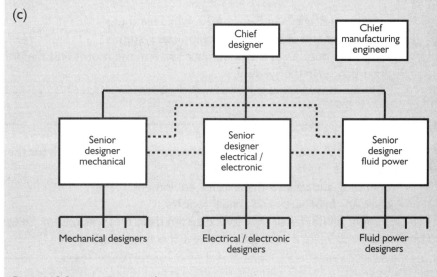

Figure 4.19 Organisational chart for an engineering company

143

Graphical communication in engineering

Another type of block diagram is used to show relationships in organisations, either between people or between different parts of the organisation. Figure 4.19 shows an organisational chart for an engineering company; the job title is in the block and the lines show a direct relationship.

Case study

The drilling-head system can be represented by a block diagram (Figure 4.20), which shows the inputs, outputs and disturbances in the system. The diagram shows that the drilling-head system is made up of the subsystems of PLC, electrical motor, pneumatic system and drill system. The block diagram shows the relationships between these subsystems.

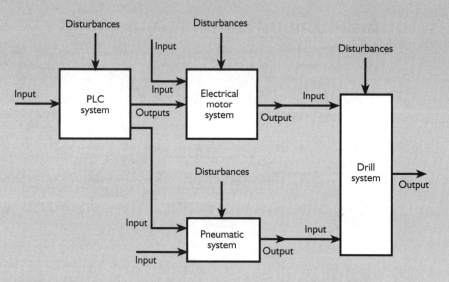

Figure 4.20 Block diagram of drilling-head system

Activity 4.7

For the block diagram in Figure 4.20, what are the:
1. main inputs and outputs of the complete system
2. inputs and outputs of the pneumatic system, the motor and the drill system
3. disturbances in the system?

Activity 4.8

Find, and copy, four examples of block diagrams. You can look for these in:
- textbooks
- operating and service manuals for equipment
- company brochures and annual reports.

Comment on how well the block diagram describes the system. Select one diagram and write a description of how the system operates.

144

Graphical methods for communicating in engineering

Flow diagrams

These are used to represent a sequence of events. A particular kind of flow chart is the flow process chart, which shows the sequence of operations in manufacturing. Symbols are used to represent different types of activity. Figure 4.21 shows the standard symbols used in flow process charts for:
- operation (processing or assembly of parts or products)
- inspection (checking or testing of parts or products)
- transportation (moving of parts or products between operations or to storage)
- storage (planned safekeeping of parts or products between operations or until required)
- delay (parts or products in queue waiting to be processed, inspected, transported or stored).

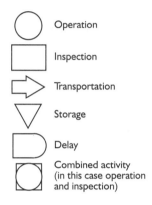

Figure 4.21 Standard symbols used in flow process charts

Case study

The parts of the pneumatic cylinder and piston assembly will need to have their manufacture planned. One way of doing this is with a flow process chart. An actual flow process chart used to produce the pneumatic cylinder for the drilling head is shown in Figure 4.22. Notice that there is written information giving some details of the part and the processes as well as distance travelled and time taken.

Another type of flow diagram is that used to show the sequential flow of operations in a computer program or in any activity that requires decisions to be taken. Figure 4.23 shows the standard symbols used in flow diagrams for:
- starting or stopping the sequence
- performing an action (process or operation)
- making a decision
- an input or output operation
- a connection between two parts of the chart
- direction of flow of data
- production of a document (print or plot).

Graphical communication in engineering

Flow process chart
Part — pneumatic cylinder (100 Off)

Distance (m)	Time (min)	Symbol	Description
		1	Storage of bars
7	1	2	Bars loaded onto forklift
60	2	3	Bars moved to lathe
2	0.5	4	Bars unloaded and put in rack
	10	5	Waiting for operation
1	5	6	Bars machined to make cylinder and inspected
	1	7	Clean and de-burr
	1.5	8	Final inspection
50	2	9	To assembly
		10	Stored until needed

Event	Number	Time (min)	Distance (m)
Operation	4	7.5	10
Transportation	2	4	110
Inspection	1	1.5	
Delay	1	10	

Figure 4.22 Flow process chart for producing the pneumatic cylinder for the drilling head

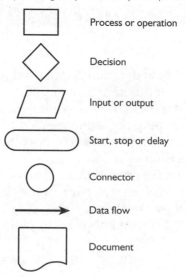

- Process or operation
- Decision
- Input or output
- Start, stop or delay
- Connector
- Data flow
- Document

Figure 4.23 Standard symbols used in flow diagrams

Graphical methods for communicating in engineering

Case study

This case study shows flow charts for (a) a simple computer program, (b) a general manufacturing process and (c) drilling a hole using the drilling head.

(a)

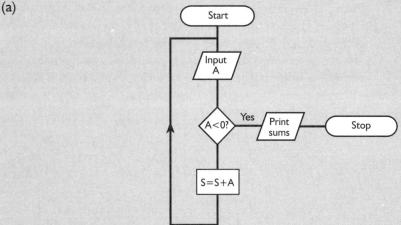

Figure 4.24 Flow chart for a computer program

Figure 4.24 shows a flow chart for a computer program to sum a series of numbers, A, which is input. The sum is printed out and the program stops if a negative value is input. Notice how the program creates a sum, S. The first time through the loop $S = A$, with A being added each time a new value of A is entered.

(b)

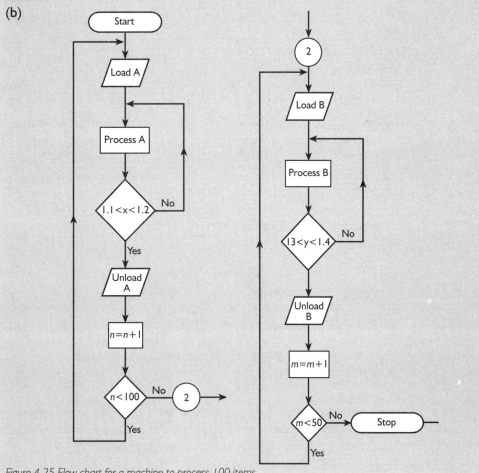

Figure 4.25 Flow chart for a machine to process 100 items

147

Figure 4.25 shows a flow chart for a machine to process 100 items, A, to a dimension, X, between 1.1 and 1.2, followed by 50 items, B, to a dimension, Y, between 1.3 and 1.4. In this program the part is loaded, processed and checked for the correct value. The part is then unloaded and the counter n increased by 1 (it will be set to 1 the first time through). When n equals 100 the program jumps to the second part, B, where the program operates in the same way as for part A. After processing the 50 parts B the machine stops.

(c)

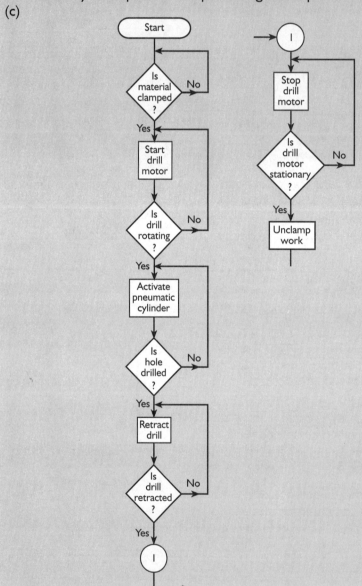

Figure 4.26 Flow chart for drilling a piece of metal using the drilling head

Flow charts can be used to describe the drilling head being used to perform actual operations such as drilling holes. Figure 4.26 shows the flow chart for drilling a piece of metal using the drilling head.

Graphical methods for communicating in engineering

Activity 4.9

1. In Figure 4.26 identify all the:
   - input and output operations
   - decision-making processes
   - actions performed.
2. Find and copy two flow process charts and two flow charts. These can be found in:
   - textbooks
   - computer manuals
   - programming manuals
   - process planning documents.

   Comment on how well the diagrams describe the process. Choose one of the flow charts and give a written description of the process.

Schematic diagrams

The word 'schematic' originates from the Greek *schema*, which means shape or figure. A schematic diagram uses simple geometry such as lines and circles to represent the way something is put together and operates. Figure 4.27 is a schematic diagram of an adjustable table lamp. The diagram shows the lamp in two different positions only, but there are a huge number of different positions for the lamp mechanism. Notice that the schematic diagram only shows the operation of the lamp. Detail and assembly drawings would be necessary to make it.

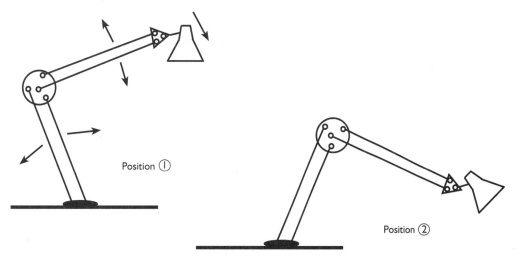

Figure 4.27 Schematic diagram of an adjustable table lamp

Case study

This case study shows schematic diagrams for (a) an electric motor and (b) the drilling head.

(a) A typical electrical schematic diagram is the electric motor/generator shown in Figure 4.28. The principle of operation and method of construction of the motor/generator are clearly shown. There is just enough detail to show the principles, but not enough to actually make a motor/generator.

Graphical communication in engineering

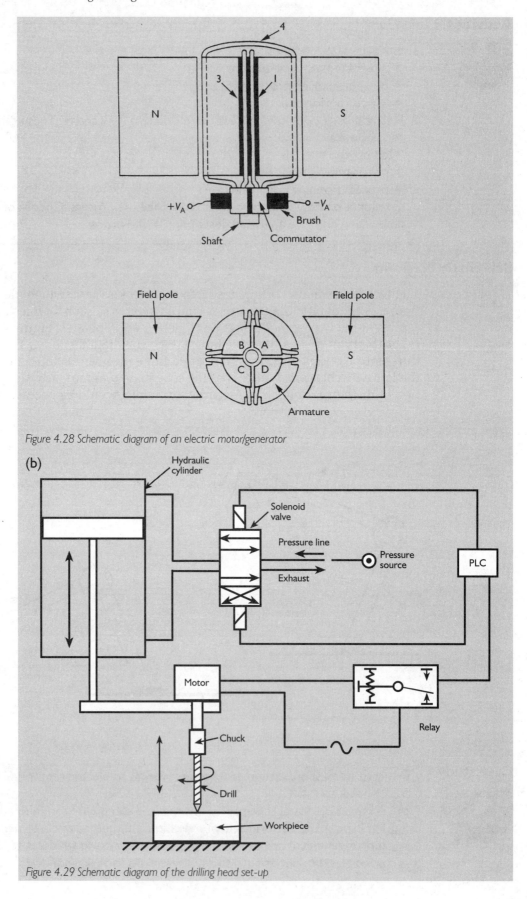

Figure 4.28 Schematic diagram of an electric motor/generator

Figure 4.29 Schematic diagram of the drilling head set-up

Graphical methods for communicating in engineering

> A schematic diagram of the drilling head set-up is shown in Figure 4.29. Notice how only the essentials of the design are given. This type of diagram would probably be done by a systems or design engineer and given to a detail designer to complete as detail and assembly drawings.

Activity 4.10

1. From the schematic diagram in Figure 4.29 comment on:
   - the type of information given
   - the effectiveness of the schematic diagram in explaining how the drilling head operates.
2. Find, and copy, a schematic diagram showing:
   - a mechanical engineering example
   - an electrical engineering example
   - a fluid power example.

   You may be able to find a diagram containing all three! Possible sources of schematic diagrams are:
   - textbooks
   - engineering journals
   - handbooks and operating manuals
   - catalogues and brochures
   - wallcharts and posters.

   Comment on how well the schematic diagram shows the operation of the device shown. Choose one of the examples and try to describe, in words, how it works.

Photographs

Photographs are used extensively in engineering. The chief applications are:
- marketing and publicity – photographs form a large part of advertising in newspapers, trade journals, brochures, etc.
- assembly and servicing – photographs can show clearly assembly/disassembly sequence and methods as well as illustrating servicing details
- analysis – 'fast-frame' photography can be used to analyse precise movements of high-speed mechanisms, etc.
- quality control – photographs can provide a permanent record for later reference or analysis.

Case study

This case study uses photographs of (a) a robot, (b) a car engine and (c) the drilling head.
(a) Figure 4.30 shows a photograph of an industrial robot used in a brochure giving details of the product. Apart from the use of film or video, this is a very good way to show the actual product.
(b) Figure 4.31 is a photograph of a car engine compartment with different parts or subassemblies being identified by the use of numbers. These numbers are then referenced in the accompanying service manual.
(c) Figure 4.32 shows a photograph of the pneumatic drilling head. The photograph is taken after the complete system has been manufactured and assembled. It can then be used in sales brochures and technical manuals.

Graphical communication in engineering

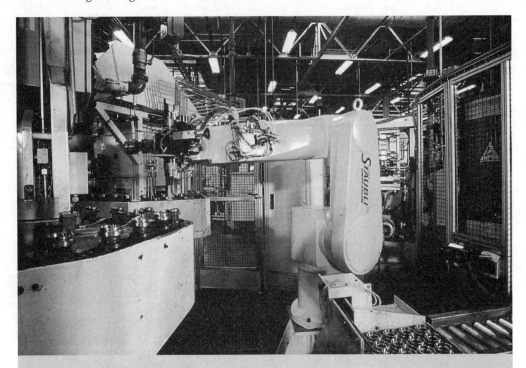

Figure 4.30 Industrial robot

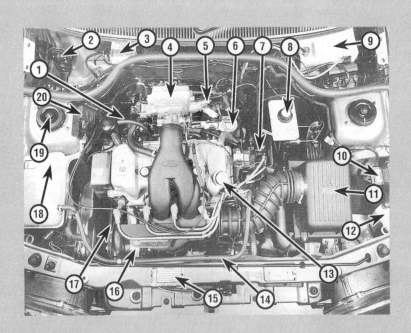

Figure 4.31 Engine compartment. Locations – 1.6-litre CVH EFi fuel-injected engine. 1, Oil level dipstick location; 2, anti-theft alarm horn; 3, windscreen wiper motor; 4, throttle housing; 5, intake air temperature sensor; 6, fuel pressure regulator; 7, EDIS ignition coil; 8, brake master cylinder reservoir; 9, battery; 10, ignition module; 11, air cleaner housing; 12, washer fluid reservoir; 13, engine oil filler cap; 14, cooling fan; 15, vehicle identification plate (VIN); 16, starter motor; 17, auxiliary drivebelt; 18, coolant expansion tank; 19, suspension upper mounting; 20, MAP sensor.

Graphical methods for communicating in engineering

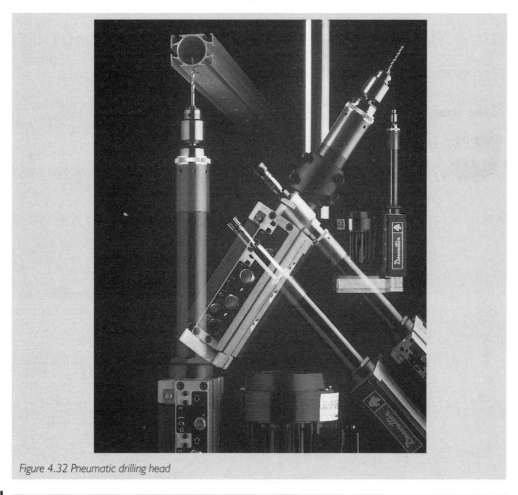

Figure 4.32 Pneumatic drilling head

Activity 4.11

1. For Figure 4.32 comment on:
   - how well this photograph represents the object
   - what types of information are contained in the photograph.
2. Find and copy a photograph from each of the following four categories:
   - marketing and publicity
   - assembly and/or servicing
   - analysis
   - quality control.

   Comment on the suitability of each photograph and how well the information is presented.

Spreadsheets, charts and graphs

Spreadsheets are an extremely useful tool for engineers. They are computer programs which work like a programmable calculator to do calculations on any kind of data that are presented. They are particularly useful for design or experimental calculations. Values can be changed quickly in a spreadsheet, which enables the results of these changes to be calculated almost instantly. The results can then be presented graphically, if required, in the form of line graphs, bar charts or pie charts. A spreadsheet is a matrix of lettered rows and numbered columns. This enables specific 'cells' to be referenced, e.g. C1, D9. Into these cells can be

Graphical communication in engineering

typed letters, words, numbers or formulae. A formula can be something very simple, e.g. C1*D9, which means multiply the number in cell C1 by the number in cell D9. Once the formula has been typed in, the cell will display not the formula but the result of the calculation.

Case study

Cases are shown here of spreadsheets used to make calculations and then present the results graphically.

1. The spreadsheet shown in Figure 4.33a shows the results of a test to find the resistance of different electrical devices when connected across a 12-V supply. The current is measured by an ammeter. The formulae (from Ohm's law $R = V/I$) put in were:

   Cell B4 = B2/B3 = 12/1 = 12 Ω
   Cell C4 = C2/C3 = 12/2 = 6 Ω
   Cell D4 = D2/D3 = 12/3 = 4 Ω
   Cell E4 = E2/E3 = 12/4 = 3 Ω

   Once the calculations are done, the spreadsheet program will convert them to graphical form if required. The results are shown in bar chart and line graph format in Figure 4.33b and c.

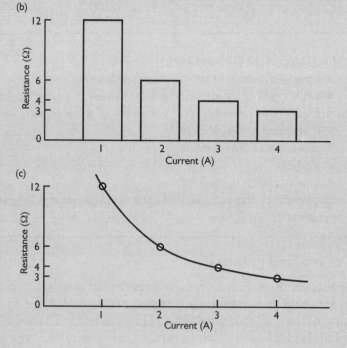

Figure 4.33 Results of a resistance test of various electrical devices: (a) as a spreadsheet; (b) as a bar chart; (c) as a line graph

Graphical methods for communicating in engineering

2. The spreadsheet shown in Figure 4.34a shows some engineering data on exports of engineering products. These data can be converted by the spreadsheet model into either bar chart or pie chart form. Both methods show the relative amounts of each product. The bar chart is shown in Figure 4.34b and has diagrammatic representations of each type of product. This kind of bar chart is called a pictogram or ideograph. The pie chart is shown in Figure 4.34c. Figure 4.34d is a stacked bar chart. This shows the different values of each type of product as 'stacked' values – that is on top of one another. It shows the sum of all the values as well as the individual values.

(a)

	A	B	C	D	E
1	Product	Vehicles	Electronic products	Machine tools	Medical equipment
2	Value of exports	£2000 m	£1000 m	£800 m	£600 m
3	Percentage	45	23	18	14

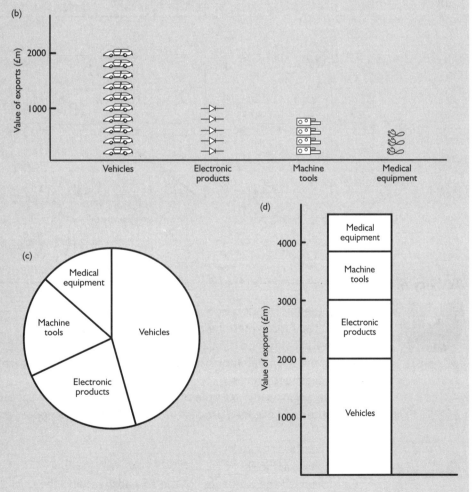

Figure 4.34 (a) Data on exports of engineering products; (b) the data as a bar chart (ideograph); (c) the data as a pie chart; (d) the data as a stacked bar chart

Graphical communication in engineering

3. It may be necessary to calculate the force that can be applied by the pneumatic cylinder. The spreadsheet in Figure 4.35a shows the variation in force applied by the pneumatic cylinder for different air pressures used. Figure 4.35b shows these spreadsheet figures presented as a graph. The formula for force is $F = PA$ (newtons) where
F = force applied (newtons)
P = pressure of compressed air supply (newtons/square millimetre)
A = area of piston (square millimetres).
In the spreadsheet
Cell C2 = A2 × B2 = 2000 × 1 = 2000 N
Cell C3 = A3 × B3 = 2000 × 2 = 4000 N
Cell C4 = A4 × B4 = 2000 × 3 = 6000 N
Cell C5 = A5 × B5 = 2000 × 4 = 8000 N.

(a)

	a	b	c	d
1	Area	Pressure	Force	
2	2000	1	2000	
3	2000	2	4000	
4	2000	3	6000	
5	2000	4	8000	

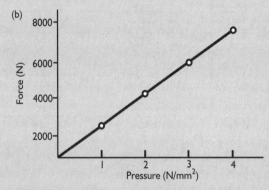

Figure 4.35 (a) Variation in force applied by pneumatic cylinder for different air pressures: (a) as a spreadsheet; (b) as a graph

Activity 4.12

1. Using the information from Figure 4.35:
   - find the force when the pressure is 2.5 N/mm^2
   - say what this type of information would be used for
   - comment on the information conveyed by the graph as compared with the spreadsheet figures.
2. Find an engineering example of a spreadsheet. Load the spreadsheet model into a computer and use it to process the data. Use the computer printer to print out the:
   - spreadsheet model
   - results graphically.
   Comment on the usefulness of the spreadsheet and the quality of the graphical output.

Graphical methods for communicating in engineering

Progress check

1. (a) Describe the advantages of graphical communication as compared with the written or spoken word. (b) State the differences between iconic and analogue models and give one example of each. (c) Give three everyday examples of the use of graphical communication.
2. (a) List the various types of engineering information. (b) Which engineering activities use graphical communication? (c) Describe, briefly, the main methods of graphical communication.
3. In which engineering activities are sketches used? Outline their uses as compared with other methods.
4. (a) What is the function of (i) detail drawings and (ii) assembly drawings? (b) Which methods of projection are used in detail and assembly drawings?
5. (a) What type of model is a circuit diagram? (b) What information does a circuit diagram contain? (c) Which types of circuit diagram are used in engineering?
6. What information does a block diagram communicate?
7. (a) What are flow diagrams used for? (b) Name the different operations in flow diagrams and draw the appropriate symbol.
8. (a) What is a schematic diagram? (b) Give two different examples of schematic diagrams.
9. State four different areas of engineering activity where photographs can be used.
10. (a) What is a spreadsheet? (b) What are the different ways in which the results of spreadsheet modelling can be presented graphically?

Specification

A complete technical definition of a product or service, including full written, numerical and graphical information.

How graphical methods relate to types of information

The types of information that are communicated include:
- qualitative – information that is not quantifiable in numbers
- quantitative – information in numerical form
- macro – information giving the overall picture without too much detail
- detail – precise and complete information giving very specific details.

Qualitative information

The term qualitative comes from the word quality. Quality, in this context, is defined as a property or characteristic. Therefore qualitative information seeks to describe properties and characteristics without using numbers. Colour, shape, layouts and relationships between objects are all examples of qualitative information.

Case study

There is qualitative information in the different graphical methods used to represent the drilling head. Typical examples here of qualitative information are the:
- connections between the pneumatic valve and cylinder (Figure 4.15b)
- colour of the electric motor casing (Figure 4.32)
- shape of the pneumatic valve (Figure 4.29).

No numbers are involved here so the information is qualitative. However, it is possible to bring numbers in as a colour can be described in terms of a specification number and a shape can be described geometrically.

Graphical communication in engineering

Activity 4.13

Find, and describe, three more examples of qualitative information in the case study graphics of the drilling head.

Quantitative information

Information is conveyed here by using numbers and calculations. Thus, any graphical communication that needs to be quantitative must include numbers. Typical examples of quantitative information are:
- variables such as size, voltage, pressure, stress
- limits and tolerances within which variables can vary.

Case study

There is much quantitative information in the graphics used to represent the drilling head. Examples are:
- length of pneumatic cylinder (Figure 4.8)
- tolerance on diameter of pneumatic piston (Figure 4.8)
- voltage of supply to electric motor (Figure 4.15b).

These examples all involve the use of numbers and so are classified as quantitative information.

Activity 4.14

Find and describe three more examples of quantitative information from the case study graphics of the drilling head.

Macro information

'Macro' is derived from a Greek word meaning large. Thus, macro information gives the 'big picture' as opposed to the smaller details. It should not be confused with the actual size of the graphics. A very large drawing can be of a very small detail and vice versa. Typical examples of macro information are:
- factory layout drawings
- photographs of an object
- schematic diagrams showing principles of operation.

Case study

The drilling head graphics contain macro information. The photograph of the drilling head exterior (Figure 4.32) is an example. This is classified as macro information because it shows an overall view of the size, shape, colour and method of construction of the drilling head.

Activity 4.15

Find three other examples of macro information in the case study graphics of the drilling head and explain why they would be classified as macro information.

Detail information

Surface roughness
The amount by which a surface deviates from being absolutely smooth

Logically this should be called 'micro' information as it is the opposite of macro information. Detail information seeks to define exactly, so that an object or process can be described precisely. Thus, it would include the totality of information on things such as:
- materials and material properties
- size, shape, finish, accuracy
- performance characteristics, e.g. velocity, force.

Case study

The drilling head graphics would have numerous examples of detail information including:
- speed of drill spindle rotation
- purity of copper wire used in electric motor windings
- surface roughness of internal diameter of pneumatic cylinder.

All these examples are classed as detail information because they give very specific information about a particular detail of the design.

Activity 4.16

Take the following three examples of detail information in the case study graphics of the drilling head:
- time taken to machine the bars for the pneumatic cylinder (Figure 4.22)
- material used for the piston (Figure 4.8)
- the size of the tapped holes in the pneumatic cylinder (Figure 4.8).

Describe them and state whether they detail the requirements fully.

Progress check

1. Explain what qualitative information is and give three general examples.
2. Give three specific examples of the use of qualitative information in engineering graphical communication.
3. Explain what quantitative information is and give three general examples.
4. Give three specific examples of the use of quantitative information in engineering graphical communication.
5. Define 'macro' and give three general examples of macro information.
6. Give three specific examples of the use of macro information in engineering graphics.
7. What is detail information? Give three general examples of its use.
8. Give three specific examples of the use of detail information in engineering graphics.

Choosing graphical methods for different purposes

The range of different purposes in engineering covers:
- conceptualising – the initial design phase when the creative ideas are developed
- designing – taking the concepts and turning them into a feasible design
- planning – the preparation of schedules and operations so that the design can be realised

Graphical communication in engineering

Conceptualising

The forming of ideas and theories by individual thinking

- estimating – preparing cost estimates for implementation of designs
- costing – evaluating actual cost after design is realised
- manufacturing and installation – the processes by which the design is realised
- marketing – covering all aspects of publicity, sales and promotion.

Clearly, communication methods vary in their suitability for each of the different purposes. For example, a complex and detailed assembly drawing of a computer circuit board is essential for manufacture but is not suitable for marketing the computer to potential purchasers.

Conceptualising

The design process is essentially one of problem solving. The designer's task is to move from a vaguely defined problem to a well-defined solution. This initial stage of design is known as conceptualising.

Case study

(a) An age-old problem in war is that of getting soldiers and materials rapidly to where they are required. From this vaguely defined problem came the well-defined solution of the helicopter. Figure 4.36 shows the problem and the range of possible concepts to solve the problem. There was much conceptualising and testing of concepts before the solution could be realised.

(b) The selection of a pneumatically fed, electrically powered drilling head came after other concepts were evaluated and tested. The ideas for the different concepts would be evaluated with the help of suitable graphical methods. For example, designers' rough sketches would probably be used to select alternative ways of joining the drilling head to the pneumatic piston rod. This is shown in Figure 4.37.

Figure 4.36 Some possible solutions to the problem of how to transport soldiers and materials

Graphical methods for communicating in engineering

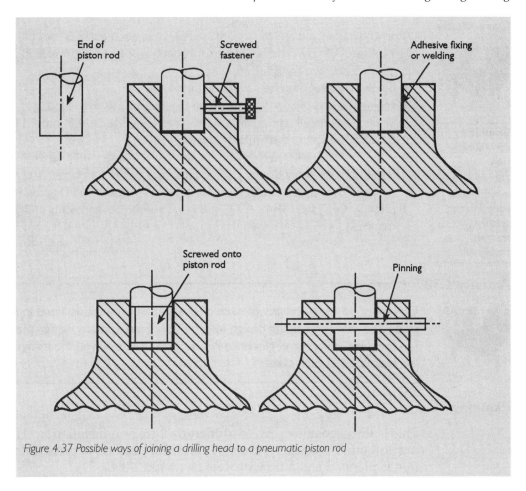

Figure 4.37 Possible ways of joining a drilling head to a pneumatic piston rod

Activity 4.17

With respect to the case studies of the drilling head, which graphical methods might be used for:
- comparing alternative connections between pneumatic cylinder, valve and air supply
- presenting the results of passing different currents through the electric motor?

Designing

Prototype

The first full-size working hardware of a new design, which is used for testing and evaluation

Once solutions have been found they have to be turned into reality. The first stage of this is the design realisation. The solution will set specific criteria, e.g. cost, strength, performance, and the design will be realised to these criteria. In the initial stages there will be much testing and refinement.

After a prototype has been tested the design can be finalised. Assembly and detail drawings can then be prepared. A block diagram of the design process is shown in Figure 4.38.

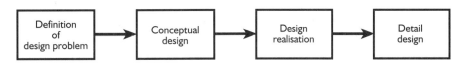

Figure 4.38 Block diagram of the design process

Graphical communication in engineering

Case study

Graphical communication

The transmission of information using drawings, paintings, sketches, photographs, etc.

After the initial concept of the drilling head has been established, the design will be realised. A likely order of events involving graphical communication is:
- flow charts (see Figure 4.26)
- block diagrams of systems (see Figure 4.20)
- schematic and circuit diagrams (see Figures 4.29 and 4.15)
- preliminary detail and assembly drawings (see Figures 4.8 and 4.11)
- building, testing and evaluating prototypes
- modifications to initial detail and assembly drawings and preparation of final detail and assembly drawings (see Figures 4.8 and 4.11)
- schematics and circuit diagrams for service manuals (see Figures 4.29 and 4.15)
- photographs (see Figure 4.32, and illustrations for marketing and publicity material).

Activity 4.18

Comment on the suitability of each of the graphical methods listed in the case study above for conveying design information. Find as many design graphics as you can for three different engineered products. State how well the design information is transmitted in each case.

Planning

This is the preparatory stage of manufacture and installation. The manufacturing and other operations must be carefully planned in great detail. Each operation is planned and a flow process chart specifies:
- processes
- machines
- tooling
- time schedules
- manufacturing routes.

Case study

Flow process charts can be used to describe the way in which the different parts of the drilling head assembly are made. The flow process chart includes information on time, distances and processes. This is shown in Figure 4.22, which is a flow process chart for producing the pneumatic cylinder.

Activity 4.19

Use the chart in Figure 4.22 to find:
- the processes used
- the machines and equipment used
- the manufacturing route
- the total distance the part travels
- the total time to process a part.

Graphical methods for communicating in engineering

Another type of planning is used for complex projects such as building ships, petrochemical plants and aircraft, or installing complex systems such as telephone networks. If there is not too much complexity a simple linear Gantt chart can be used. Figure 4.39 shows a Gantt chart for a project.

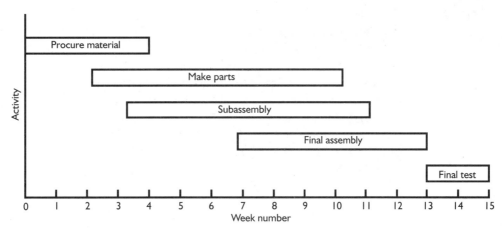

Figure 4.39 A Gantt chart

Interconnections

The actual joining together of different components in a circuit

For planning projects of great complexity, the programme review and evaluation technique (PERT) is used. This method allows interconnections between different activities, unlike the simple Gantt chart. Figure 4.40 shows a PERT chart for a simple project. The numbers indicate time, in this case weeks.

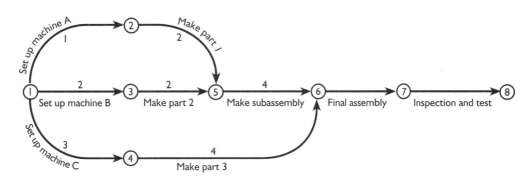

Figure 4.40 A PERT chart

Activity 4.20

Find examples of graphics used for planning the manufacture or installation of three different engineered products or services.

Estimating

From the detailed information in the planning documents an estimate of cost can be prepared. This is essential before any project can be approved and proceed. Customers will also need fast and accurate estimates so that purchasing decisions can be made correctly.

Graphical communication in engineering

Case study

Figure 4.41 shows a chart with estimated costs for producing the pneumatic cylinder. This is using information from the flow process chart for the cylinder (Figure 4.22). Times are multiplied by cost/hour and the cost of material and transport is added. The costs are:
- material £1.50
- machining and inspection £20/hour
- transportation £10/hour
- waiting £5/hour.

The estimated costs are then calculated by multiplying the time for each stage multiplied by the hourly rate. The estimated costs are:

machining and inspection = £(9/60 × 20) = £3
transportation = £(4/60 × 10) = £0.67
delay = £(10/60 × 5) = £0.83
material = £1.50
total cost = £6

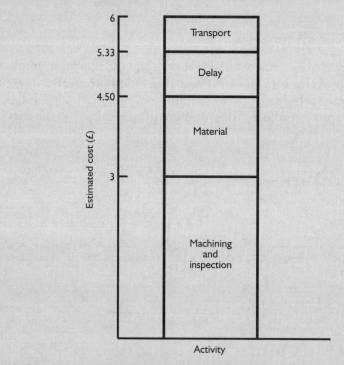

Figure 4.41 *Estimated costs for production of the pneumatic cylinder*

Costing

This has much in common with estimating. It usually refers to the situation after the product has been made or the project completed. As is well known, costs quite often exceed estimates. The Channel Tunnel is a well-known example: an estimated building cost of around £2 billion turned out to be over £8 billion. All costs incurred have to be accounted for, and these include:

Graphical methods for communicating in engineering

- management and administration
- national and local taxes
- research, development and design
- buildings and equipment
- direct and indirect labour costs
- services such as energy, water etc.
- marketing and publicity
- materials.

To simplify matters, costs are classified into direct costs and indirect costs (also called overheads). Direct costs are those costs actually incurred in manufacturing or installing, with indirect costs being any of the other costs of running the business.

Case study

Spreadsheets can be used to construct costing models, and this can be done to get a cost for producing the pneumatic cylinder. Figure 4.42a shows a spreadsheet costing model for the pneumatic cylinder. Figure 4.42b is the stacked bar chart showing graphically the direct, indirect and total costs.

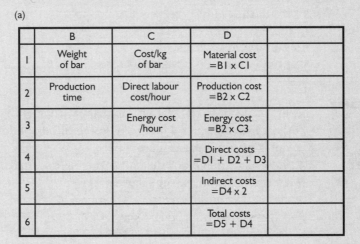

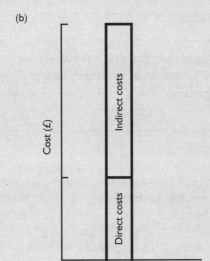

Figure 4.42 (a) Spreadsheet costing model for the pneumatic cylinder; (b) stacked bar chart of direct, indirect and total costs

Graphical communication in engineering

Activity 4.21

Using the basic model in Figure 4.42, build your own spreadsheet model and graph for cylinder costs if:
- material cost = £2.50/kg
- direct labour cost = £15/hour
- energy cost = £0.06/kWh
- indirect cost = 2 × direct costs.

Manufacturing and installation

This covers all aspects of producing the hardware and installing it, if required, for the customer. All graphical communication will have to be very detailed, precise and to agreed international standards. This is because manufacturing today is a worldwide business. The design may be made anywhere in the world. The main graphical methods in manufacturing and installation are:
- flow charts
- system diagrams
- schematic diagrams
- circuit diagrams
- detail drawings
- assembly drawings
- spreadsheets, charts and graphs.

Case study

The manufacture, installation and maintenance of the drilling head will require correct use of graphical methods. Some of the methods used are:
- detail drawings (see Figure 4.8)
- flow charts and block diagrams (see Figures 4.20 and 4.22)
- assembly drawings (see Figure 4.11)
- circuit diagrams (see Figure 4.15a and b)
- schematic diagrams (see Figure 4.29).

Activity 4.22

1. Describe the uses in manufacture and installation to which each of the methods listed in the case study above will be put.
2. Find examples of manufacturing and installation graphics for at least three different engineered products. List the methods used and comment on their effectiveness in conveying information.

Marketing

Commercial
Relating to products and services that are competing with others in a free market situation

This is the process by which the product is promoted to prospective customers. It covers the entire range of methods for promotion and selling, which include:
- advertising in journals and newspapers and on video, television and radio
- production and distribution of brochures, catalogues, leaflets and videos
- lectures, talks and demonstrations.

In today's competitive conditions, any marketing graphics will have to be of high quality, eye catching and usually in full colour. Graphical techniques for marketing and sales promotion include:

Graphical methods for communicating in engineering

- photographs
- artist's illustrations
- exploded diagrams
- schematic diagrams.

Case study

Graphics suitable for use in marketing have been covered in this chapter. Figure 4.5 shows an artist's impression of a car for use in advertising. This type of graphic can often be more attractive than photographs. The drilling head will have to be promoted in the markets where it is to be sold and this activity will be vital for commercial success. Figure 4.32 shows a photograph of the drilling head that could be used for marketing purposes.

Activity 4.23

1. What type of promotional material might the photograph in Figure 4.32 be included in?
2. Find examples of promotional and marketing material for at least three different engineered products. List the types of graphics used and comment on their effectiveness.

Progress check

1. (a) Define the process of conceptualising. (b) Which graphical methods are the most useful for conceptualising?
2. (a) Draw a block diagram of the design process. (b) Select three graphical methods used in design. State the purpose of each method and the type of information transmitted.
3. Name three different types of graphics used in planning and describe the types of information conveyed.
4. (a) What is the difference between costing and estimating? (b) What types of graphical model are most useful in costing and estimating?
5. (a) List the range of graphical techniques used in manufacturing and installation. (b) Which graphical technique is most suitable for:
   - giving dimensional information about a part
   - showing all the parts put together
   - giving interconnections between electronic components?
6. (a) Give three graphical techniques used for marketing an engineering product or service. (b) What features of graphical methods would be likely to attract customers?

Selection criteria for graphical methods

The guru of twentieth century communications, Marshall McLuhan, said, 'the medium is the message'. What this statement emphasises is the primary importance of selecting the correct method and medium by which to convey information. When selecting graphical methods, clearly defined criteria must be used. Engineering covers an enormous range of types and sizes of artefacts, from the very small scale of microelectronics to the large scale of supertankers and jumbo jets, and everything in between. Despite the obvious differences, the principles

Graphical communication in engineering

of good communication remain the same. In all cases methods must be matched to the purpose and to whoever is to receive the communication.

The selection criteria are:
- clarity – is the information clear, complete and unambiguous?
- cost – can the chosen method be delivered within a fixed budget?
- audience and user – is the method matched correctly to the audience or user?

Clarity

Clarity implies that the sender of the message is clear about what the message should be. Provided this is true, the method chosen should, if it is used correctly, be capable of transmitting that message with only one possible interpretation. Whatever graphical method is used it should be open to only one interpretation.

Figure 4.43a shows a wire-frame model of a cube. Figure 4.43b and c show two possible interpretations of this wire-frame model. The ambiguity is removed by removing the hidden lines (in this case shown dotted).

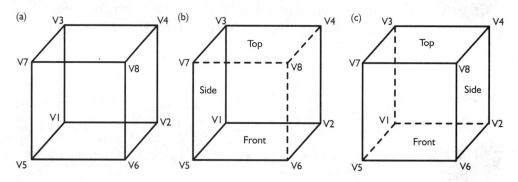

Figure 4.43 (a) Wire frame model of a cube; (b and c) two possible interpretations of (a)

Any graphical method used should have clarity. It cannot be said that any one method is less clear than any other. When testing a particular piece of graphics for clarity the following questions should be asked.
- Is there only one possible interpretation?
- Is the solution fully defined?
- Are all the numbers consistent, e.g. do all individual values contributing to an overall value add up to the overall value?

Figure 4.44 shows a drawing with some errors and omissions, leading to a lack of clarity.

Activity 4.24

Make a list of the errors and omissions in the drawing in Figure 4.44.

Cost

All graphical methods will incur costs in their production and distribution. These costs will include:
- labour – time spent by people producing and distributing, etc.

Graphical methods for communicating in engineering

Accuracy

The same as error, which is the difference between a specified value and the actual value of the outcome.

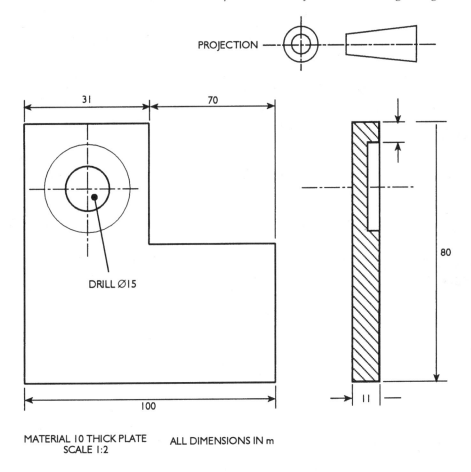

Figure 4.44 A drawing with some errors and omissions

- materials – inks, pencils, pens, paints, paper, disks, etc.
- energy and services – heating, electricity, water, etc.
- overheads – management and administration, buildings, taxes, etc.

Although costing of graphics has to be considered, their costs are generally low compared with the overall cost of designing and manufacture. Although it is important to control costs, other factors may be more important. Accuracy and freedom from misinterpretation are most important. Any mistakes arising from graphical errors are likely to be extremely costly and cause time delays.

Audience and user

Clearly, there has to be a good fit between the method used and the audience and user being addressed, for example a glossy coloured photograph of a machine would not be of any use to a fitter sent out to a remote location to repair this machine.

In engineering the types of audience/user for which graphics are produced include:

- prospective customers – marketing and advertising graphics
- actual customers – schematics, circuit diagrams and exploded diagrams for installation and servicing
- designers – conceptualisation sketches, block diagrams of systems, spreadsheet calculations and graphs

Graphical communication in engineering

- manufacturing, installation and maintenance engineers – detail and assembly drawings, circuit diagrams, schematic diagrams, flow charts.

Case study

For the drilling head, the graphics selected would be for:
- prospective customers – Figures 4.11, 4.15, 4.32
- actual customers – Figures 4.8, 4.11, 4.15, 4.29
- designers – Figures 4.3, 4.29, 4.35, 4.37.
- manufacturing, installation and maintenance engineers – Figures 4.8, 4.11, 4.15, 4.22, 4.26, 4.41.

Progress check

1. What is meant by clarity when applied as a criterion to assess engineering graphics.
2. Give an example of unclear or ambiguous graphics.
3. What types of cost are incurred when producing engineering graphics?
4. Comment on the statement that minimum cost is the most important criterion when selecting and producing engineering graphics.
5. Name the typical audiences and users for which engineering graphics are produced and list the types of graphics used by each audience or user.

Assignment 4
Identification of graphical methods

This assignment provides evidence for:
Element 2.1: Select graphical methods for communicating engineering information
and the following key skills:
Communication 2.2: Produce written material
Communication 2.3: Use images
Communication 2.4: Read and respond to written material
Information technology 2.1: Prepare information
Information technology 2.2: Process information
Information technology 2.3: Present information
Application of number 2.1: Collect and record data
Application of number 2.2: Tackle problems
Application of number 2.3: Interpret and present data

Ideally, your notes and diagrams will be produced using word processor and spreadsheet so that credit can be given for the key skills of communication and information technology.

Your tasks

Produce a set of notes that identify, and give an example of, each of the following graphical methods:

Graphical methods for communicating in engineering

- sketches
- block diagrams
- flow diagrams
- schematic diagrams
- circuit diagrams
- detail drawings
- assembly drawings.

From these notes three tables should be produced as follows.

- Table 1. This identifies which graphical methods are most suited to communicating the different types of engineering information, including electrical, electronic, mechanical, hydraulic and pneumatic.
- Table 2. This identifies which graphical methods are most suited to communicating the different types of information, including qualitative, quantitative, macro and detail.
- Table 3. This gives brief comments on each type of graphical method with respect to clarity, cost and type of audience for which it is suitable.

The setting out of the tables can be done manually or by using a word processor or spreadsheet.

Table 4.1 shows a template suitable for Table 1, and you will need to construct other templates for Tables 2 and 3. These templates need to be constructed and then the relevant blank 'cells' filled in. All the information to fill in the blank cells can be obtained from this chapter or from your own Activity and Case study work.

Table 4.1 Template

	Mechanical	**Electrical**	**Electronic**	**Pneumatic**	**Hydraulic**
Sketches					
Flow diagrams					
Block diagrams					
Schematic diagrams					
Circuit diagrams					
Detail drawings					
Assembly drawings					

Chapter 5: Scale and schematic drawings in engineering

> **This chapter covers:**
> Element 2.2: Produce scale and schematic drawings for engineering applications.
> **...and is divided into the following sections:**
> - Selecting media to produce drawings
> - Projections, conventions and standards
> - Producing scale drawings for engineering applications
> - Producing schematic drawings for engineering applications
> - Produce a folder of scale and schematic drawings.

Producing drawings, whether manually or by using a computer, is all about communicating information clearly and accurately. Drawing is an essential aid for the design engineer. It is a universally understood modelling tool for communicating ideas and solutions. The solid object to be represented is modelled by drawing it on two-dimensional paper using conventions to represent the true three-dimensional situation. Any drawing will:
- show the properties being modelled, e.g. shape, form, function, size
- be a particular model type, e.g. iconic or analogue (see the introduction to Chapter 4)
- have a particular use, e.g. for simulation, manufacture, investigation, publicity
- have a user or audience, e.g. designer, manufacturing engineer, test engineer, customer

Plotter
A graphical output device that uses a pen to draw lines on paper in response to programmed instructions

Figure 5.1 *The purpose of an engineering drawing is to communicate information so the design can be made. Leonardo da Vinci produced some wonderful drawings of helicopter design in the 15th century, but these designs could not be made!*

- use universally understood standards, e.g. BS 308 Engineering drawing practice, BS 3939 Electrical and electronic graphical symbols, BS 499 Welding symbols, etc. These cover aspects such as projection systems, symbols, types of line, conventions.

Selecting media to produce drawings

Manually
By hand, without the use of any powered devices

Drawings can be done in two basic media:
- manual – either freehand or using appropriate equipment such as rules, tee squares, set squares, draughting machines, etc.
- computer based – the image is constructed on a computer screen using a computerised modelling package; hard copy, i.e. a drawing on paper, can be obtained from a printer or plotter connected to the computer.

Manual medium

The technique of producing drawings manually is one requiring skill and care. It is an essential tool for any engineer and can be learned only by actually doing different kinds of drawing. Table 5.1 shows some of the advantages and disadvantages of manual draughting.

Table 5.1 Some advantages and disadvantages of manual draughting

Advantages	Disadvantages
Low initial cost	Manual storage and slow retrieval of drawings
Initially fast and flexible	Modifications are slow and expensive
Can be done anywhere without electrical power	No linkages to manufacturing systems, etc.
Very suitable for initial sketches of concepts	Manipulation of images and data is difficult
Good for artistic impressions of designs	Animation is not really practical

Tools used in manual drawing

Tools used in manual draughting are:
- line-producing tools such as pencils and pens. Pencils are normally used initially, with pens being used for making a more permanent drawing. A drawing should normally be done entirely in pencil or pen. Pencil lines and pen lines should normally never be mixed on the same drawing.
- equipment to assist line production and measurement, e.g. rules, tee squares, set squares, protractors, pairs of compasses, templates, draughting boards, draughting machines, etc.

Pencils The lines on the paper are made by the pencil lead, which is actually a mixture of graphite and clay. Draughting pencils are available from the softest grade, HB, to the hardest, 10H. Artists' pencils range from the softest grade, 8B, to the hardest, F. Writing pencils are graded from a soft 1 to a hard 4. In practice, most engineering draughting is done with HB, F, H, 2H and 3H pencils. Choosing a grade of pencil lead is often a matter of personal preference. Generally the harder grades are used for the thinner and fainter construction lines that may be erased later. The softer grades are used for the thicker and bolder outlines and

Graphical communication in engineering

Indenting
Permanently deforming a surface

for lettering. Whatever grade is used, the pencil must never be pressed hard. Too much pencil pressure will indent the paper and this indentation cannot be removed. Choice of the correct lead and a minimal pressure will produce a satisfactory line without indenting the paper. Sharpening the pencil correctly is very important. This should be done with a blade and a glasspaper block and not with a normal pencil sharpener. There are three main types of pencil point:
- chisel edge, for thin straight lines and thin curves of large radius
- sharp point, for thin curves of small radius
- round point, for thick straight lines and radii.

These pencil points are shown in Figure 5.2a–c.

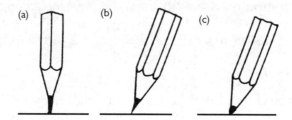

Figure 5.2 Pencil points: (a) chisel edge; (b) sharp point; (c) round point

Lines drawn in pencil can be removed easily with a rubber (also called an eraser). Rubbers come in various grades of hardness. The harder grades are used with harder pencils and harder papers. Too hard a rubber will destroy the paper surface, whereas too soft a rubber will not rub out. A very soft rubber can be used to remove surface dirt from a drawing without affecting any drawn lines.

Pens These are used for producing a permanent line in manual draughting (although it is possible to erase ink lines) and also for plotting when using computer draughting packages. Draughting or plotting pens differ from ball point and writing pens. There is no ball or nib. Instead the ink flows through a circular orifice of controlled size to produce a line of a known and uniform width. The two most used pen sizes are 0.7 mm and 0.35 mm. These give the line widths as recommended in BS 308.

Equipment to assist line production and measurement
Drawing boards Drawing boards are available to suit the standard paper sizes (in mm) of:
- A0 (841 × 1189)
- A1 (594 × 841)
- A2 (420 × 594)
- A3 (297 × 420)
- A4 (210 × 297).

It can be seen that A1 has half the area of A0, A2 half the area of A1, and so on. Figure 5.3 shows the relationship between the recommended sizes of drawing paper/drawing boards.

Drawing boards will always have two accurate straight edges at right angles to enable a tee square to be used with them. Figure 5.4 shows a tee square and drawing board. The surface of a drawing board must be flat and smooth. If it is not, it must be faced with a sheet of flat plastic or other smooth material.

Drawing instruments Figure 5.5 shows a selection of drawing instruments as follows.

Scale and schematic drawings in engineering

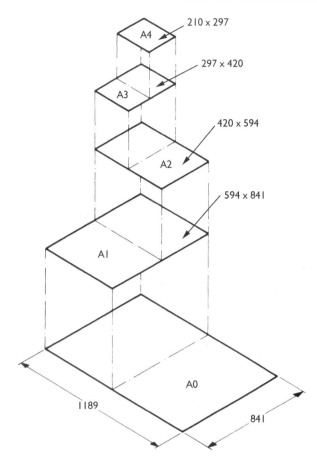

Figure 5.3 Relationship between recommended sizes of drawing sheets

Figure 5.4 Drawing board and tee square

- Compasses/dividers (Figure 5.5a). A pair with one steel point and one lead point or pen is used for drawing circles and radii. A pair with two steel points is used for dividing up lines. A good set for engineering drawing will consist of a small springbow for radii up to 50 mm, a large springbow for radii up to 150 mm and a beam compass for radii over 150 mm.

Graphical communication in engineering

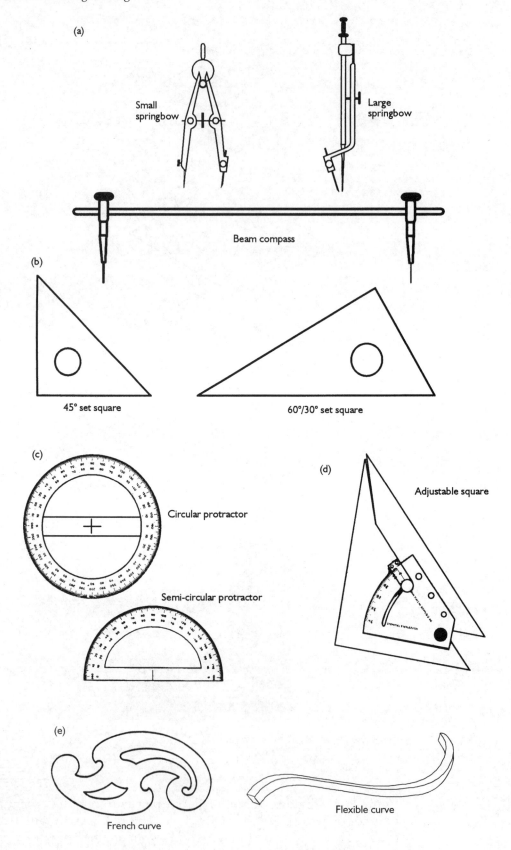

Figure 5.5 Selection of drawing instruments

Scale and schematic drawings in engineering

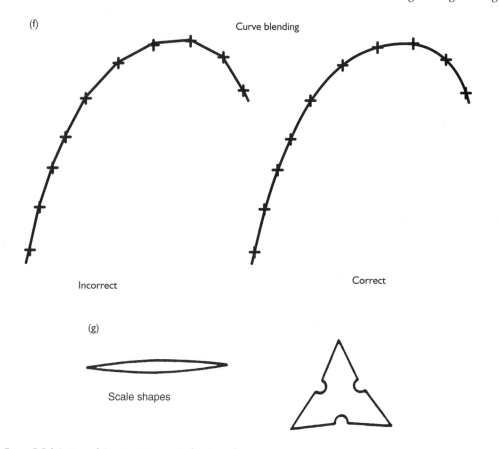

Figure 5.5 Selection of drawing instruments (continued)

- Set square (Figure 5.5b). These are available as 30°/60° and 45°/45°. Used together they can set out angles on multiples of 15°.
- Protractor (Figure 5.5c). Protractors are used for measuring and drawing at any angle.
- Adjustable square (Figure 5.5d). An adjustable square is a versatile instrument that functions as a protractor and set square.
- French curves and flexible curves (Figure 5.5e). These are used for drawing non-circular curves. It is important to blend curves smoothly when using these curves, as shown in Figure 5.5f.
- Scale (Figure 5.5g). This is used for measuring line lengths and linear distances to various scales. The recommended multipliers and divisors for scale drawings are 2, 5 and 10.

Reduction scale factors used are 2:5:10:20:50:100:200:500:1000:2000:5000:10 000.

1:2 is a drawing half the size of the object being drawn.

1:50 is a drawing one fiftieth the size of the object being drawn.

Enlargement scale factors used are 2:5:10:20:50.

2:1 is a drawing two times larger than the object being drawn

50:1 is a drawing 50 times larger than the object being drawn

Graphical communication in engineering

These scale multipliers and divisors refer to a scaling of the linear dimensions. This means, for example, that a scale drawing of 5:1 has an area 25 times (5 × 5) larger than the object being drawn. Similarly, a scale drawing of 1:50 has an area 2500 times (50 × 50) smaller than the object being drawn.

Stencils or templates Stencils or templates are available to enable rapid drawing of features such as fillet radii, small holes, and electrical and electronic components. Figure 5.6a shows a stencil for small holes/fillet radii and Figure 5.6b shows one for electronic components.

Eraser shield An eraser shield is a piece of thin steel with holes and slots that help in selectively erasing portions of a drawing.

Draughting machines These combine the functions of drawing board, tee square, set squares, protractor and scale rule. Their use allows for maximum productivity when doing manual draughting. They are available for all the standard paper/drawing board sizes, i.e. A0, A1, A2, A3 and A4. Figure 5.7 shows a typical draughting machine.

(a)
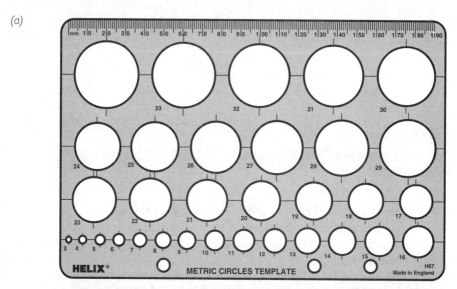

(b)
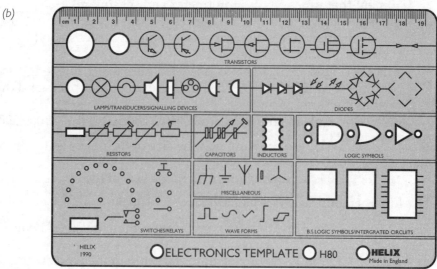

Figure 5.6 (a) Stencil for small holes and fillet radii; (b) stencil for electronic components

Scale and schematic drawings in engineering

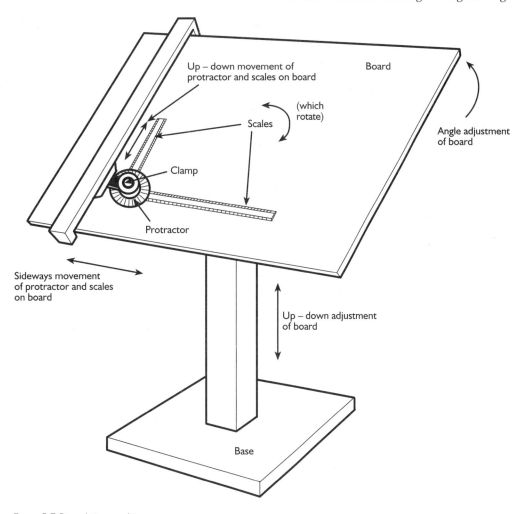

Figure 5.7 Draughting machine

Line types

Pencils or pens can be used to produce a variety of line types. This section is equally applicable to computer-based methods as appropriate size pens will be used when plotting out drawings. Table 5.2 shows the thickness and type of line and gives examples of applications.

The line thicknesses used are from the range (all in mm) of :

$$0.25:0.35:0.5:0.7:1.0:2.0$$

These lines should be of consistent thickness and density, with the thick and thin lines in the ratio 2:1. Line thicknesses of 0.35 mm and 0.7 mm are frequently used in engineering drawings. The application of these lines in a typical drawing is shown in Figure 5.8.

Activity 5.1

In Figure 5.8 identify examples of:
- visible outline
- hidden outline
- hidden edge

Table 5.2 Thickness and types of line

Line	Description	Application
A ▬▬▬▬▬▬	Continuous thick	A1 Visible outlines A2 Visible edges
B ───────	Continuous thin	B1 Imaginary lines of intersection B2 Dimension lines B3 Projection lines B4 Leader lines B5 Hatching B6 Outlines of revolved sections B7 Short centre lines
C ∼∼∼∼	Continuous thin irregular	C1 Limits of partial or interrupted views and sections, if the limit is not an axis
D ─⋀─⋀─	Continuous thin straight with zigzags	D1 Limits of partial or interrupted views and sections, if the limit is not an axis[a]
E ▬ ▬ ▬ ▬ ▬	Dashed thick	E1 Hidden outlines E2 Hidden edges
F ─ ─ ─ ─ ─	Dashed thin[b]	F1 Hidden outlines F2 Hidden edges
G ──·──·──	Chain thin	G1 Centre lines G2 Lines of symmetry G3 Trajectories and loci G4 Pitch lines and pitch circles
H ▬·──·──·▬	Chain thin, thick at end and changes of direction	H1 Cutting planes
J ▬·▬·▬·▬	Chain thick	J1 Indication of lines or surfaces to which a special requirement applies (drawn adjacent to surface)
K ·──··──··──	Chain thin double dashed	K1 Outlines and edges of adjacent part K2 Outlines and edges of alternative and extreme positions of movable parts K3 Centroidal lines K4 Initial outlines prior to forming K5 Parts situated in front of a cutting plane[c] K6 Bend lines on developed blanks or patterns

Note: The lengths of the long dashes shown for lines G, H, J and K are not necessarily typical owing to the confines of the space available. [a]This type of line is suited for production of drawings by machines. [b]The thin F-type line is more common in the UK, but on any one drawing only one type of dashed line should be used. [c]Included in ISO 128-1962 and used mainly in the building industry.

Scale and schematic drawings in engineering

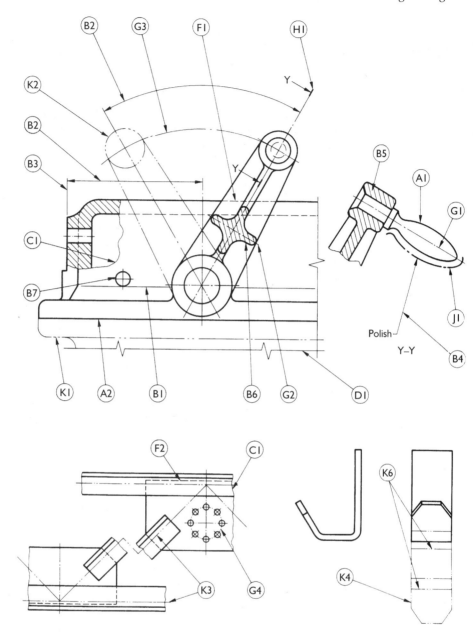

Figure 5.8 Application of various line types: BS 308; part 1; 1993

- cutting plane
- leader line
- limits of partial view
- centre line
- cross-hatching
- surface with special requirements
- extreme position of movable part
- dimension line
- trajectory/loci
- imaginary line of intersection
- initial outline prior to forming
- bend line.

Trajectory

The path of an object; a locus (plural – loci) is the path of an object constrained by a mechanism or a control system

Graphical communication in engineering

Activity 5.2

Draw Figure 5.8 in both pen and pencil using the correct line types and widths.

Lettering

Lettering on engineering drawings is a matter of individual preference. However, clear and neat lettering is very important for high-quality draughting. The main points to try to achieve are:
- uniformity of size and style
- legibility
- correct spacing
- use of upper case whenever possible
- emphasis of importance of lettering by size rather than by underlining.

Figure 5.9 shows the recommended British Standard lettering/numbering and is for guidance only.

Note that both straight and sloping styles are shown, but these should not be mixed on one drawing. The minimum height for characters in lettering and numbering is shown.

Computer-based medium

Computers are being used increasingly for engineering drawing and design. Compared with manual methods they can be expensive to start with. At 1997 prices a personal computer with a high-resolution screen and a colour plotter for hard copy can cost up to £7000, although cheaper versions with lower resolution and an A4 plotter can be bought for around £1200. CAD (computer-aided de-

Minimum character heights are
- For drawing numbers, etc. A0,A1,A2 and A3   7mm
  A4   5mm
- For dimensions and notes A0   3.5mm
  A1,A2,A3 and A4   2.5mm

Figure 5.9 Recommended British Standard lettering and numbering

Scale and schematic drawings in engineering

sign) software packages vary in price according to speed and capability. A typical cost to an industrial user would be in the region of £2000, but educational versions are available for about £200. However, this cost can often be recovered by the extra productivity gained compared with manual methods. In terms of actually producing a drawing, there may be little to choose between manual and computerised methods. Table 5.3 shows some of the advantages/disadvantages of computerised methods.

Table 5.3 Advantages and disadvantages of computerised methods

Advantages	Disadvantages
Modifications are easy to make	Initially not as fast as manual methods
All types of view easily displayed/plotted	Difficult to use for making rough sketches or artistic impressions
Multiple copies can be produced rapidly	
Libraries of parts/assemblies can be stored on disk and retrieved easily	Possible loss of data if the system goes down (which can be prevented by making frequent back-up copies)
Complex assemblies are easily managed by keeping a database of subassemblies and parts	
Manufacturing programs can be produced directly from the drawing	
Parts can be tested for stressing using finite element meshes produced from the part drawing	
Simulations of circuits, mechanisms, etc. can be carried out	
Files of standard parts such as springs, fasteners, electronic/electrical parts can be held on disk	

As computer graphics are displayed on a screen a coordinate system is used. This uses the normal graphical convention of x and y coordinates, with x values increasing from left to right and y values increasing from bottom to top. The origin $x = 0$, $y = 0$ is normally taken as the bottom left-hand corner of the screen, or as the middle of the screen as shown in Figure 5.10.

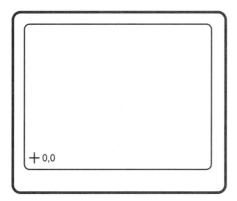
Computer screen – origin bottom L.H. corner

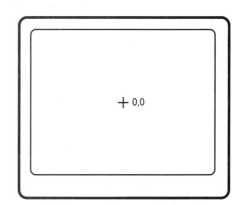
Computer screen – origin middle of screen

Figure 5.10 Origins on computer screens

Graphical communication in engineering

Vector-based computer graphics

The majority of computer-aided draughting and design systems, e.g. Autocad, Silverscreen, fastCAD, etc., use vector graphics. A vector is a line representing both magnitude and direction. Figure 5.11 shows a vector **v**, joining points P_1 and P_2, of length 90 mm at a direction of 30° from the horizontal. On the computer this vector would be generated by an electron beam being 'steered' across the VDU (visual display unit).

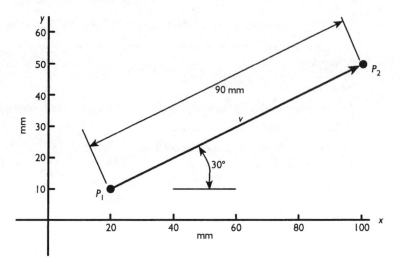

Figure 5.11 *Vector joining points P_1 and P_2*

One big advantage of vectors is that very little information has to be stored in the computer, as a line is defined only by position (of point P_1), length and direction. It is also very easy to change this line to alter the drawing. The line can be:
- rotated
- scaled up or down
- moved to another position (translated).

Figure 5.12 shows the original vector (a) rotated clockwise by 90°, (b) scaled by 1.5:1 and (c) translated by –10 mm in x and +10 mm in y.

Because vectors are straight lines, they give good line quality when output to a plotter, and so are ideal for engineering drawings. Vectors can be combined to give any shape required. Even circles and other curves can be closely approximated if enough vectors are generated. Figure 5.13 shows vectors being combined to form (a) an octagon and (b) a circle.

Raster-based graphics

This is a system often used for drawing and 'painting' packages such as Microsoft Paint. A raster-scan display works by an electron beam continuously scanning a series of horizontal lines. This scan line is subdivided across the screen into a number of individual elements. This means that the whole screen is a grid pattern of picture elements, which are called pels or pixels. This grid pattern is ideal for displaying lettering and for filling in areas with colour, but it is less suitable for computer-aided draughting. Typical screen resolutions are 1024 by 1024, giving a total of 1048576 pixels. There are two main problems with raster graphics for use in engineering drawing:
- the 'staircasing' effect on lines (shown in Figure 5.14)
- huge amounts of memory are required – one simple vector has to be represented by possibly thousands of pixels.

Scale and schematic drawings in engineering

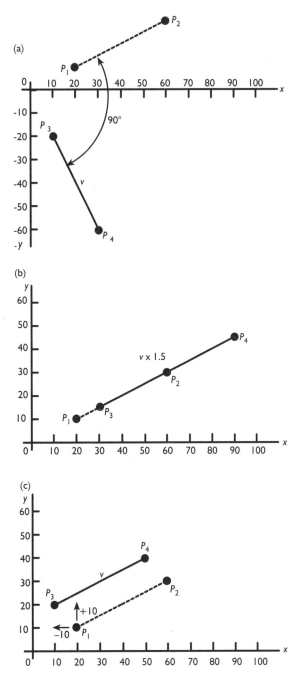

Figure 5.12 The vector, v, from Figure 5.11 (a) rotated clockwise by 90°, (b) scaled by 1.5:1 and (c) translated by −1 in x and +1 in y

Thus, the system is generally not suited for use in draughting packages.

Activity 5.3

Produce the shape in Figure 5.15 on:
- a vector-based computer graphics system
- a raster-based computer graphics system.

In each case:

Graphical communication in engineering

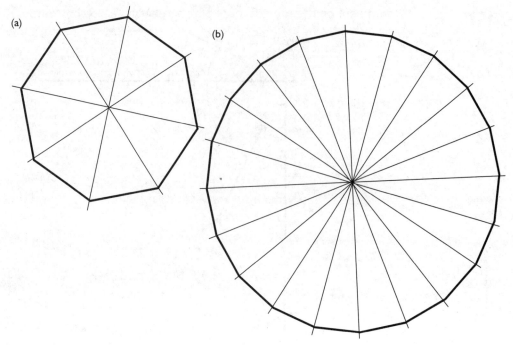

Figure 5.13 Vectors combined to form (a) an octagon and (b) a circle

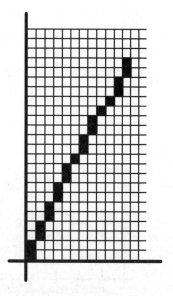

Figure 5.14 'Staircasing' effect with raster graphics

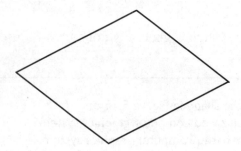

Figure 5.15 Shape for activity 5.3

Scale and schematic drawings in engineering

- comment on the line quality of both screen and hard-copy output
- save to disk and state the amount of memory required to store the image
- state the time taken.

Do the drawing manually and comment on the difference in time taken.

Equipment used in computer graphics

Unlike manual draughting, no special tools or craft skills are required to produce high-quality drawings. The basic equipment consists of:

- a 'stand-alone' personal computer or networked intelligent terminal
- output devices including a high-resolution VDU, a plotter to give hard copy for all required sizes of drawings, a printer for lists, schedules, results of calculations, etc.
- input devices including keyboard, mouse, graphics tablet, scanner
- storage and input/output devices including floppy disk, hard disk, compact disk (CD).

Figure 5.16 shows a typical set-up for producing drawings by computer-aided methods.

Techniques for producing drawings in computer-aided draughting systems

Software will vary as to how the lines and drawings are created. However, there are similarities, and most systems will use menus to create the different entities such as lines, circles, polygons, etc. To draw a line, this option is selected from the menu. The user is then prompted for a start point, which is selected by keyboard, mouse or tablet pen (collectively called pointing devices). The line can then be extended from the start point by the pointing device to the chosen length and in the chosen direction. The system will then ask for an end point, which is

Figure 5.16 CAD set-up showing the Silverscreen Solid Modeller being used

(Leonardo Computer Systems, Reading)

Graphical communication in engineering

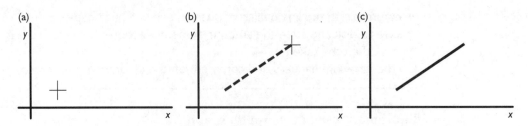

Figure 5.17 Procedure for creating a line: (a) point created; (b) line extended from point; (c) line confirmed

entered by the pointing device. Similarly, for circles and polygons, the user will be prompted for centre position, radius, number of sides, etc. Figure 5.17 shows the procedure for creating a line.

Activity 5.4

Use a computer-aided draughting package to draw Figure 5.8 using the correct line types and widths.

Comparison of manual and computer-aided media

These can be compared with respect to:
- speed
- flexibility
- cost
- presentation
- storage
- retrieval.

Activity 5.5

All the comparison factors listed above have been referred to in this section. Present a comparison table that would assist anyone making a choice between manual and computer-aided draughting.

Progress check

1. What are the purposes of engineering drawings?
2. Name the essential characteristics of any engineering drawing.
3. What are the advantages and disadvantages of manual draughting?
4. List the tools used to produce drawings manually.
5. State (a) the recommended sizes for engineering drawings and (b) the line widths recommended by BS 308 and give their uses.
6. What are the advantages and disadvantages of computer-aided draughting?
7. Describe vector-based graphics and give the reasons why they are the main method used for computer-aided draughting.
8. Describe raster-based graphics and their main uses. State why they are not entirely suitable for computer-aided draughting.
9. What equipment is required for a computer-aided draughting system that needs to produce a hard copy.
10. Describe how a line may be drawn on a typical computer-aided draughting package.

Scale and schematic drawings in engineering

Projections, conventions and standards

Orthographic
Using straight lines and right angles

As has been said before, the language of engineering is the engineering drawing. In order that information may be transmitted correctly and at minimum cost, it is vital that the drawings are produced according to universally accepted rules and conventions. The rules and conventions are detailed in BS 308 Engineering drawing practice. The abridged version for schools and colleges, PP7308, is adequate for most courses in engineering drawing, including GNVQ. These standards should be referred to when doing drawings and activities.

Projections and sectioning

This topic was covered briefly in Chapter 4 but it will be looked at in greater detail here. All drawings have to represent a three-dimensional object on two-dimensional paper. In engineering drawings this is done by using orthographic projection, in which an object is usually represented by three or more views (although in some cases one or two views are sufficient) projected from the X, Y and Z planes of the three-dimensional object. Pictorial projection attempts to give a three-dimensional impression of the object by combining the X, Y and Z projections into a single view.

Orthographic views: first- and third-angle projections

Figure 5.18 shows a car drawn in third-angle orthographic projection. Starting with the side view of the car:

- the rear view is projected from the left-hand side and drawn on the left-hand side
- the front view is projected from the right-hand side and drawn on the right-hand side

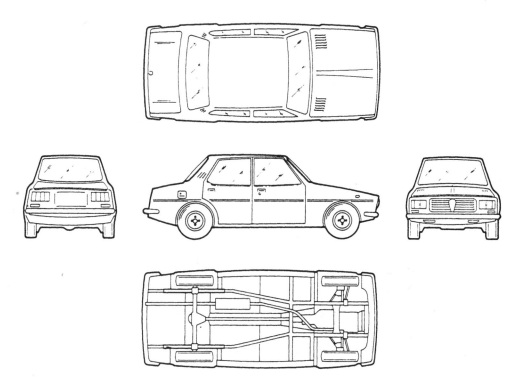

Figure 5.18 Orthographic projection of a car

Graphical communication in engineering

- the top plan is projected from the top and drawn above
- the bottom plan is projected from the underneath and drawn below.
  In first-angle projection the views would look the same but:
- the rear view would be projected from the left-hand side and drawn on the right-hand side
- the front view would be projected form the right-hand side and drawn on the left-hand side
- the top plan would be projected from the top and drawn below
- the bottom plan would be projected from the underneath and drawn above.

Activity 5.6

Sketch the five orthographic views of the car in first-angle projection.

Case study

Figure 5.19a is a pictorial view of a bracket. Figure 5.19b and d shows how first- and third-angle orthographic views are projected, using one view as a starting point. The completed drawings, with projection lines removed, are shown in Figure 5.19c and e. Note the truncated cone symbols for first- and third-angle projections.

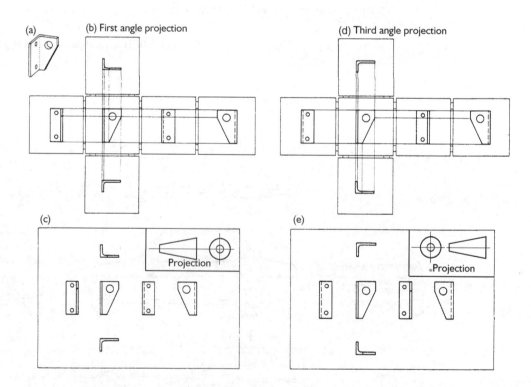

Figure 5.19 (a) Pictorial view of a bracket; (b) and (d) how first and third angle views are projected; (c) and (e) completed drawings, with projection lines removed

Scale and schematic drawings in engineering

Hidden detail

When drawing views in orthographic projection some lines will be invisible because they are:

- internal
- hidden by other features.

Such lines are referred to as hidden detail and are conventionally shown by using dashed thin lines as shown in Table 5.2. Commonsense should be applied to hidden detail lines. If they are liable to cause confusion or misinterpretation they can be omitted. Also, if lines coincide, visible lines take precedence over hidden lines, and hidden lines take precedence over centre lines. Hidden detail is often best shown by a sectional view. Sectional views are described later in this chapter.

Figure 5.20 shows the use of hidden detail in various situations.

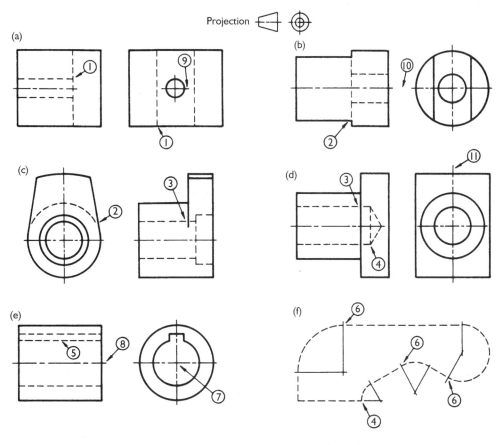

1. Hidden lines touch outlines and other hidden lines
2. Gap where hidden line continues outline
3. Where a hidden line crosses an outline a gap is left each side of the outline
4. Hidden lines meet at a point
5. Gaps staggered in parallel hidden lines
6. Hidden arcs stop at tangent points
7. Centre lines cross at long dashes
8. Long dash at ends of centre line extends a short distance past the outline
9. For small circles (up to about 12 mm diameter) centre lines are continuous thin lines
10. Centre lines do not extend across the space between views
11. Line of symmetry of the rectangular base of the part

Figure 5.20 Treatment of hidden lines and centre lines

Graphical communication in engineering

Activity 5.7

Make an orthographic drawing, showing hidden lines, of the shaped block in Figure 5.21, using both manual and computer-aided methods.

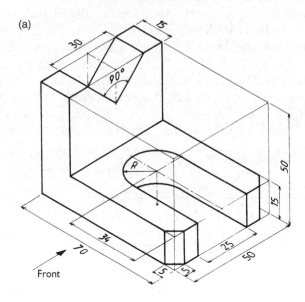

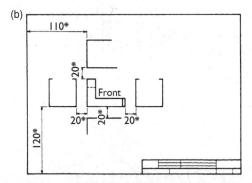

Figure 5.21 Shaped block (a) and initial stages of its orthographic drawing (b)

Auxiliary projection

As well as using the three mutually perpendicular planes for projection, planes at any angle can also be used. This is necessary on parts that have inclined faces, in order to show the true lengths and shapes of the inclined face.

Figure 5.22a and b shows two parts with inclined faces. It can be seen, in each case, that an auxiliary plane that is parallel to the inclined face has been used. This auxiliary view has the same widths as the normal view, and this is shown in Figure 5.22.

Scale and schematic drawings in engineering

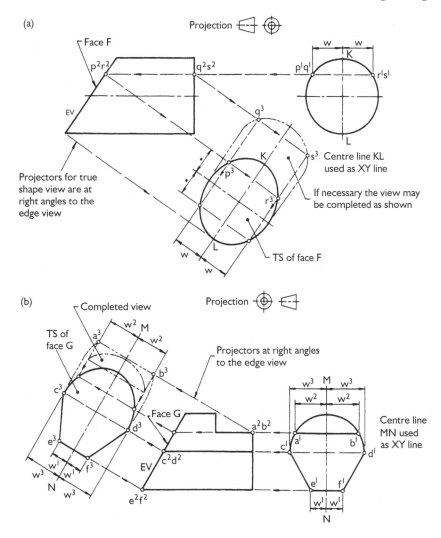

Figure 5.22 Projection of curves in true shape views

Activity 5.8

Draw the orthographic views as indicated in Figure 5.23 and add an auxiliary view projected from the long-angled face. Do the drawing both manually and using computer-aided draughting.

Sectioning

Ambiguity

When two or more meanings are communicated

As discussed above, hidden interior detail of drawings is difficult to show. Complicated interior detail is easily shown if the part is cut along a suitable plane. The portion of the part in front of the cutting plane is then removed, thus revealing the interior detail. A sectional view is included on a drawing to:
- reduce ambiguity and confusion by having fewer hidden detail lines
- improve representation of internal features
- make dimensioning of internal features easier
- show how hidden parts are assembled.

193

Graphical communication in engineering

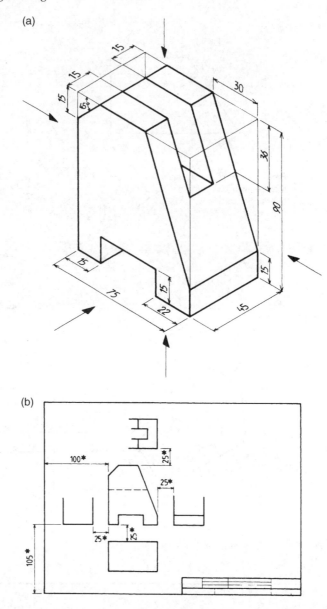

Figure 5.23 Drawing of part for activity 5.8

Counterbored

A larger hole drilled on the same axis as a smaller hole

The sectioning procedure is shown in Figure 5.24 and consists of:
(a) selection of the part, in this case a cylindrical part with a counterbored hole
(b) initial drawing of two orthographic views
(c) cutting the part along the cutting plane with an imaginary cutting tool and separating the two halves
(d) a final drawing of the sectional view of the part with hatching at 45° using thin lines to indicate the cut surface.

Case study

This case study indicates how to section various parts and assemblies. Figure 5.25 shows the sectioning of an assembly. The application of sectioning to different types of parts is shown in Figure 5.26.

In some cases, features or parts in an assembly are left unsectioned. This is done to keep the drawing as clear and uncluttered as possible.

194

Scale and schematic drawings in engineering

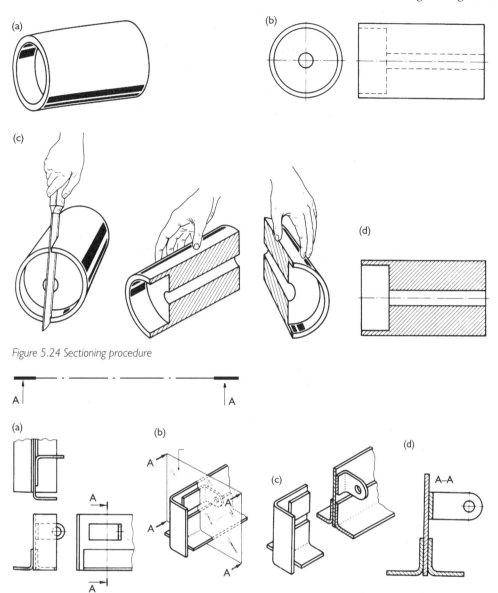

Figure 5.24 Sectioning procedure

Figure 5.25 Sectioning of an assembly. (a) Orthographic views and indication of cutting plane as shown by plane A–A. Note the use of the thick ends of lines joined with a chain dotted line, plus arrows to show the direction of viewing. (b) An isometric view of the assembly with the cutting plane A–A. (c) An isometric view of the assembly after sectioning. (d) The resulting sectional view of the assembly when viewed in the direction of the arrows A–A; note the use of hatching in different directions to indicate the different parts

> Examples of unsectioned features and parts are:
> - ribs, webs and spokes
> - shafts and small solid cylindrical parts
> - screw threads and fasteners including nuts, bolts, washers and rivets
> - keys and cotters
> - plugs and studs
> - balls and rollers in bearings
> - split pins and taper pins
> - gear teeth.
>
> Figure 5.27 is an assembly showing typical unsectioned features and parts.

Graphical communication in engineering

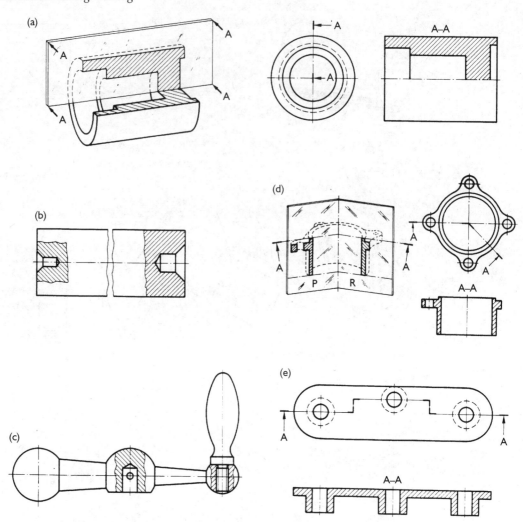

Figure 5.26 Application of sectioning to different types of parts. (a) A half-section used for a symmetrical part to reduce the amount of drawing. (b) and (c) Scrap sections are used to section only the areas where an internal feature is to be shown. (d) and (e) Off-set sections are used to section the desired parts of the drawing. The part in (d) has an off-set section to show both the flange with the hole and the flange without the hole. The part in (e) has an off-set section to show the details of all three circular bosses

Activity 5.9

For Figures 5.28 to 5.30 complete the drawings as instructed.

Pictorial projections

These were mentioned in Chapter 4 when working on the identification and selection of graphical methods. They will now be covered so that the actual techniques of producing pictorial views are shown. For many engineering drawings, the three orthographic views are sufficient for design and manufacturing purposes. However, there are some drawings used for publicity, modelling, assembly, servicing, etc. that need a three-dimensional representation. These three-dimensional views, usually isometric or oblique, are called pictorial projections.

The advantages of pictorial projections include:
- clear presentation of information to customers and clients
- easier understanding of complex designs

Scale and schematic drawings in engineering

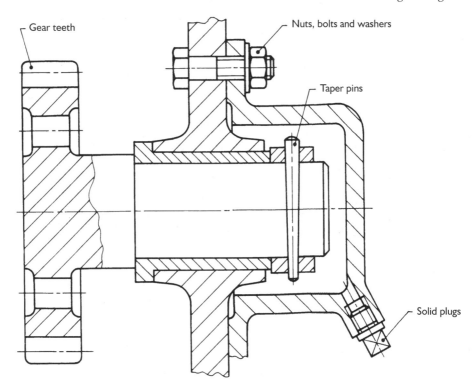

Figure 5.27 Details not sectioned on assemblies

- better presentation of ideas at initial design stage
- possible avoidance of the necessity of building an actual model; this is particularly true for computer graphics where 'solid' computer models are used to speed up design and development work in engineering.

On the other hand, pictorial projections do have disadvantages, which include:

- excessive time required to produce realistic views
- difficulty of fully dimensioning
- some lines cannot be measured
- circles, curves and other shapes can be distorted and take a long time to draw manually (this is not such a problem in a good computer-aided draughting package)
- production of isometric and oblique pictorial views.

It is a useful skill to be able to make pictorial views easily and quickly. Figure 5.31 shows the basic principles of producing isometric and oblique views of a cube.

Case study

Figure 5.32 shows orthographic views of two blocks. This case study illustrates the methods of producing drawings of the two blocks in both isometric and oblique projection.

The basis of the method, in all cases, is to construct (in faint construction lines) a rectangular prism of the same size as the block. The parts to be cut out are then drawn in as faint construction lines. The cut-out lines are then erased and the finished drawing lines drawn in bold.

An alternative method is to use grids, as shown in Figure 5.33.

The isometric grid (a) uses vertical lines and lines at 30° to the horizontal. Thus, instead of using the outline prism, the part can be drawn directly on the

Graphical communication in engineering

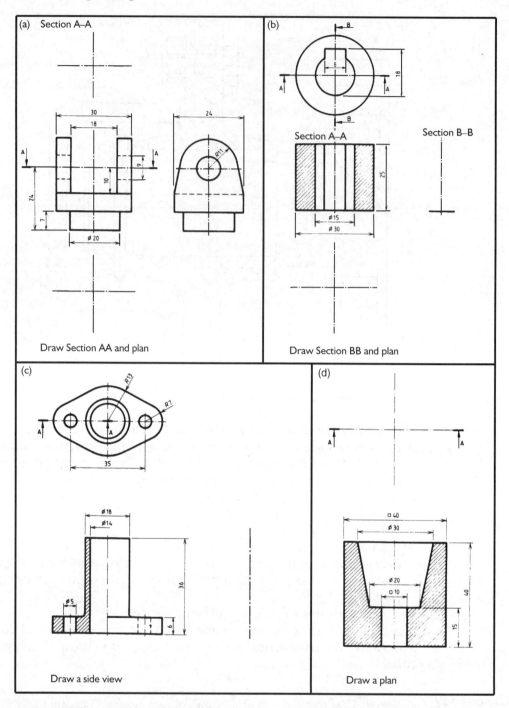

Figure 5.28 Drawings for activity 5.9

grid. In one of the examples used there is a circle. Note that a circle in isometric projection is drawn as an ellipse. The use of a grid enables the ellipse to be constructed by the use of ordinates. The construction of ellipses by use of ordinates is shown in Figure 5.34.

The oblique grid in Figure 5.33b uses vertical lines and lines at 45°. The oblique views can now be drawn directly on to the grid.

Scale and schematic drawings in engineering

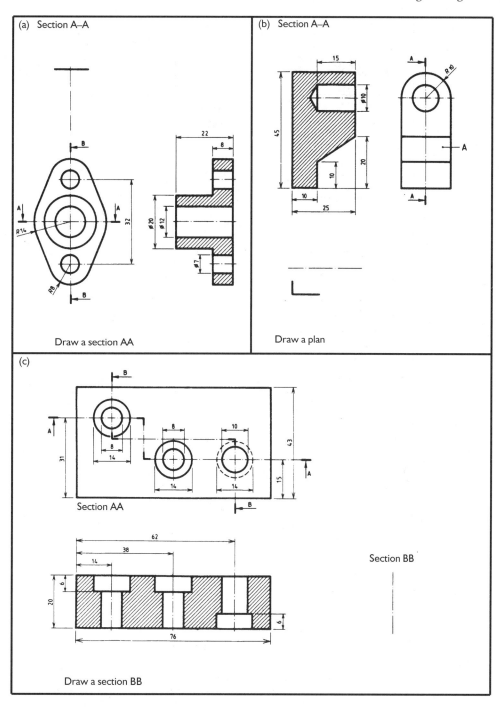

Figure 5.29 Drawings for activity 5.9

Activity 5.10

Using the grids in Figure 5.33 draw isometric and oblique views of the parts shown in Figure 5.20a–e.

199

Graphical communication in engineering

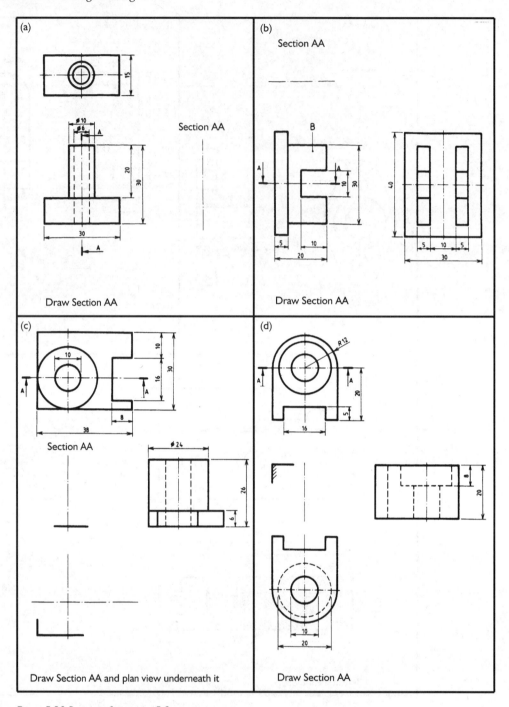

Figure 5.30 Drawings for activity 5.9

Scale and schematic drawings in engineering

(a)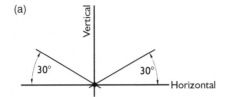

(b)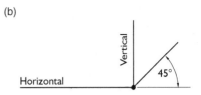

For isometric sketches vertical lines remain vertical whilst receding lines are shown to the same scale at 30° to the horizontal. So, to sketch a simple cube of side 20 mm in isometric projection proceed as below

For oblique sketches the front face of a component is shown as it is; the receding lines are shown at an angle of 45° and to a reduced scale, to avoid a foreshortening effect. There are no fixed rules for the reduced scale of the receding lines, but sketches will appear in good proportion, if half scale is used. So, to sketch a simple cube of side 20 mm in oblique projection, proceed as below.

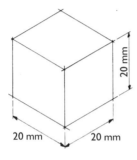

Stage 1: sketch, showing the simple basic shape (note the receding lines at 30° to the horizontal)

Stage 1: sketch, showing the simple basic shape (note the receding lines at 45° and 1/2 scale)

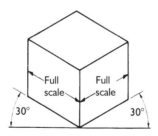

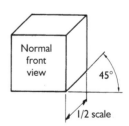

Stage 2: line it in

Stage 2: line it in

Figure 5.31 (a) Isometric sketches; (b) oblique sketches

Graphical communication in engineering

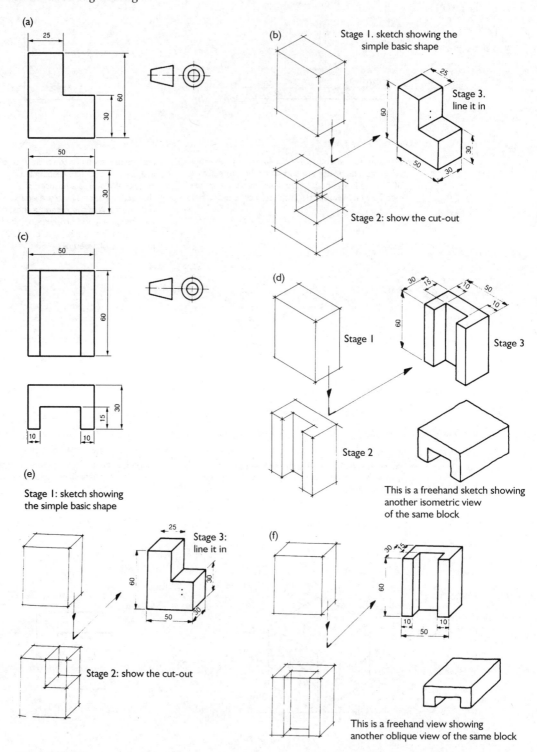

Figure 5.32 (b) Isometric projection of the blocks in (a). (d) Isometric projection of the blocks in (c). (e) Oblique projection of the blocks in (a). (f) Oblique projection of the blocks in (c)

Scale and schematic drawings in engineering

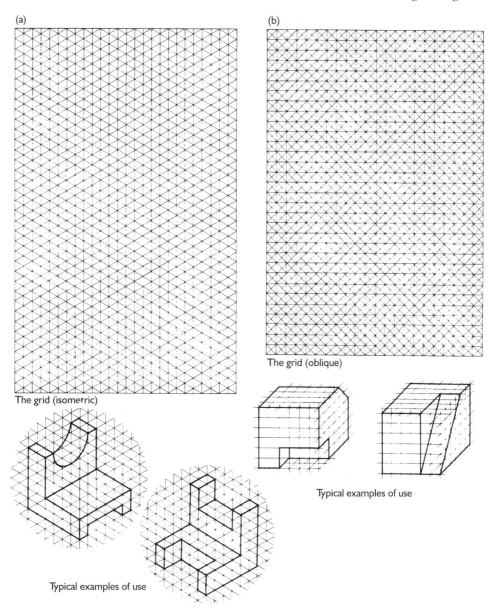

Figure 5.33 (a) A grid for guided sketching (isometric); (b) a grid for guided sketching (oblique)

203

Graphical communication in engineering

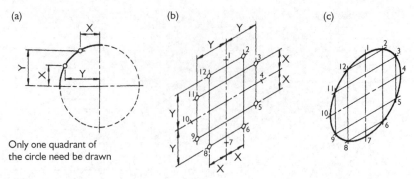

Figure 5.34 Circle construction by ordinates

Conventions

Symbolises
Conveys a meaning by using something to represent something else, e.g. 2 symbolises a quantity of two

A convention, in graphical terms, is defined as a representation that simplifies, symbolises or substitutes for the natural form. It is, if you like, a form of graphical shorthand. Most of these conventions have been standardised and are illustrated in the appropriate British or International standard. The use of conventions will:

- speed up the drawing process
- reduce the cost of drawing
- save space on the drawing
- improve the look of the drawing
- enable the drawing to be easily understood by engineers anywhere in the world
- allow the use of templates, stencils and standardised computer graphics.

Most of the conventions used are specified in the BS 308 Engineering drawing practice. Reference should be made to BS 308 and PP 7308, which is an abridged version of BS 308 for schools and colleges.

Standard drawing conventions and representations

Table 5.4 shows a selection of standard representations for features as recommended in BS 308, including screw threads, interrupted views, repeated parts and features, plane faces, knurling, rolling bearings. It can be seen that the use of these conventions will make drawings understandable to anyone who has a knowledge of BS 308.

Standard abbreviations are used to save space and reduce drawing effort. Table 5.4 shows a selection of commonly used standard abbreviations.

Dimensioning

Functional
Performing a desired task as required

Dimensions are of extreme importance in engineering drawing. Use of correct dimensions and tolerances ensures that:

- parts can be made accurately
- parts can be made economically
- parts can be assembled
- finished parts and assemblies function properly.

To help meet the above requirements, the general principles of dimensioning are as follows:

- each dimension necessary for the complete definition of a finished product should be on the drawing and should appear once only
- calculation of dimensions from other dimensions should not be necessary
- scaling of drawings to obtain dimensions should not be necessary

Scale and schematic drawings in engineering

- dimensions relating to a single feature should generally be next to that feature on a single view
- surfaces should also be dimensioned, if required, for features such as flatness and smoothness.

Drawings must be dimensioned using standards for:
- lettering
- numbering
- tolerancing
- projection lines
- dimension lines
- leader lines
- arrowheads and dots
- 'balloons' for letters and numbers.

Types of dimensions

Depending on their function and purpose dimensions are of three types:
- functional dimensions, which specify the size, shape or location of various features
- non-functional dimensions, which may assist in manufacturing or quality control but have no direct effect on function
- auxiliary dimensions, which may give useful information for things such as raw material size, but are not necessary for manufacture or quality control.

The assembly and components in Figure 5.35 show all three types of dimensions. Functional dimensions are marked F, non-functional dimensions NF and an auxiliary dimension AUX.

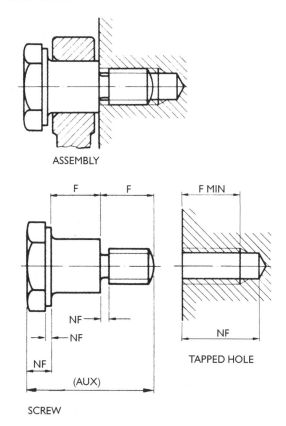

Figure 5.35 Showing functional (F), non-functional (NF) and auxiliary dimensions (AUX; given without tolerances, for information only)

Graphical communication in engineering

Table 5.4 Conventions for common features (BS 308)

Representation	Notes
Screw threads (visible, hidden; insection and assembled)	
(a) (b) (c) (d) (e) (f) (g) (h) (i) (j) (k) (l) (m) 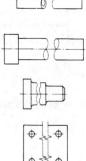	The following applies *independently* of the type of screw threads **For visible threads** Crests are shown by a type A line Roots are shown by a type B line On an end view the line showing the root diameter extends for not fewer than three quadrants (b and J) **For hidden threads** Crests and roots are shown by type F lines (f and i) On end views the line showing the root diameter is also type F and extends for not fewer than three quadrants (e) **For sectioned views** Hatching extends to the crest diameter (c, d, g, h and k) Taping hole sizes are shown by a type A line (f) To show the limit of useful length of thread, a transverse line extending to the major diameters is used: type A for visible threads (a and h) type B for hidden threads (c and f) On sectional views where parts are shown assembled, external parts are shown covering the internally threaded parts (l and m)
Interrupted views	In order to save space, only those parts of a large or long object that are sufficient for its definition are shown Breaklines are shown using a type B line and revealing the circular shape The type D breakline continues for a short distance beyond the outline

206

Scale and schematic drawings in engineering

Repeated parts and features

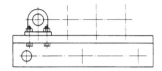

In order to avoid repeated illustrations of identical features or parts, they are shown only once, the position of others being indicated by their centre line

Patterns of repeated features

When several holes, bolts, rivets, slots, etc. are required in a regular pattern, only the locating centre lines to establish the pattern are shown

Other information is given in a note

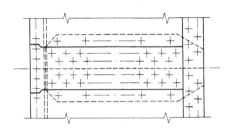

Plane faces on cylindrical parts

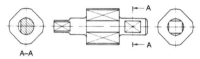

Flat surfaces, e.g. squares, tapered squares and local flats, are shown by thin, crossed, diagonal lines

Knurling

Straight knurl   Diamond knurl

The type of knurling is shown by type B lines on the surface to be knurled

Rolling bearings

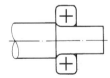

Rolling bearings are shown in this manner, without taking into account the type of bearing (e.g. roller or ball, radial or thrust)

Rules of dimensioning
The 'rules' of dimensioning are fully covered in BS 308 and some of them are shown in Figure 5.36.

Tolerances
Tolerances are the amounts by which variables such as sizes can vary. This means that any dimension will have:
- an upper limit and a lower limit

Graphical communication in engineering

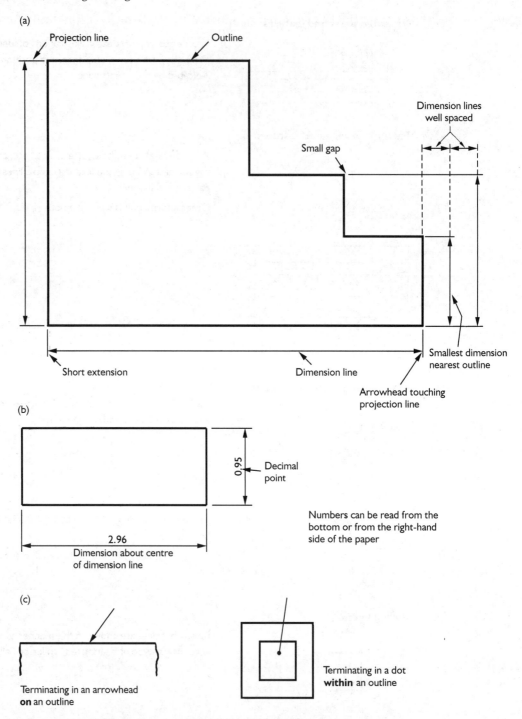

Figure 5.36 (a) Projection and dimension lines. (b) Positioning of numerals and decimal points. (c) Termination of leaders from a note, symbol or numeral to a surface or within an outline. (a)–(c) Arrowheads, which should be between 3 mm and 5 mm long

- a tolerance that is equal to the difference between the upper limit and the lower limit, i.e. tolerance = upper limit − lower limit.

Tolerances are necessary because all manufacturing processes produce parts of different sizes, even when set to produce the same size. Designers have to work within the limits that can be produced economically, while still ensuring that the parts function properly. Figure 5.37 shows some of the principles applied when tolerancing drawings. These are:

Scale and schematic drawings in engineering

(a) use the largest tolerance that still allows correct function; the graph shows the relationship between tolerance and production costs.
(b) either specify a size with limits above and/or below (i) or specify both limits of size (ii)
(c) specify angular tolerances using the same principles as for linear tolerances
(d) put a general tolerance on a drawing, to apply to all dimensions unless otherwise stated.

Datums and dimensioning

A datum is a reference point. For use in dimensioning and measuring it may be defined as a point, edge, surface or line from which a feature is dimensioned or measured. The application of datums and dimensioning methods are shown in Figure 5.38, which shows:

(a) the surface of a part used as a datum line from which all the features are dimensioned
(b) a hole axis line used as datum line
(c) a method of dimensioning from a datum, where space is limited
(d) a common corner point used as a datum for both directions
(e) features individually dimensioned and toleranced from a common datum (progressive dimensioning – sometimes called parallel dimensioning)
(f) features dimensioned in series, with no common datum (chain dimensioning)
(g) (and h) the effect of using chain dimensioning; the drawings show the dimensions produced when all individual tolerances are at their maximum positive or negative tolerances

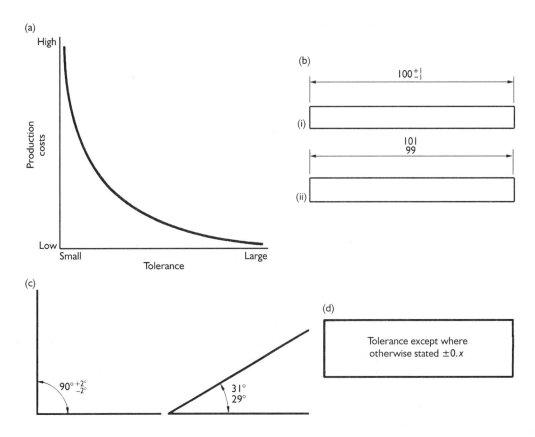

Figure 5.37 Principles applied when tolerancing drawings

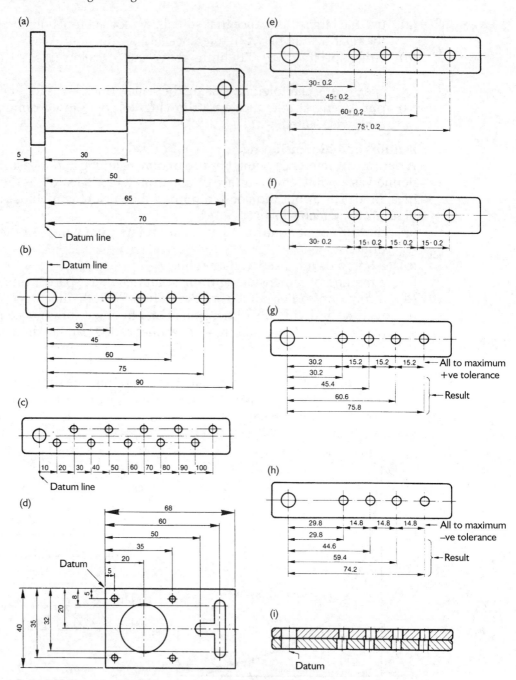

Figure 5.38 The application of datums and dimensioning methods

(h) two assembled plates; the bottom one is progressively dimensioned, with the top one being chain dimensioned; it can be seen that the accumulated errors in the top plate have made assembly impossible.

The basic message of datums, dimensioning and tolerancing can be summarised as:
- progressive dimensioning from a common datum avoids the cumulative build-up of tolerances
- chain dimensioning can result in excessive cumulative build-up of tolerances.

Scale and schematic drawings in engineering

Case study

Figure 5.39 shows an example of an assembly in which the cumulative build-up of tolerances cannot be avoided. This assembly consists of three rollers of different diameters, which are assembled on to a pin. If all the rollers are made to the maximum positive tolerance, the overall length of the assembly will be: 20.2 + 20.2 + 20.2 = 60.6 mm. If all the rollers are made to the maximum negative tolerance, the overall length of the assembly will be: 19.8 + 19.8 + 19.8 = 59.4 mm. This gives a possible total variation of 60.6 − 59.4 = 1.2 mm. If this is unacceptably large then the individual tolerances will have to be reduced, e.g. a reduction of tolerance to 0.1 mm on each roller will give a maximum total variation of 6 × 0.1 = 0.6 mm.

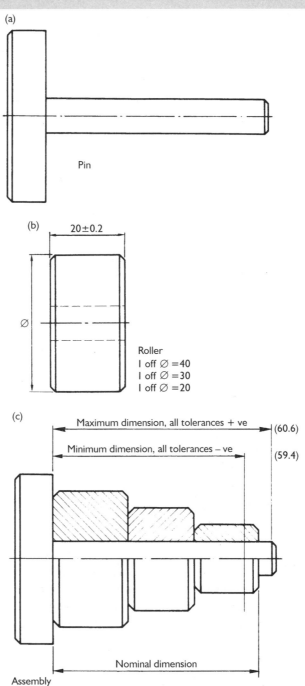

Figure 5.39 Assembly of rollers on a pin

Graphical communication in engineering

Case study

Figure 5.40 shows a dimensioned drawing of a pin. This drawing illustrates some of the principles of good practice in dimensioning, which include:
- use of a common datum from which are dimensioned the 10, 30, 40 and 50 lengths
- use of specific tolerances only where required, in this case for the 15 and 30 diameters; all other dimensions will be to the general tolerance
- correct use of projection lines from the part to the dimension lines
- use of notes and leader lines to the two chamfer notes (in this case as there is plenty of space, chamfer is not abbreviated to CHAM as shown in Table 5.5)
- use of symbol for diameter (⌀) as shown in Table 5.5.

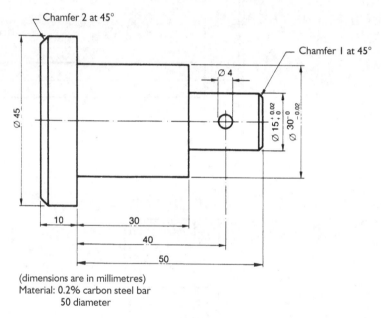

(dimensions are in millimetres)
Material: 0.2% carbon steel bar
50 diameter

Figure 5.40 Pin

Activity 5.11

1. A shim of thickness 2.5 has an outside diameter of 120 and a bore of 80. The shim has eight equally spaced holes of diameter 10, on a pitch circle diameter of 100. Draw and fully dimension the shim (all dimensions in millimetres).
2. A component is 50 square by 100 long. It has a through hole of 25 diameter drilled centrally in the 50 square. This hole has a counterbore of 40 diameter by 10 deep. Make a dimensioned drawing of the part, fully sectioned across the hole and counterbore (all dimensions in millimetres).
3. A circular pin is 50.00 mm diameter at its upper limit, and has to fit in a hole that is 50.00 mm diameter at its lower limit. Both pin and hole have a tolerance of 0.05 mm in diameter. Produce toleranced drawings of the hole and pin diameters.
4. Figure 5.41 shows a dimensioned drawing that has some errors in the method of dimensioning. Redraw the part and redimension it correctly.

Table 5.5 Abbreviations BS 308 as referred to in case study

Attribute	Abbreviation
Across flats	AF
Assembly	ASSY
Centres	CRS
Centre line	
on a view	℄
on a note	CL
Chamfered	CHAM
Cheese head	CH HD
Countersunk	CSK
Countersunk head	CS HD
Counter bore	C BORE
Diameter: in a note	DIA
Diameter (preceding a dimension)	∅
Drawing	DRG
External	EXT
Figure	FIG.
Hexagon	HEX
Internal	INT
Left hand	LH
Long	LG
Material	MATL
Maximum	MAX
Minimum	MIN
Number	NO.
Pitch circle diameter	PCD
Radius	
in a note	RAD
preceding a dimension	R
Required	REQD
Right hand	RH
Round head	RD HD
Sheet	SH
Specification	SPEC
Spherical diameter (preceding a dimension)	SPHERE ∅
Spherical radius (preceding a dimension)	SPHERE R
Spotface	S'FACE
Square: in a note	SQ
Square (preceding a dimension)	□
Standard	STD
Undercut	U'CUT

Graphical communication in engineering

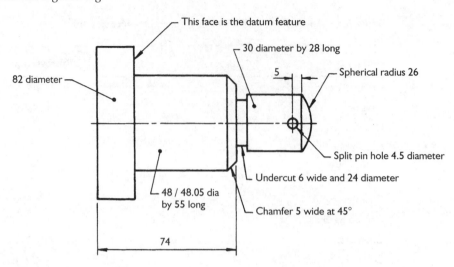

Figure 5.41 Drawing for activity 5.11

Dimensioning of screw threads and threaded fasteners

Table 5.4a shows the conventional representations of screw threads on drawings. Once drawn in the correct form the screw threads will need to be dimensioned and toleranced. Thread sizes and tolerances are always indicated by a note. The most commonly used thread in Europe is the ISO metric thread and is typically given as:

$$M12 \times 1.75 - 6H \text{ for an external thread}$$

$$M12 \times 1.75 - 6g \text{ for an internal thread}$$

In this system:
- M is the symbol for an ISO metric screw thread
- 12 is the nominal major diameter in millimetres
- 1.75 is the pitch in millimetres
- 6 is the international tolerance (IT) number that specifies the thread tolerance
- H and g are the deviations from nominal size of the threads.

Some examples of screw thread dimensioning are shown in Figure 5.42. In particular note that: (a) and (b) show the use of dimension notes to the circular view of the thread; (c)–(e) show dimensioning methods for internal threads; (f) and (g) show how thread lengths are shown.

Case study

Figure 5.43 shows a drawing of the component parts of an assembly. The drawings of the fork, bolt and strut end show correct methods of dimensioning and tolerancing screw threads.

The most common types of threaded fasteners are:
- hexagonal headed bolts and screws (hexagonal headed screws are threaded right up to the head, whereas bolts have an unthreaded section)
- screws with other heads such as countersunk, pan, cheese or socket cap head (socket cap head screws are also called allen screws)
- studs (basically threaded shafts without any head).

All these fasteners are shown in Table 5.6, together with their recommended preferred diameters and lengths.

Scale and schematic drawings in engineering

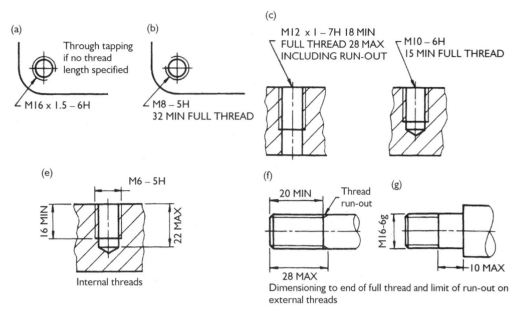

Figure 5.42 Dimensioning of screw threads

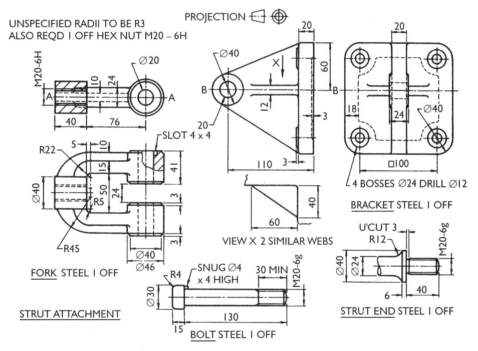

Figure 5.43 Drawing of the component parts of an assembly

Case study

Applications of threaded fasteners in assemblies are shown in Figure 5.44. Particular points to note here are:
(a) use of a nut, bolt and washer for a joint that can be easily disassembled
(b) use of a stud when a bolt cannot be used
(c) use of a cheese head (or cap head) screw when a flush finish is required.

215

Graphical communication in engineering

Table 5.6a ISO bolt specifications

Material	Head shape	Thread diameter (mm)	Length (mm)
Brass	Hexagon	M.14	40
Steel	Hexagon	M.12	50
Steel	Hexagon	M.10	30

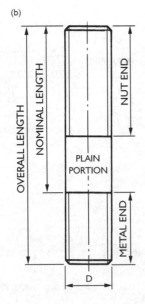

Table 5.6b Hexagon-headed bolts

Recommended diameters for metric bolts (mm)

M3	M10	M30	M64
M4	M12	M36	M72
M5	M16	M42	M80
M6	M20	M48	M90
M8	M24	M56	M100

L to be stated

The length of the threaded portion varies according to the length of the bolt. The following is nomally accepted:

Bolts up to 125 mm long = 2D + 6 mm

Bolts up to 200 mm long = 2D + 12 mm

Bolts over 200 mm long = 2D + 25 mm

Table 5.6c Screwed studs

Preferred nominal lengths (mm)

12	35	65	100
14	40	70	110
16	45	75	120
20	55	85	140
30	60	90	150

Length of thread on 'nut' end = 2D + 6 mm.

Length of thread on 'metal' end = D or 1.5D.

Recommended diameters as for bolts.

Scale and schematic drawings in engineering

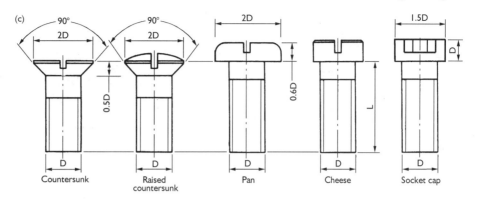

Table 5.6d Machine screws

Preferred diameters and minimum length of thread for machine screws (mm)

Diameter	M1.6	M2	M2.5	M3
Thread length	15	16	18	19
Diameter	M4	M5	M6	M8
Thread length	22	25	28	34
Diameter	M10	M12	M16	M20
Thread length	40	46	58	70

Activity 5.12

1. Referring to Table 5.6 draw the following threaded fasteners: (a) an M20 hexagon-headed bolt of length 100 mm, (b) an M12 stud of nominal length 75 mm, (c) an M10 countersunk-head machine screw of thread length 40 mm and (d) an M8 socket cap-head screw of thread length 34 mm.
2. Draw an assembly drawing of the clamp parts in Figure 5.45. Show the assembled clamp screwed to a vertical surface with two countersunk head screws.

Conventions for non-threaded fasteners

The other main ways of holding parts together in assemblies, apart from welding or adhesives, are by using:
- rivets
- keys
- dowel pins
- circlips (also called spring retaining rings).

Rivets These are used to make permanent joints. Unlike a joint made with a threaded fastener, a riveted joint cannot be dismantled easily. Figure 5.46a shows the main types of rivet used for thick plate and Figure 5.46b shows rivets for lighter gauge work.

The different types of riveted joint are shown in Figure 5.47.

Some of the lighter types of rivet as used in vehicles, consumer products, electronics, aircraft, etc. are shown in Figure 5.48.

Graphical communication in engineering

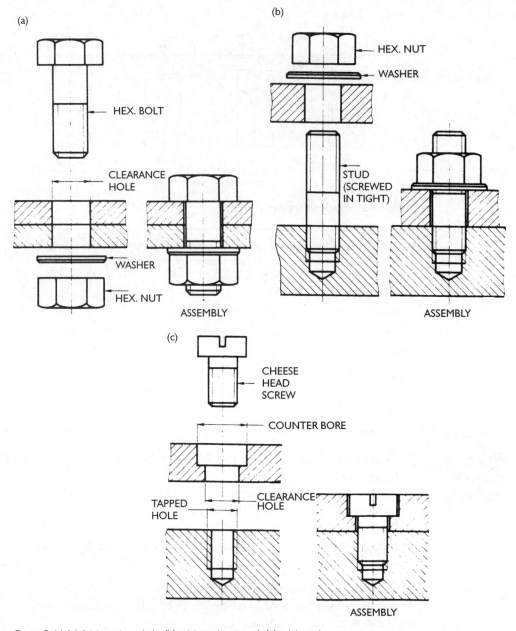

Figure 5.44 (a) A joint using a bolt: (b) a joint using a stud; (c) a joint using a screw

Case study

The particular advantages of tubular rivets, which can be fixed from one side only, are shown in the assemblies in Figure 5.49. Two types of rivet are shown, together with the fixing methods. These fixing methods are widely used in many assemblies because of their low cost and speed of assembly.

Keys Keys are used to prevent rotation between a shaft and an attached hub. Figure 5.50 shows the correct methods for the dimensioning of keyways.

Scale and schematic drawings in engineering

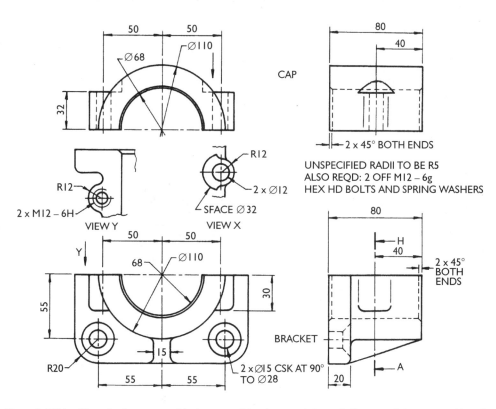

Figure 5.45 Use this projection to assemble the parts of the clamp and draw a half-sectional elevation on AA and a plan view

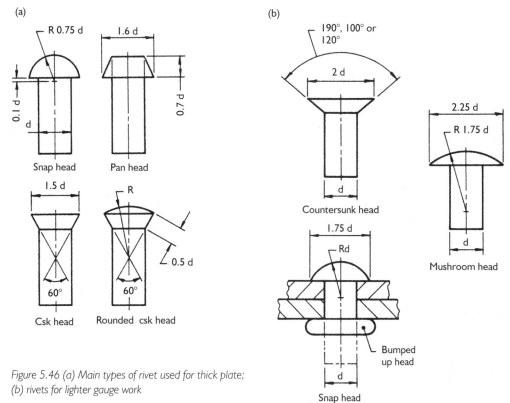

Figure 5.46 (a) Main types of rivet used for thick plate; (b) rivets for lighter gauge work

Graphical communication in engineering

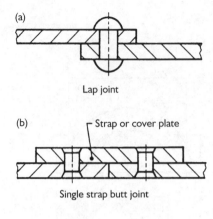

Figure 5.47 *Different types of riveted joint*

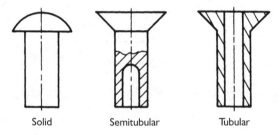

Figure 5.48 *Types of rivet for light-gauge work*

Case study

The application of keys in different assemblies is shown in Figure 5.51.
In all cases the key is used to prevent radial movement between the shaft and the hub. Of special interest are: (c) the feather key, which also allows axial movement; (d) the woodruff key, which can self-adjust to any taper.

Dowel pins These are used to locate parts together accurately as well as being able to prevent sideways movement when the parts are subject to shear stress. Figure 5.52 shows a drawing of a dowel pin assembly.

Circlips Circlips are used to allow rapid assembly/disassembly of parts on a shaft or in a bore (internal diameter). These two applications of circlips are shown in Figure 5.53.

Activity 5.13

A solid wheel of 100 mm diameter by 25 mm wide is to be assembled on to one end of a shaft of diameter 30 mm by 200 mm long. Draw two views (one sectioned) showing the parts assembled with a 5-mm square key.

Scale and schematic drawings in engineering

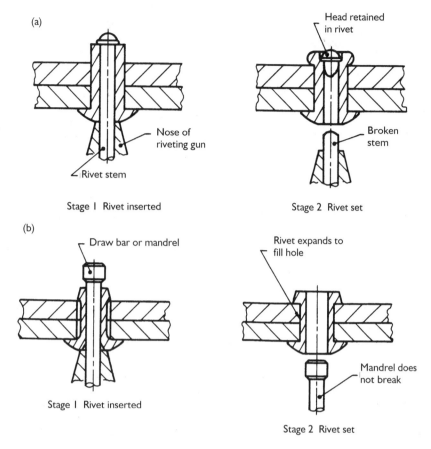

Figure 5.49 Examples of tubular rivets: (a) Tucker' Pop' rivet; (b) Avdel 'Briv' rivet

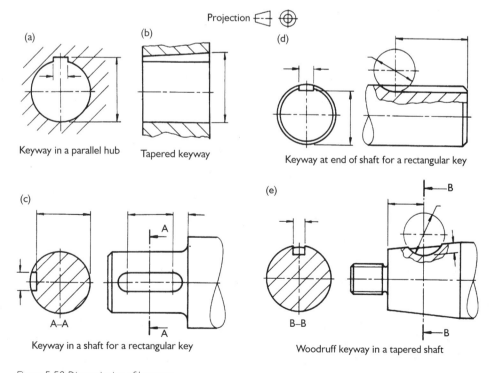

Figure 5.50 Dimensioning of keyways

221

Graphical communication in engineering

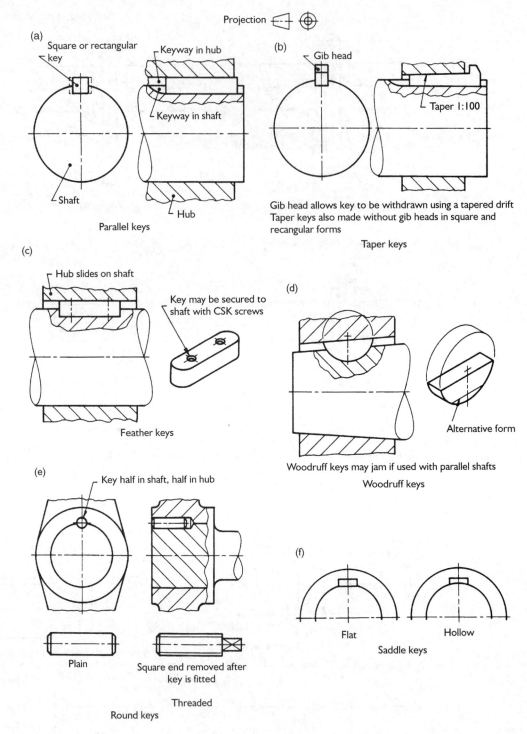

Figure 5.51 Types of keys

Gears

Figure 5.54 shows the methods of representation for the different kinds of gears. Note that the gear tooth profile is not drawn, except in the case of the rack. For the rack just the two teeth profiles at each end are drawn.

Scale and schematic drawings in engineering

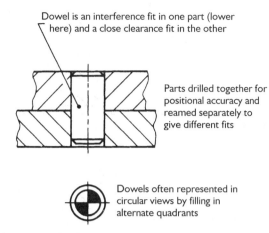

Figure 5.52 *Location of parts using dowels*

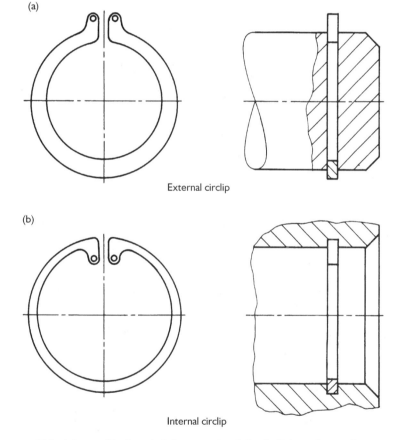

Figure 5.53 *Circlips (spring retaining rings)*

Case study

Gears are used to transmit motion and are always used in assemblies with mating gears. Figure 5.55 shows assembly drawings of three types of gears using the correct conventional representations. Of particular interest are:
(a) views of spur gears in mesh; note that the axes of the two gears are parallel
(b) views of bevel gears in mesh; in this case the axes are at 90° but they could be at other angles
(c) views of worm and wormwheel in mesh (here the axes are at 90° to each other); worm and wormwheel are used to give very large speed reductions.

Graphical communication in engineering

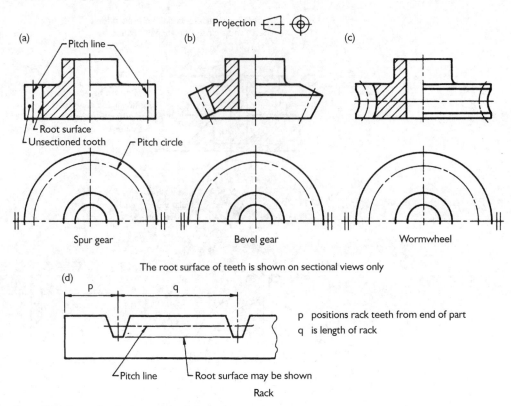

Figure 5.54 Conventional representation for gears

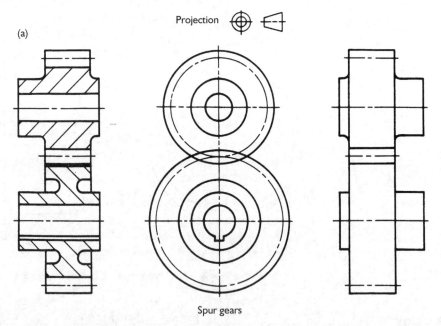

Figure 5.55 Conventions for gears in mesh

Activity 5.14

A spur gear has a pitch circle diameter of 100 mm. The outside diameter of the gear is 110 mm and the tooth depth is 10.75 mm. The bore is 30 mm. Using conventional representation draw two views of the gear, including a half-section.

Scale and schematic drawings in engineering

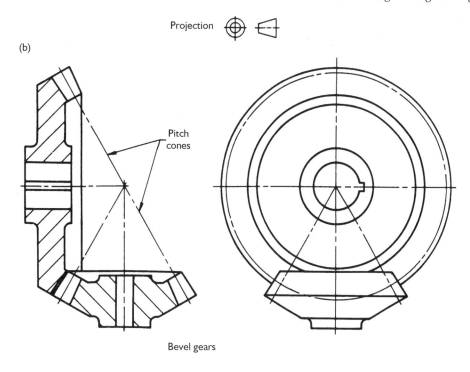

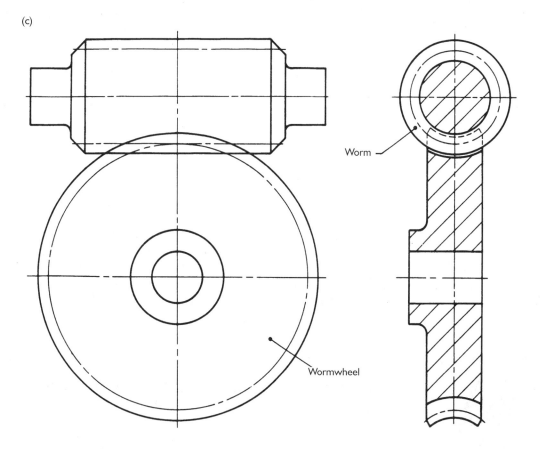

Figure 5.55 Conventions for gears in mesh (continued)

Graphical communication in engineering

Springs

Table 5.7 shows the correct conventional representations for springs.

Table 5.7 Conventions for common features (BS 308)

Subject	Representation		
	View	Section	Simplified
Cylindrical helical *compression spring* of wire of circular cross-section. If necessary, an indication 'wound left' or 'wound right' is included in an adjacent note			
Cylindrical helical *compression spring* of wire of rectangular cross-section			
Cylindrical helical *tension spring* of wire of circular cross-section			
Cylindrical helical *torsion spring* of wire of circular cross-section (wound right)			
Cup spring			
	View		Simplified
Semielliptic *leaf spring*			
Semielliptic *leaf spring* with eyes			

The convention for representing coil springs simplifies the helix to a straight line. Only two or three coils need to be shown at each end of the spring. The rest of the spring is shown as a thin chain line through the centre of the spring cross-section. For diagrams and schematic views the spring can be shown just as a simplified helix drawn in a single thick line. The representations for cup and leaf springs are also shown.

Activity 5.15

Figure 5.56 shows the parts of a door catch, which include a compression spring and a steel ball. Draw the assembly as instructed.

Surface finish and machining symbols

The use of surface finish and machining symbols is shown in Figure 5.57.

226

Scale and schematic drawings in engineering

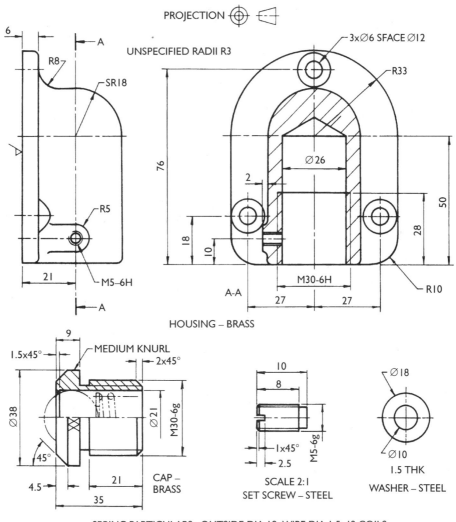

Figure 5.56 Assemble the parts of the door catch and draw twice full size a sectional view replacing the given left-hand view of the housing and an outside view replacing section AA

Case study

The application of surface finish and machining symbols to a component is shown in Figure 5.58.
  It can be seen that:
- symbols are shown once only on each surface
- symbols are preferably shown on the same view as the feature's dimensions
- if space is limited the symbol may be put on an extension line or a leader line
- numerical values should be positioned to read from the bottom or the right-hand side.

Activity 5.16

A 100-mm cube has a hole of 25 mm bored centrally. Draw and dimension the cube, showing a machined external surface of between 0.8 and 1.6 mm and a maximum value of 0.4 mm for the machined bore.

Graphical communication in engineering

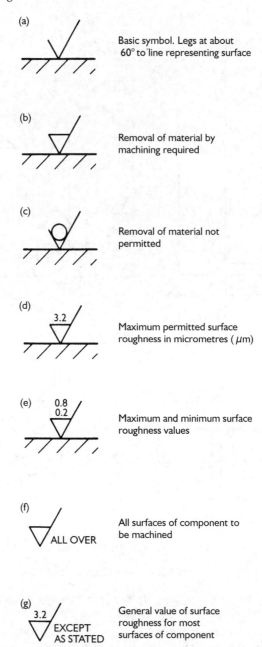

Figure 5.57 Indications of machining and surface texture

Welding symbols to BS 499
Welding is used very frequently to make permanent joints in all kinds of fabrications and assemblies. A quality weld should be as strong as the parent material. Welding can be used with confidence to join most kinds of materials. Figure 5.59 shows the most common types of weld and their symbolic representation.

Scale and schematic drawings in engineering

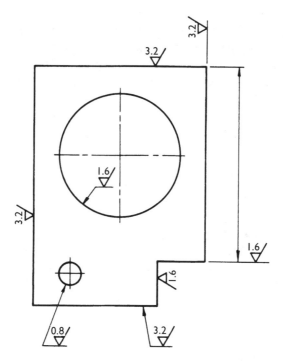

Figure 5.58 Application of machining symbols and surface texture values

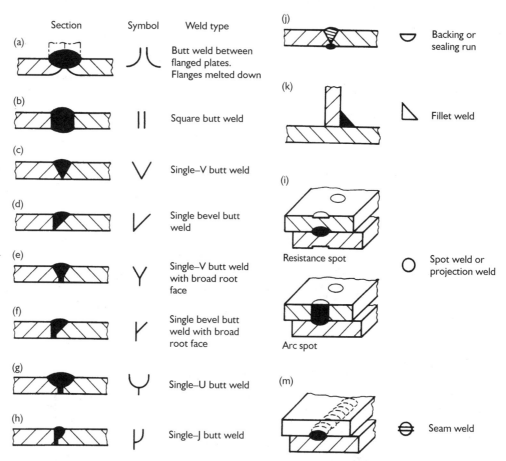

Figure 5.59 Welded joints and their symbols

229

Graphical communication in engineering

Case study

Clearly it would be too time consuming to draw the full graphical representation on drawings. Instead the symbols are used as shown in Figure 5.60. Note the use of a leader line (at an angle) and the horizontal reference line. Although not as vivid as the full graphical representation, the symbols do convey the type of weld to be used very well.

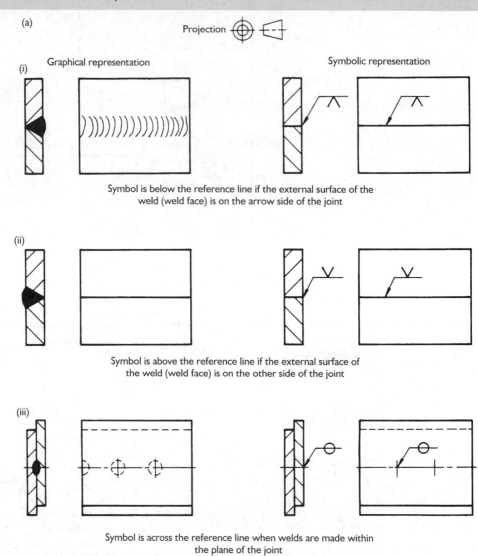

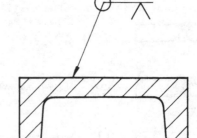

Figure 5.60 (a) Position of the welding symbol; (b) symbol for a peripheral weld

Scale and schematic drawings in engineering

Activity 5.17

Figure 5.61 shows a welded fabrication, with parts A, B and C being welded together. Draw two views of the completed fabrication, showing all weld symbols.

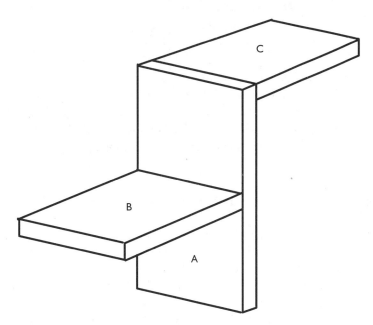

Figure 5.61 A welded fabrication

Standards

The section on conventions made use of the following standards:
- BS 308 Engineering drawing practice
- BS 499 Welding terms and symbols.

To complete the two scale and two schematic drawings, reference will also have to be made to:
- BS 3939 Electrical and electronic graphical symbols
- BS 2917 Symbols for fluid power systems
- BS 1553 Graphical symbols for general engineering (this covers all process plant as well as heating and ventilating)
- BS 5070 Drawing practice for engineering diagrams (this covers the general principles to be followed in all types of engineering circuit and installation drawings)
- BS 4058 Specification for data processing, flow chart symbols, rules and conventions.

Chapter 6 covers all the details of, and symbols for, many of the components for use in the following types of engineering applications:
- mechanical
- electrical
- electronic
- fluid power systems (hydraulic and pneumatic).

When producing the drawings for the evidence indicators it will be necessary to refer to Chapter 6.

Graphical communication in engineering

Progress check

1. (a) What is orthographic projection? (b) Explain the difference between first-angle and third-angle projection.
2. (a) What is pictorial projection? (b) What are the two main types of pictorial projection?
3. (a) Why are parts and assemblies sectioned? (b) Name the components that are not normally sectioned.
4. Name the three types of dimension.
5. Define a datum and explain its importance in dimensioning.
6. Define tolerance and limits.
7. Which features of a screw thread must be specified on a drawing?
8. What is the function of a surface finish symbol?
9. What is the function of a welding symbol?
10. Explain why standards are important when producing engineering drawings.

Producing scale drawings for engineering applications

Copyright
A legal right to any written, graphical or musical work

Everything an engineer produces must be of high quality. This is because all engineered products are expected to perform correctly and safely for their specified life. As all products start life on the drawing board (or its equivalent, the CAD system), it is very important that any drawing is of high quality. A high-quality scale drawing will:

- be drawn or plotted on good-quality clean paper with a border
- be drawn with correct-width pencils or pens
- not mix pencil and pen lines on the same drawing
- use correct standards and conventions throughout the drawing
- be drawn to the appropriate scale (see the section on equipment for line production and measurement earlier in the chapter for information on scales)
- be fully dimensioned and use correct arrowheads for dimension lines
- use notes as required to explain requirements, state general tolerances, etc.
- have just enough views to define it fully for the purposes required
- have the views correctly spaced on the paper
- have a title block stating:
  - name of company, draughtsperson and person approving the drawing
  - part or assembly name and drawing reference number (also repeated at top left-hand corner of drawing)
  - issue number with date drawn and approved
  - scale and projection (first angle/third angle/isometric/oblique)
  - material(s) and any surface or heat treatments, etc.
  - copyright details
  - modifications and date.

Preparatory stages of producing scale drawings

Figure 5.62 shows a typical drawing sheet, with title block.

Tables 5.8 and 5.9 show basic and additional information respectively, as used when producing drawings.

Scale and schematic drawings in engineering

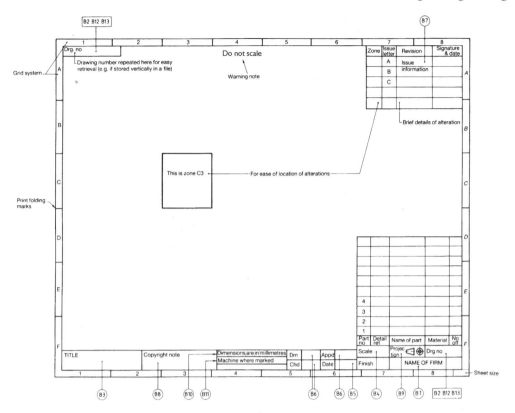

Figure 5.62 A typical drawing sheet

Case study

The numbers B1 to B12 are referenced in Figure 5.62. The information contained in typical entries in title blocks is shown in Table 5.10. The details vary according to whether the parts are made by the company or bought from an outside supplier. Examples of this are:
- made by the company – the location pin has details of material specification and treatment
- bought from a supplier – the cylindrical tension spring does not need a separate specification as this will be in the manufacturer's catalogue details.

The stages in starting a drawing are:
- selection of paper size from those available: A0:A1:A2:A3:A4 (see Figure 5.3)
- fixing paper in plotter or on drawing board/draughting machine
- choosing suitably prepared pens, pencils and instruments
- choosing correct scale for the drawing to fit the paper
- drawing the border and putting in title block (see Figure 5.62)
- positioning views correctly allowing space between and around views for dimensioning – faint outlines can be drawn at this stage. Note that the minimum number of views required to fully describe the part or assembly should be drawn. If this is two or three, then they should be evenly spaced on the paper.

Graphical communication in engineering

Table 5.8 Basic information on drawing sheets

Reference no.	Basic information	What the user needs to know	Comment
B1	Name of firm	Who/where is this from?	Useful outside the place of origin
B2	Drawing number	Where the drawing can be kept and retrieved easily	Useful inside the place of origin
B3	Descriptive title of depicted part or assembly	What does it show?	This may be important if components are to be grouped for economic production
B4	Original scale	What is the original size of drawing/size of feature?	The size may be altered by the subsequent reprographic process
B5	Date of the drawing	When was it drawn?	
B6	Signature(s)	Who drew it? Who checked it? Who approved it?	
B7	'Issue' information	Is this the latest issue of this drawing?	Drawings are reissued when they are modified or updated
B8	Copyright clause	Is this information confidential? Secret? May it be copied?	Original ideas and patents need to be safeguarded
B9	Projection symbol	What type of projection has been used? First angle? Third angle?	
B10	Unit of measurement?	Metric? Imperial?	
B11	Reference to drawing practice standards	Which standards have been used? Machining marks? Position tolerances?	
B12	Sheet number Number of sheets	How many sheets are there related to this drawing?	Often in the same space as the drawing number. A simple statement '3 of 6' means that this sheet number 3 from a total of six sheets

Producing schematic drawings for engineering applications

Schematic drawings do not need to be drawn to scale but do need to be absolutely correct and drawn with care and quality. This is because an electrical/electronic or fluid power circuit is put together on the basis of this diagram. If a scale drawing is prepared as well, it will be based on the schematic or the circuit diagram. Chapter 6 gives full details of how schematic and circuit diagrams are interpreted and should be referred to when producing schematic diagrams. The section on conventions earlier in this chapter should also be referred to for details of the appropriate conventions and standards to be consulted for use of correct methods and symbols.

Table 5.9 Additional information on drawing sheets

Additional information	What the user needs to know	Comment
Material and specification	What is it made from?	Alloy steel? Concrete mix?
Related specifications	Identity of special requirements	
Treatment/hardness	Is the part acceptable as machined? Does it need any special treatment before it is used?	
Finish	Plated? Painted?	
Surface texture	Is it acceptable rough? Smooth?	If so, how rough? How smooth?
General tolerances	How much variation from the general dimensions given can be tolerated?	This information will have a considerable influence on the manufacturing mehods to be used
Screw thread forms	BS Whitworth? British Association? Metric? ISO?	
Sheet size	How big was the original drawing sheet?	This may have been increased or reduced when copied
First used on	Has this component/idea been used before?	
Similar to	Does this component belong to a similar group of components?	These items help in the location of tools and equipment which aid manufacture
Equivalent part	Does an equivalent part exist?	
Supersedes or is superseded by	Is this drawing up to date?	Out-of-date reproductions are not always withdrawn from circulation quickly, but the fact that they are retained should be shown clearly
Tool reference	Are there any special tools to help with this job?	
Gauge references	Do I need any special gauges to help me to assess the quality of this job?	
Grid systems or zoning	How to locate particular features quickly	For instance, if a drawing is being discussed by telephone
Warning notes, e.g. DO NOT SCALE	Any special requirement to be noted	Drawings should never be scaled. Some are deliberately increased or reduced in size
Print folding marks	Where to fold drawings so that important details do not become obscured	Drawings accumulate dirt along creases. Folds which obscure important features must be avoided

Table 5.10 Typical entries in title blocks: (a)–(c) manufactured items; (d)–(f) bought-in items

Part no. (Location of part)	No. off (How many?)	Detail ref. (Detail or code)	Name of part	Raw material	Treatment required after machining
(a) 1	6	—	ANGLE BRACKET	CAST IRON CASTING	NONE
(b) 17	32	—	LOCATION PIN	20-MM-DIAMETER COLD-ROLLED D2% CARBON STEEL	CASE HARDEN AND GRIND
(c) 2	4	XYZ SPEC NO. 20 SPEC.	BEARING BUSH BEARING	40-MM-DIAMETER PHOSPHOR BRONZE BAR	FOR DETAILS, SEE MAKER'S SPECIFICATION
(d) 32	108	GKN 16	HEX HEAD SCREWS	Not required as diameter, type of thread, length, material, etc. will be included in maker's detail or code	
(e) 8	20	ABC 16/2	CYLINDRICAL TENSION SPRING	Not required as wire diameter, coil diameter, shape of ends and heat treatment will be included in maker's detail or code	
(f) 6	24	PQR 100	100-W LAMP BAYONET FITTING	—	—

Scale and schematic drawings in engineering

Mechanical scale drawing

Case study

Figure 5.63 shows a drawing of three parts that will form an assembly. Note that all three parts are drawn on one drawing.

To produce the drawing manually the stages are:
- preparation stage as detailed above (in this case: select A4 paper, choose the scale of 1:2, select first-angle projection, draw border, put in title block, space views equally on the paper – two views are sufficient to describe the parts, draw in faint outlines of views)
- draw in centre lines and use these as basic datums
- draw all lines in faintly
- when all lines are correct, draw in firmly with suitable pencil/pen to correct line width (0.7 mm for outlines: 0.35 mm for centre lines, cross-hatching and break lines)
- erase all construction lines
- draw in projection and dimension lines
- dimension and tolerance the drawing
- add machining symbols as required
- if the drawing is done with a CAD package on computer then the same basic principles apply, except that all lines will be drawn once only.

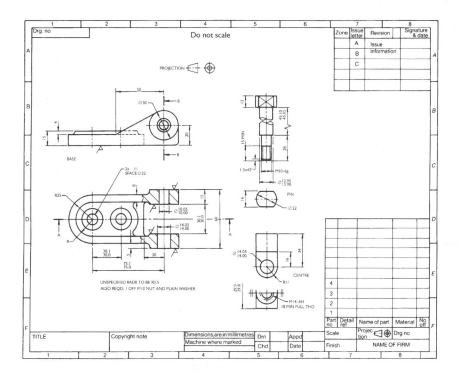

Figure 5.63 Use the diagram to assemble the parts and draw a sectional elevation on AA and a sectional end view on BB

Graphical communication in engineering

Activity 5.18

Using similar principles to those given in the case study, draw an assembly of the parts shown in Figure 5.63. Draw a sectional elevation on A–A and a sectional end view on B–B.

Mechanical schematic drawings

Case study

Figure 5.64 shows the schematic for the assembled parts shown in Figure 5.63. This schematic can be drawn freehand, or with the aid of a rule. It is not to scale and its purpose is just to show the function of this pivot. Some CAD packages will have these symbols in their library, which simplifies the production of schematic drawings.

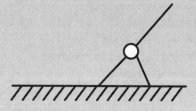

Figure 5.64 Schematic for the assembled parts shown in Figure 5.63

Activity 5.19

Draw a copy of Figure 5.64 and add a screw pair to the pivot arm (see Chapter 6, Figure 6.3).

Electrical/electronic scale drawings

Case study

Figure 5.65 is a scale drawing of a fluid-level detector, the circuit diagram of which is shown in Figure 5.66. Note that the same basic principles of draughting are followed as for mechanical scale drawings.

Activity 5.20

Draw a copy of Figure 5.66 and also produce a scale drawing of the PCB showing tracks and pin holes. Notice that the integrated circuit (IC) has 14 pins, but there would only be tracks to the pins used (1, 7, 10, 11, 12, 13 and 14).

Electrical/electronic circuit diagrams

Case study

Figure 5.66 shows the circuit diagram for the fluid-level detector. It is purely schematic, but note the clear way in which all the components are drawn, identified and correctly connected. The general principles to be followed when drawing circuit diagrams are given in BS 3939 and can be summarised as:

Scale and schematic drawings in engineering

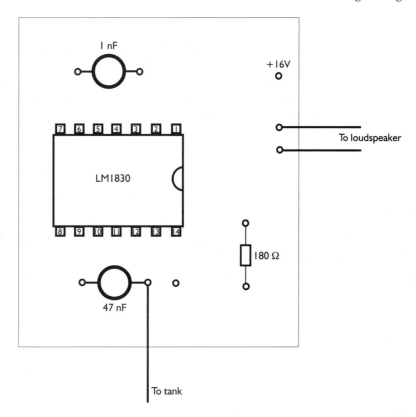

Figure 5.65 Scale drawing of a fluid-level detector

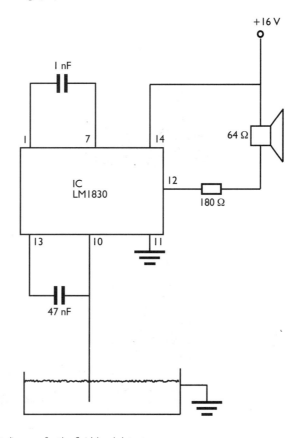

Figure 5.66 Circuit diagram for the fluid-level detector

Graphical communication in engineering

- the meaning of any circuit must be able to be easily and quickly understood
- correct symbols must always be used
- symbols should be suitably spaced and orientated
- interconnections must be carefully routed to minimise track and cable lengths
- if the circuit has a clear sequence from cause to effect it should be from left to right and/or top to bottom
- conductors should be drawn as a continuous horizontal or vertical line
- switches or relays should normally be drawn in the state they would be in if the circuit is not connected to the supply.

Activity 5.21

Using sources such as catalogues, electronics magazines, textbooks, etc. draw the circuit diagram for a circuit to control the speed of an electric drill. If time permits, also draw a scale drawing and a PCB layout.

Fluid power scale drawings

Case study

Figure 5.67 shows a typical hydraulic system that could be used to operate a car lift in a garage. Again, the same basic principles are followed as for mechanical scale drawings. The circuit diagram is shown in Figure 5.68.

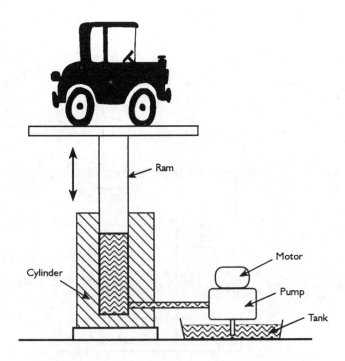

Figure 5.67 Typical hydraulic system

Scale and schematic drawings in engineering

Fluid power circuit diagrams

Case study

As stated above, Figure 5.68 is the circuit diagram for the car lifter. The basic principles of the drawing of fluid power circuits is quite similar to those for electrical/electronic circuits. The introduction to BS 2917 (fluid power symbols) states that:

- symbols are functional, not to scale and not orientated in any particular direction
- hydraulic and pneumatic units are normally shown in the unoperated position, e.g. no pressure in the circuit
- symbols show connections, flow paths and functions but do not include constructional details.

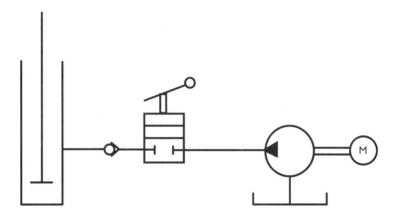

Figure 5.68 Circuit diagram for the car lifter

Activity 5.22

Draw the circuit diagram and scale drawing for the hydraulic braking system of a vehicle.

Progress check

1. State why quality and accuracy are essential when producing scale or schematic drawings for engineering applications.
2. List the preparatory stages of producing an engineering drawing.
3. What information is contained in the title block of a drawing?
4. Give two examples of the information contained in a general note on a drawing.
5. How do schematic and circuit diagrams differ from scale drawings?
6. List at least five principles to be followed when drawing electrical/electronic circuit diagrams.
7. State what scale drawings of electrical/electronic circuits are used for.
8. Give the basic principles to be followed when producing fluid power circuit diagrams.
9. What is the purpose of a scale drawing of a fluid power system?

Assignment 5
Produce a folder of scale and schematic drawings

This assignment provides evidence for:
Element 2.2: Produce scale and schematic drawings for engineering applications
and the following key skills:
Communication 2.3: Use images
Information Technology 2.1: Prepare information
Information Technology 2.2: Process information
Information Technology 2.3: Present information
Information Technology 2.4: Evaluate the use of information technology

This assignment is designed to provide evidence for assessment of the unit. As drawing is communication, it can also be used as evidence for key skills. It is expected that a CAD system will be used (a good idea is to do half the drawings manually and the other half on a CAD system) and this will also provide evidence for key skills in IT.

Your tasks

Quality work is essential. Good-quality work should be presented in a folder.
The drawings required are:
- A scale drawing of a mechanical engineering application, e.g. bearings and shafts, brackets, couplings, gears, linkages, chain drive, screw and nut, piston and cylinder, valve, etc.
- A schematic circuit diagram of a mechanical engineering application. Rather than use purely mechanical schematics, the unit requires the use of fluid power circuits. Therefore, the circuit will need to be pneumatic/hydraulic, although it could also contain purely mechanical elements. Suitable examples are: hydraulic clutch control on a vehicle, air drill, reciprocating cylinders on machines, etc.
- A scale drawing of an electrical application that also has electronic components mounted on a PCB, e.g. motor-control circuit, lighting-control circuit, burglar-alarm circuit, temperature controller of electric heating, etc.
- A schematic circuit diagram of an electrical/electronic application. It clearly makes sense to use the same application for both scale drawing and circuit diagrams.

Chapter 6: Interpreting engineering drawings

This chapter covers:
Element 2.3: Interpret information presented in engineering drawings.
...and is divided into the following sections:
- The use of symbols to represent components
- The representation of physical situations
- Extracting information from engineering drawings
- Interpretation of information from engineering drawings.

As engineering drawings are about communicating information it is important that this information is correctly interpreted by the person reading the drawing. Chapters 4 and 5 have been about selecting graphical methods and actually producing the drawing. Certainly, selection of the right method and correctly produced drawings, using the right conventions and standards, will help ensure that drawings are correctly interpreted. However, it is possible that mistakes can be made at the design and draughting stage. It is, therefore, vital that anyone reading the drawing understands exactly what they are reading and whether it is correct.

Case study

To illustrate the sort of thing that can happen because of incomplete or misleading information, I will tell a story of what happened in an engineering company where I was working. One of my colleagues was given the job of doing a new plant layout. He did this and presented it to the chief engineer, who checked and approved it. The new layout was done over one weekend. On Monday morning all hell had broken loose. The new layout was fine but my colleague had omitted to draw in one of the girders holding up the factory roof. The millwrights doing the job had assumed the new layout needed the space and so removed the girder. The result, as you would expect, was that the roof had caved in. I think the moral of the story is, when interpreting a drawing and something does not seem right, check it out.

Gudgeon pins
The pins holding a piston to the connecting rod that connects the piston to the crankshaft

This chapter will cover the engineering activities of:
- manufacture
- installation
- maintenance

in the various types of engineering including:
- mechanical
- electrical

Graphical communication in engineering

Assembled systems

Complete functioning devices such as internal combustion engines assembled from cylinder block, cylinder head, pistons, gudgeon pins, connecting rod, crankshaft, bearings, starter motor, alternator, etc., or fluid power components such as cylinders, valves, etc., connected to make a functioning circuit such as the case study drilling machine example in Chapter 4

Figure 6.1 A case of a misinterpreted drawing

- electronic
- fluid power (pneumatic and hydraulic).

The types of information will include:
- dimensions
- locations
- functions

and will cover individual components and their use together in situations including:
- assemblies
- circuits
- layouts.

The use of symbols to represent components

All engineered products are made from individual components. Sometimes the individual components are complete systems in themselves, for example:
- mechanical components such as couplings and ball bearings, etc., which are made up of several parts and thus are subassemblies, but which function as a single component
- electronic semiconductor components such as operational amplifiers, voltage regulators, etc., which are complete circuits on a single chip
- fluid power components such as pumps and cylinders, etc., which are subassemblies of several parts, but which function as a single component.

At other times the individual components are used together to make a complete system or subsystem. Examples of components assembled into systems are:
- electronic components such as ICs, capacitors, resistors, etc. assembled into products such as computers, television and videos.

The components covered will be:
- mechanical
- electrical and electronic
- fluid power (pneumatic and hydraulic).

Mechanical components

Although many mechanical parts have to be designed individually there are many standard components available. Designers will frequently use these parts, so anyone reading a drawing needs to be able to recognise them and know something about them. For details of the standard mechanical components available, reference should be made to manufacturers' catalogues, e.g. Renold (belt and chain drives), Ratcliffe (springs), RHP (Ransome Hoffmann Pollard) (ball and roller bearings), David Brown (gears), GKN (Guest Keen Nettlefold) (couplings, drives, clutches, etc.), SPS (Standard Pressed Steel) (fasteners), etc.

Springs
The function of springs is to deform under load and to return to the original length when the load is removed. While deformed they are storing energy, which can be released when the load is removed. The different types of spring are detailed in Table 6.1 and their conventional representations are shown in Table 5.7.

Table 6.1 Different types of spring

Component	Purpose/function	Applications
Compression spring	To compress under load and return to original length when load is removed	Valve springs for internal combustion engines; suspension springs for motor vehicles
Tension spring	To extend under load and return to original length when load is removed	Return springs for doors, flaps, table lamps, etc.
Torsion springs	To rotate when a torque is applied and return to original angle when torque is removed	In devices when return rotation is required, e.g. clockwork mechanism
Cup spring	As for compression spring	Where space is limited
Leaf spring	To bend under load and return to original shape wehn load is removed	Vehicle suspension systems

Gears
Gears function as a means of taking an input rotational or linear force and converting it to an output rotational or linear force. The use of gears enables the output speeds, forces and directions to be changed. Table 6.2 details the main types of gears, with their conventional representations being shown in Figure 5.54. The schematic representations of gears and chain drives are shown in Figure 6.2.

Screw threads
Screw threads produce a linear movement when rotated. Their uses include:
- as fasteners to hold parts together in assemblies to allow dismantling
- to transmit motion, e.g. leadscrew on a lathe

Graphical communication in engineering

Table 6.2 Types of gears

Component	Purpose/function	Applications
Spur gear	To transmit parallel rotation and torque	Gearboxes in vehicles, machine tools, etc.
Bevel gear	To transmit rotation and torque at different angles	Where direction of shaft rotation is changed, e.g. a vehicle driveshaft is turned through 90 degrees to drive wheels
Worm and wheel	To increase torque, reduce speed and change direction of rotation by 90 degrees	Drives for slow-speed, high-torque applications such as winches

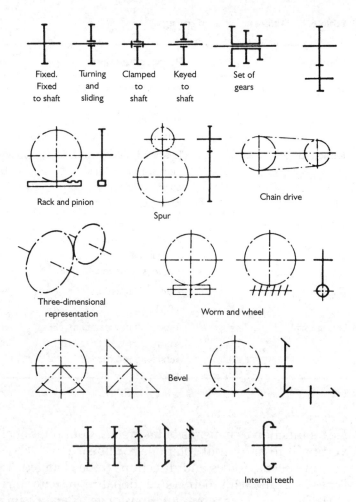

Figure 6.2 Schematic representations of gears and chain drives

- as adjusters, e.g. levelling screws for instruments, machines etc.

The scale drawing of screw threads was dealt with in Chapter 5 and their conventional representations are shown in Table 5.4a.

Interpreting engineering drawings

In schematic drawings threaded fasteners are not normally shown, except by the use of centre lines. A screwed pair such as leadscrew and nut are as shown in Figure 6.3.

Figure 6.3 Leadscrew and nut

Bearings
Bearings can be rotary or linear, with the main types being:
- plain; a typical application is the big-end bearing on an engine connecting rod
- ball/roller; a typical application is for wheel and shaft bearings in vehicles.

The representation of rolling bearings in scale drawings is shown in Table 5.4c. The schematic representations for various types of bearings are shown in Figure 6.4.

Notice the differences between the bearings that:
- can rotate and move axially
- are fixed pivots
- are just a sliding pair (linear bearing), e.g. piston and cylinder.

Note the symbol for a shaft is used here (a straight line).

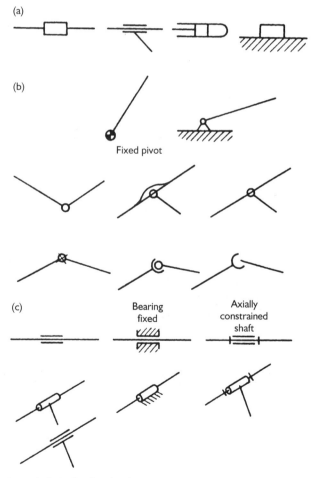

Figure 6.4 Schematic representations of various bearings

247

Splines, serrations and keys

These are used to prevent rotation between a hub and a shaft. Splines also allow axial movement. A typical application for a spline is on a gearbox shaft, which allows axial movement of the gears.

Serrations allow precise angular positioning but not axial movement and are used where small adjustments to angular positioning of a hub may be necessary.

Keys (see Figure 5.51) prevent rotation between hub and shaft but do not usually allow axial movement, except for the feather key shown in Figure 5.51c. Typical applications for keys are also shown in Figure 5.51.

The conventional representations for splines and serrations in scale drawings are shown in Figure 6.5, but there are no recognised schematic symbols.

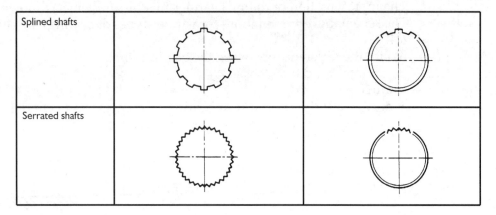

Figure 6.5 Conventional representations for splines and serrations

Couplings

Couplings
Mechanical or fluid connections between two or more devices

Couplings allow two rotating shafts to be connected together. There is no standard conventional representation for scale drawings, because there are so many types. Typical applications are:
- the coupling between the power take-off at the rear of a tractor and an implement such as a rotovator or a mower
- a vehicle clutch (friction coupling).

The schematic symbols are shown in Figure 6.6.

Dampers

These are also called shock absorbers and are used mainly to damp out spring oscillations. A typical application is a vehicle suspension shock absorber. The schematic symbols are shown in Figure 6.7.

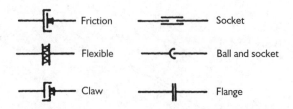

Figure 6.6 Schematic coupling symbols

Interpreting engineering drawings

Linear   Torsional

Figure 6.7 Schematic symbols for dampers

Electrical and electronic components

The full range of symbols for these components is given in BS 3939 Graphical symbols for electrical power, telecommunications and electronic diagrams. A good selection is also given in PP7307 Graphical symbols for schools and colleges. The symbols are used for circuit diagrams. For the scale drawings of the actual layouts on circuit boards the components are drawn to scale. Reference should also be made to the excellent catalogues of electrical and electronic components produced by companies such as Maplin Electronics and Farnell Electronics. Electrical components are normally joined to each other by copper wire or copper tracks on a circuit board. Electronic components can be connected in the same way but they can also be connected on an integrated circuit. In the latter case the integrated circuit is a single component but performs the function of several interconnected components. Soldering is the usual method for connecting components by wire or track. Terminals can also be used for connections and three types are shown in Figure 6.8.

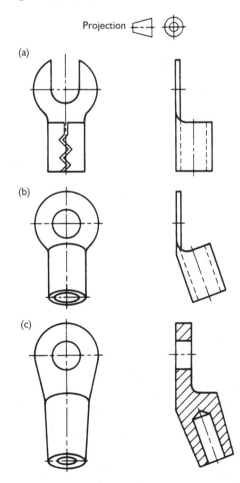

Figure 6.8 Conductor terminals

Graphical communication in engineering

Figure 6.8a shows a fork end terminal that is crimped on to the wire. Figure 6.8b and c shows thin-gauge and thick-gauge washer end terminals, with soldered connections to the terminals.

Electromotive force (EMF) components and systems

These represent the sources of electrical power in a system and are detailed in Table 6.3.

Table 6.3 Electromotive force components

Component	Purpose/function	Applications	Symbol
Primary cell	Storage and provision of direct current	Assembly in series to make batteries	—∣⊢—
Battery	Storage and provision of direct current	Provision of electrical power to circuits and devices or back-up in case of current failure	—∣⊢-----∣⊢— or —∣∣∣∣⊢—
Rectifier	Conversion of alternating current to direct current	Power supply to direct current circuits	—▷∣—
DC generator	Provision of direct curent	Power supply to direct current circuits	Ⓖ
AC generator (alternator)	Provision of alternating current	Power supply to direct current circuits	Ⓖ∼
Motor	Conversion of electrical energy to kinetic energy	Using alternating or direct curent to provide power for machine tools, domestic appliances, trains, etc.	Ⓜ
Transformer	Reduction of alternating current voltages	Provision of power to devices requiring a lower voltage	(transformer symbol) or (circle symbol)

Fuses and circuit breakers

These give protection in a circuit and are shown in Table 6.4.

Circuit components

Circuit components are used to make the circuit function. A range of the most commonly occurring components is shown in Table 6.5.

Switches, contact makers/breakers and relays

All these devices are used to allow current to flow or not to flow, and are shown in Table 6.6.

Interpreting engineering drawings

Table 6.4 Fuses and circuit breakers

Component	Purpose/function	Applications	Symbol
Fuse	To break a circuit by melting of internal wire if circuit is overloaded	Protection of circuits, people and property	—[ ]— or —⊳⊲—
Circuit breaker	To break a circuit under overload by switching off power	Protection as for fuses	—o ⁄o—

Table 6.5 Circuit components

Component	Purpose/function	Applications	Symbol
Resistor	Provide resistance to a current	Control of current and voltage in circuits	—[ ]— or —/\/\/—
Variable resistor (general symbol)	Provides a variable resistance to a current	Control of current and voltage in circuits	⌿[ ]
Potentiometer	Provides a variable resistance to a current	Control of current and voltage in circuits	↓—[ ]—
Thermistor	To decrease resistance as temperature changes	Temperature measurement, circuit control, circuit protection	TH
Capacitor	To store electrical energy temporarily	With resistors, to control circuits in applications (such as power supplies, radio and TV circuits, etc.)	—||—
Inductor (coil)	To induce a magnetic field	Current filters/limiters, chokes for fluorescent lights, etc.	—||—

Lamps
Lamps are devices that convert electrical energy into light energy. Table 6.7 shows two symbols for lamps.

Measuring instruments
These enable circuit values to be measured. A selection of instruments is shown in Table 6.8.

Semiconductor devices
These are solid-state components that perform a variety of functions in circuits. They are made from materials such as silicon, which are intermediate between conductors and insulators. They are doped with impurities to either add electrons, which are n (negative) carriers, or provide more holes, which are p (positive) carriers. A range of these are shown in Table 6.9.

Doped
Impregnated with an impurity to achieve a desired property

Table 6.6 Switches, contact makers/breakers and relays

Component	Purpose/function	Applications	Symbol
Switch	To turn current on or off	On–off switches for all types of circuits	
Break contact (normally closed)	To turn current off	Off switch for machines, etc.	
Make contact (normally open)	To turn current on	On switch for machines, etc.	
	As a relay to switch a large current with a small signal	Machine and fluid power control, etc.	

Table 6.7 Lamps

Component	Purpose/function	Applications	Symbol
Signal lamp	To illuminate when a current is passed	Indication of circuit function	
Filament lamp	To illuminate when a current is passed	Provision of light	

Fluid power components (hydraulic and pneumatic)

The full range of symbols for these components for use in circuit diagrams is given in BS 2917 Symbols for fluid power systems. A reasonable selection is also given in PP 7307 Graphical symbols for schools and colleges. In scale drawings of actual layouts, the components would be drawn to scale. For details of components, reference should be made to manufacturers' catalogues from companies such as Festo, Norgren, Compair Maxam, David Brown, Normalair Garrett, Robert Bosch and Denison. Fluid power components are connected by piping, which is the equivalent of wiring in the electrical and electronic circuits. However, most fluid power systems have electrical wiring, because many of the circuit functions will be controlled electronically. Switches and valves can be elec-

Interpreting engineering drawings

Table 6.8 Measuring instruments

Component	Purpose/function	Applications	Symbol
Ammeter	Measurement of current	Measurement of current flow/magnitude in circuits	Ⓐ
Voltmeter	Measurement of electromotive force	Measurement of voltages in circuits	Ⓥ
Ohmmeter	Measurement of resistance	Measurement of resistances in circuits	Ⓞ (Ω)
Oscilloscope	Measurement of circuit parameters	Analysis of circuit function	ⓥ OSC

Table 6.9 Solid-state components

Component	Purpose/function	Applications	Symbol
Diode	Acts as a semiconductor to control circuits	Rectifier, switch, voltage regulator, etc.	─▷│─ or ─▷│─
Light-emitting diode (LED)	Emits light when a current is passed	Indicating lights for circuits	(LED symbol)
Photodiode	Passes a current when exposed to light	Generation or conduction of electricity when exposed to light	(photodiode symbol)
Triac	A bidirectional triode triggered by positive or negative voltages	Full-wave rectifier	(triac symbol)
Transistor	Semiconductor n-p-n or p-n-p device	Switch, oscillator, amplifier, counter, gate, etc.	(transistor symbol)
Integrated circuit (IC)	A complete electronic circuit on a single chip	Operational amplifiers, keypad encoders, temperature controllers, etc.	(IC symbol)
Operational amplifier	A specific IC that amplifies an input to give a much larger output	Audio amplifier, control devices, etc.	(op-amp symbol)

253

tronically controlled. If computers and programmable logic controllers (PLCs) are used, very large and complex fluid power systems can be controlled. Typical applications of fluid power systems are:
- vehicle braking and transmission systems
- control and actuation of machine tools such as drilling machines, presses, etc.
- control and actuation of hoists, cranes, diggers (JCBs), etc.
- control and actuation of complex sets, stages and scenery for theatres and night clubs.

To interpret drawings of fluid power circuits, a good knowledge of the schematic symbols for circuit diagrams is required.

Energy conversion components
These components, which change the form of energy, are shown in Table 6.10.

Table 6.10 Energy conversion components

Component	Purpose/function	Applications	Symbol
Pump	Converts electrical energy to kinetic energy of fluid flow at pressures up to 100 bar and flow rates up to 700 litres/min	Actuation of hydraulic motors and cylinders	
Compressor	Converts electrical energy to kinetic energy of air flow at pressures up to 175 bar and flow rates up to 4000 litres/min	Actuation of pneumatic motors and cylinders	
Motor	Converts energy of fluid kinetic energy of rotation	Rotation for machine operation etc.	
Accumulator	Stores fluid under pressure to act like a fluid spring	Supply of fluid to meet a sudden demand	
Reservoir	Container for hydraulic fluid	Source of fluid supply and place where fluid is returned to	

Interpreting engineering drawings

Proportional
In correct relationship to a value

Cylinders

These convert fluid power into linear motion. Table 6.11 shows the common types of cylinder.

Control valves and control mechanisms

Overall control of a fluid system is often achieved by using electronic controllers such as PLCs and computers. However, any electronic signals will have to be converted into control of the fluid by control valves. Fluid comes into the valve at inlet pressure and leaves the valve at a normally lower outlet pressure. These valves themselves can be controlled manually, mechanically, by fluid pressure or electrically/electronically. Control mechanisms are shown in Table 6.12 and valves in Table 6.13.

Table 6.11 Common types of cylinder

Component	Purpose/function	Applications	Symbol
Single acting	Power actuation in one direction	Where one-directional motion is required with power actuation in one direction and return by gravity, e.g. hoist/jack	
Single acting with spring return	Power actuation in one direction	Where one-directional motion is required with spring return, e.g. clamp	
Double acting	Power actuation in both directions	Where powered motion is required in both directions, e.g. machine tool slide	
Air–oil actuator	Conversion of pneumatic pressure to hydraulic pressure or vice versa	Where change from air to oil or oil to air is required	

Activity 6.1

Using resources such British and International Standards, manufacturers' catalogues, actual components, etc., build up your own reference sources for mechanical, electrical, electronic and fluid power components and symbols.

Progress check

For each of the components listed in the following questions sketch the schematic symbol (or the conventional representation if there is no schematic symbol), describe its function and give an application.
1. (a) Compression spring, (b) tension spring and (c) torsion spring.
2. (a) Spur gear, (b) bevel gear and (c) worm and wheel.
3. (a) External screw thread and (b) internal screw thread, (c) plain bearing,

Table 6.12 Control mechanisms

Component	Purpose/function	Applications	Symbol
Push-button control mechanism	Manual valve actuation	Valve control	
Lever control mechanism	Manual valve actuation	Valve control	
Pedal control mechanism	Manual valve actuation	Valve control	
Plunger or tracer control mechanism	Mechanical valve actuation	Valve control	
Spring control mechanism	Mechanical valve actuation	Valve control	
Roller control mechanism	Mechanical valve actuation	Valve control	
Solenoid control mechanism	Electrical valve control	Valve control	
Motor control mechanism	Electrical valve control	Valve control	
Pilot control	Indirect valve control by fluid pressure	Valve control	

(d) ball and roller bearing, (e) spline, (d) serration and (e) socket coupling.
4. (a) Fuse, (b) resistor, (c) potentiometer, (d) thermistor, (e) capacitor and (f) inductor.
5. (a) Transformer, (b) filament lamps, (c) ammeter, (d) oscilloscope and (e) AC motor.
6. (a) Integrated circuit, (b) diode, (c) thryristor, (d) transistor, (e) diac and (f) triac.
7. (a) Two-direction variable-capacity hydraulic pump, (b) single-direction fixed-capacity pneumatic motor, (c) reservoir with inlet pipe, (d) accumulator and (e) variable-capacity compressor.
8. (a) Single-acting cylinder with spring return, (b) double-acting cylinder with double-ended piston rod, (c) air-oil actuator, (d) a 2/2 directional control valve (DCV) with solenoid control and (e) a 4/2 DCV with pressure control one way and electric motor control the other.
9. (a) Spring-loaded non-return valve, (b) shuttle valve, (c) rapid-exhaust valve, (d) throttle valve with plunger control, (e) proportional pressure relief valve.

Interpreting engineering drawings

Table 6.13 Valves

Component	Purpose/function	Applications	Symbol
Non-return valve	To open if inlet pressure is higher than outlet pressure	For clamps, jacks, etc. to prevent failure if inlet pressure drops	
Shut-off valve	Prevention of any fluid flow	To shut down circuits rapidly	
Pressure control/ throttle valve	To control flow and pressure	Provision of suitable pressure/ flow rates for circuits	
Pressure-relief valve (safety valve)	Opens the outlet to reservoir or atmosphere if working pressure is exceeded	Prevention of danger or damage by excessive pressure	
Two port–two position (2/2) directional control valve (DCV)	To control one inlet port and one outlet port	Control of single-acting cylinder etc.	
Three port–two position (3/2) DCV	To control two outlet ports and one inlet port or vice versa	Control of two cylinders etc.	
Five port–two position (5/2) DCV	To control two outlet ports, two inlet ports and one exhaust port to atmosphere or reservoir	Control of double-acting cylinders etc.	

257

Graphical communication in engineering

The representation of physical situations

As discussed earlier, in the section on the use of symbols to represent components, components can be used in:
- single-component form
- multicomponent form
- layouts.

Examples of single-component form are:
- mechanical components, e.g. couplings, ball and roller bearings, dampers (also called shock absorbers)
- electrical and electronic components, e.g. motors, transformers, ICs such as operational amplifiers, etc.
- fluid power components, e.g. cylinders, valves, pumps/compressors, motors, etc.

Examples of multicomponent form are:
- mechanical systems and assemblies, e.g. gearboxes, pumps, clutches
- electrical and electronic systems, e.g. burglar alarm systems, satellite receivers
- fluid power systems, e.g. 'pick and place' robots, air-operated drilling machines
- assemblies and systems using mechanical, electrical/electronic and fluid power components, e.g. CNC machine tools such as machining centres, which have:
  - mechanical systems, e.g. leadscrew and nut drives, spindle assemblies
  - electrical systems, e.g. motors for spindles and to drive leadscrew and nut
  - electronic systems, e.g. control of motor speeds and cutter positions
  - fluid power systems, e.g. clamping of workpiece, automatic tool changing, etc.

Layout drawings are used when facilities such as broadcasting and recording studios, manufacturing plants, power stations and so on have to be installed or changed. The individual outlines (to scale) of the complete machines or devices, e.g. CNC machining centres, hydraulic presses, sound and vision recording systems, etc. are shown on a floor plan of the facility.

Single-component drawings

To illustrate the principles involved one case study from each area will be covered.

Case study

Mechanical

Figure 6.9 shows a shaft with two bearings. Both the BS 308 conventional symbols and the actual cross-sectional drawing of the bearings are shown. Note that the same symbol is used for both the ball bearing and the roller bearing. This makes purely graphical interpretation difficult and so there would have to be a note on the drawing that specified the bearing. It can be seen that the bearings consist of an inner and an outer circular ring, separated by either balls or rollers.

Interpreting engineering drawings

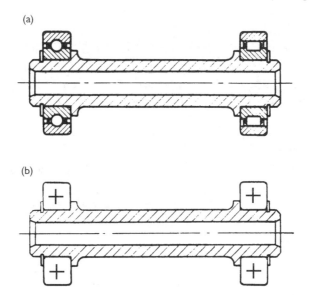

Figure 6.9 A shaft with two bearings: (a) ball and roller bearings; (b) BS conventions

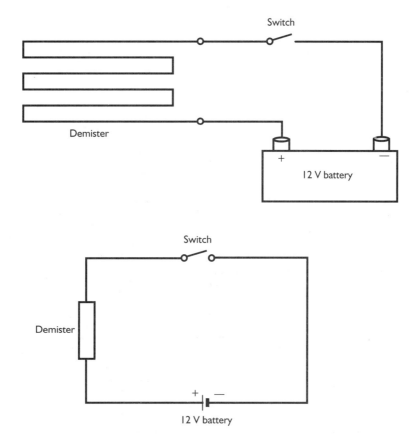

Figure 6.10 Drawing of a heating element

Graphical communication in engineering

Case study

Electrical
Figure 6.10 shows a drawing of a heating element for the rear windscreen of a car. Notice that the circuit symbol is just that of a resistance, so there has to be a note to identify it. The resistance element can be seen to be a long element shaped to cover the windscreen area.

Case study

Electronic
Figure 6.11 shows a light-emitting diode (LED). Note that the symbol lacks any dimensions or pin details, whereas the full drawing of the LED gives this information. The LED is seen to be a domed component with two protruding pins.

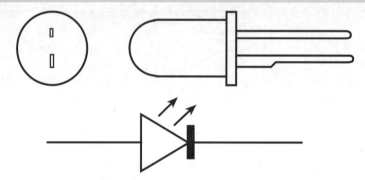

Figure 6.11 A light-emitting diode (LED)

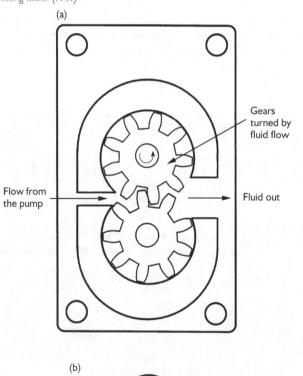

Diagrammatic

Form or method of operation shown in sketch or outline

Figure 6.12 A gear motor in (a) diagrammatic form and (b) symbol form

Interpreting engineering drawings

Case study

Fluid power
Figure 6.12 shows a gear motor drawn in both symbolic and diagrammatic form. The symbol form gives little information and would have to be supplemented by a note giving the manufacturer's reference number. The diagrammatic form illustrates the principle of operation of the motor, with fluid entering under pressure and turning the gears.

Activity 6.2

(Hint: use the preceding case studies as typical answers to this activity, but note that different components are used.)

Using resources such as British and International Standards, reference books, manufacturers' catalogues, machine manuals, circuit diagrams, etc., find and copy both the symbol and the drawing/photograph of the actual physical form of the components, and describe the actual physical situation of the following components:

Universal joint

A joint connecting two rotating devices that allows freedom of radial movement between the two devices

- motorcycle chain drive to rear wheel
- ball and socket universal joint for front wheel drive car
- relay
- domestic light switch
- operational amplifier (IC)
- capacitor.

If possible, check your interpretation of the drawings against the real component.

Multicomponent drawings

The principles involved in recognising and describing the physical situation when different components are put together in circuits and assemblies are similar to those of single components. What is different is the way the components interact when put together. However, at this level it is sufficient to be able just to recognise the components and to describe the physical layout. To illustrate the principle one case study is looked at from each area.

Case study

Mechanical assemblies
Figure 6.13 shows a schematic diagram and scale drawing of a mechanical assembly.

The schematic diagram using symbols shows:
- a shaft
- a spur gear screwed to one end of the shaft
- a coupling at the other end
- a bearing on the shaft.

The sectioned sectioned drawing, which is to scale, gives more information, including:
- diameter and length of shaft
- diameter and type of bearing
- diameter and width of spur gear
- use of spur gear and screw thread to fix bearing to shaft
- more details of the coupling and the way it is secured to the shaft with a key and to the other part with rivets.

261

Graphical communication in engineering

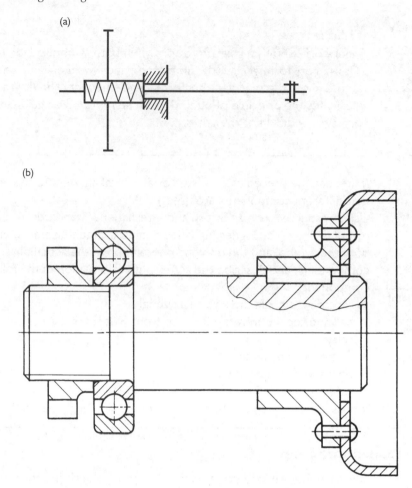

Figure 6.13 (a) Schematic diagram and (b) scale drawing of a mechanical assembly

Case study

Electrical circuits

Figure 6.14 shows the circuit and layout diagrams for the side and tail lights on a car.

The circuit diagram shows:
- four bulbs connected in parallel across a 12-V battery, which ensures that the voltage drop is the same across each bulb, and that if one fails the others carry on working
- a fuse and an on–off switch in the circuit.

The layout diagram shows:
- a plan view of the actual components in place as they are on the car
- the size of the components and their distance apart.

Interpreting engineering drawings

(a)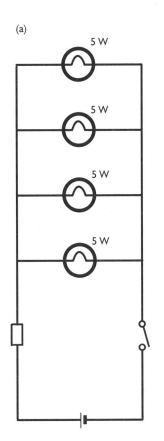

(b)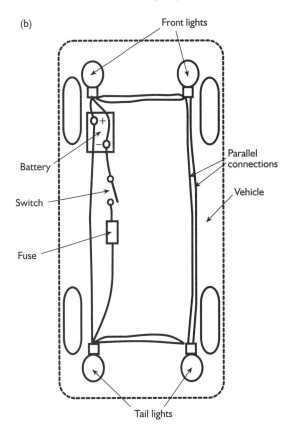

Figure 6.14 (a) Circuit diagram for car sidelights; (b) plan view of actual layout in car

Case study

Audio amplifier

An electronic device that magnifies a small electrical signal into a power signal to drive a loudspeaker

Electronic circuits

Figure 6.15 shows an audio amplifier circuit diagram and photograph of the components assembled on to a printed circuit board (PCB).

The circuit diagram gives the connections for the various components, which are:

- an IC number TBA810P (which is an audio amplifier on a single chip); this is drawn separately as an IC and in the circuit as an amplifier with the chip number stated
- resistors R1, R3 and R_f
- potentiometer R2
- capacitors C1 to C9.

The photograph shows all components on the circuit board in their correct positions. The TBA810P chip can be seen clearly in the middle.

263

Graphical communication in engineering

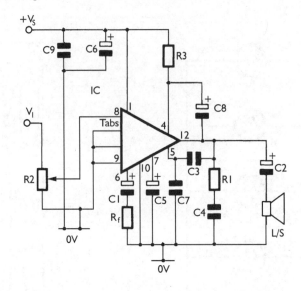

Figure 6.15 Audio amplifier circuit diagram and photograph of the components assembled on a printed circuit board

Case study

Reciprocating
Movement forwards and backwards along an axis

Fluid power circuits
Figure 6.16 shows a continuously operating reciprocating pneumatic cylinder. The circuit diagram shows:
- a double-acting pneumatic cylinder
- a lever-operated 2/2 directional control valve (DCV) – valve no. 1
- a plunger-operated 2/2 DCV – valve no. 2
- a pressure-controlled 5/2 DCV feeding air into each end of the cylinder and also exhausting air from each end; it is controlled by valves 1 and 2.

The layout drawing shows cylinders, valves and piping in their actual positions and to scale.

Interpreting engineering drawings

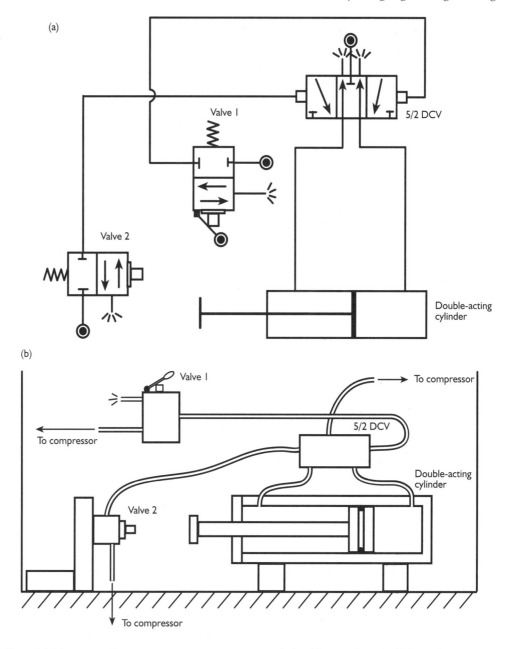

Figure 6.16 A continuously operating reciprocating pneumatic cylinder: (a) circuit diagram; (b) layout drawing

Case study

Layouts of plant and buildings

Figure 6.17 shows the layout of a manufacturing facility.

This plan view drawing with the machines in outline shape is the most usual form of layout. The machine shapes are usually cut-out shapes that can be arranged and rearranged freely on the drawing. If a three-dimensional representation is required then models are used. Computer-aided draughting is useful here as three-dimensional modelling can be done on screen and then plotted when correct. What the drawing shows is:

- position of each machine
- entrances, exits and gangways
- space taken up by machines and free space around machines
- connections to services.

265

Graphical communication in engineering

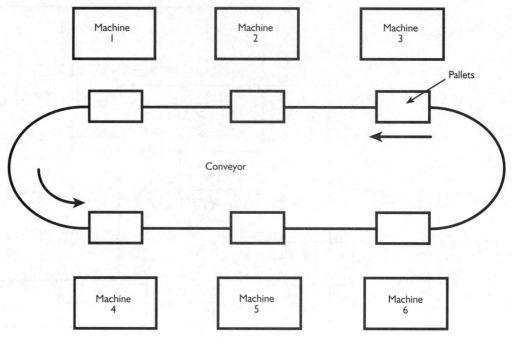

Figure 6.17 *Layout of a manufacturing facility*

Activity 6.3

Using the same principles as in the preceding case studies obtain schematic or circuit drawings and, if possible, scale drawings or photographs for a typical:
- mechanical assembly
- electrical circuit
- electronic circuit
- fluid power circuit
- plant layout.

Identify all the different component parts and describe the physical situation. If possible look at the real objects/circuits/layouts to check your interpretation of the drawings.

Case study

For questions 1–4 following, describe the physical characteristic and sketch the schematic or circuit symbol of:
1. (a) rack and pinion, (b) friction coupling (clutch), (c) damper and (d) roller bearing.
2. (a) socket, (b) push-button switch, (c) AC motor and (d) filament lamp.
3. (a) capacitor, (b) IC, (c) LED and (d) p-n-p transistor.
4. (a) single-acting cylinder with spring return, (b) hydraulic pump, (c) throttle valve and (d) 5/2 DCV.

Figures 6.18 to 6.22 are drawings of multicomponent situations. For each figure you are required to describe the physical situation.

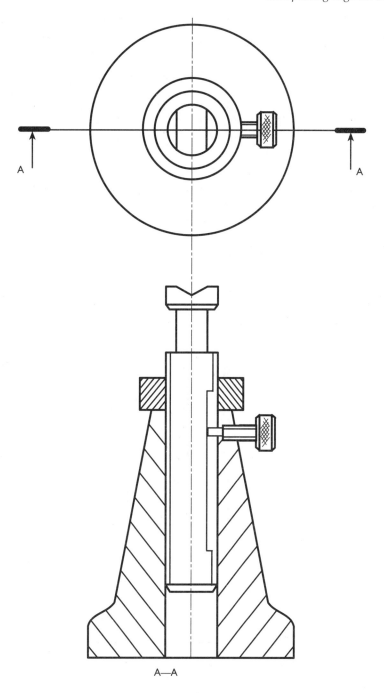

A—A

Figure 6.18

Graphical communication in engineering

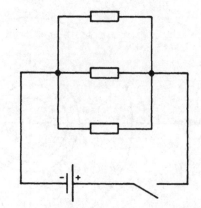

Figure 6.19

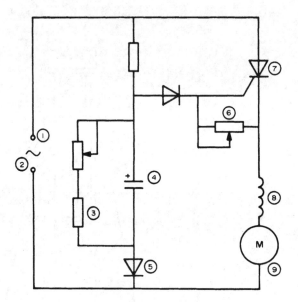

Figure 6.20

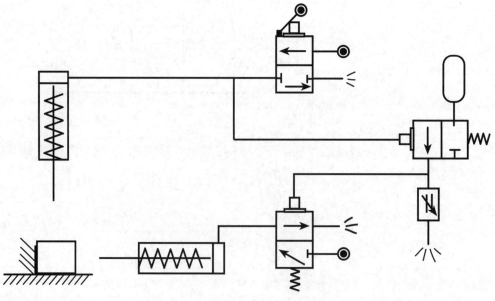

Figure 6.21

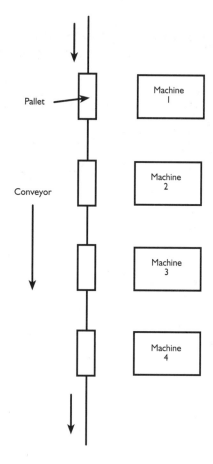

Figure 6.22

Extracting information from engineering drawings

Information about an engineered product is required at each activity and stage of:
- manufacture
- installation
- maintenance.

Although the same kinds of drawings may be used at each stage, it is clear that the user of the drawing may be looking for different kinds of information.

Manufacture

The kinds of information that have to be extracted for manufacture include:
- specification of materials
- specification of bought-out components
- all dimensions, properties and tolerances for components to be made
- instructions for assembly and test
- list of all parts used in the assembly.

The extraction of information for manufacture is now covered for each area.

Graphical communication in engineering

Case study

Mechanical components and assemblies

Referring to Figure 6.13, the information required for manufacture will include:
- dimensions, tolerances, finishes and material specification for the shaft, including the screw thread
- gear tooth size and profile, dimensions, tolerances, finishes and material specification for the spur gear
- dimensions, tolerances, finishes and material specification for the two parts of the coupling
- dimensions and specifications for the bought-out rivets, ball bearing and key
- instructions for assembly
- test specifications.

Case study

Electrical circuits

Referring to Figure 6.14, the manufacturing information required at the car assembly plant will include:
- number and length of wires for wiring harness
- specification of bulb holders, bulbs, fuse and switch
- assembly instructions
- test specifications.

Case study

Electronic circuits

Referring to Figure 6.15, the manufacturing information required will include:
- specifications for all parts
- track and pin hole details for the PCB
- insertion positions for all components
- assembly and soldering instructions
- test specification.

Case study

Fluid power circuits

Referring to Figure 6.16, the manufacturing information at the component manufacturers will be similar to the mechanical assembly. For a user, such as a machinery manufacturer, the information required will include:
- specifications of all components
- position of all components
- length and position of piping
- assembly instructions
- test specifications.

Installation

The information to be extracted for installation is that required for:
- positioning
- final assembly on site, if required
- connecting to services
- testing and troubleshooting to achieve satisfactory functioning.

Interpreting engineering drawings

Case study

Mechanical assemblies

Referring to Figure 6.13, the installation information required in this case will include:
- connecting to the rest of the assembly
- checking for concentricity, alignment, etc.
- testing by running in operating conditions.

Case study

Electrical circuits

Referring to Figure 6.14, it is clear in this case that installation and manufacture are the same. This is because this car lighting circuit is installed at the time of manufacture of the car. Therefore the installation instructions are the same as the assembly instructions.

A better example for an installation example is shown in Figure 6.23.

In this case the information to be extracted for installation in the customer's house will include:
- positions of sockets, connection box, etc.
- distances between sockets, connection box, etc.
- type and length of wiring
- specifications for sockets, fuse box, etc.
- wiring regulations to be followed
- testing for function and safety.

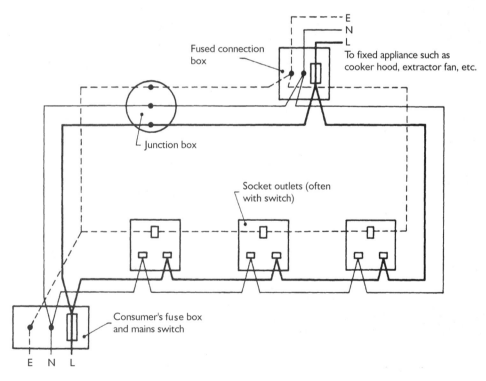

Figure 6.23 Ring-mains system for domestic distribution of electricity

Graphical communication in engineering

Case study

Electronic circuits

Referring to Figure 6.15, the installation information will include that for:
- rating and stability of input power
- testing the circuit function before installation
- connecting the amplifier assembly to the rest of the audio or video circuit
- testing the complete audio or video circuit.

Case study

Fluid power circuits

Referring to Figure 6.16, installation of fluid power circuits will be very similar to the manufacturing procedure described earlier. When the fluid power circuit is to be installed as an integral part of a machine such as a press or drilling machine the information required will be:
- services required, e.g. compressed air, electricity
- test specification under service conditions.

Maintenance

The information required for maintenance will include:
- service intervals for replacement of oils, filters, components, etc.
- repair and/or replacement procedure for components
- fault-finding/troubleshooting procedures.

Case study

Mechanical assemblies

Referring to Figure 6.13, the maintenance information required will include:
- intervals for lubricant changes for gear and bearing
- dismantling procedure for gear, bearing and coupling
- estimated service life of gear and bearing.

Case study

Electrical circuits

Referring to Figure 6.14, the maintenance information required will include:
- procedures for changing bulbs and fuses
- procedures for fault finding/troubleshooting
- specifications for replacement bulbs and fuses
- bulb life.

Case study

Electronic circuits

Referring to Figure 6.15, maintenance information to be extracted will include:
- specifications for replacement components
- test specifications
- service life of components
- fault finding and troubleshooting procedures.

Interpreting engineering drawings

Case study

Fluid power circuits
Referring to Figure 6.16, maintenance information required will include:
- service intervals for replacement of seals, etc. in valves and cylinders
- instructions for fitting of replacement seals and components
- fault finding and troubleshooting procedures.

Activity 6.4

Find examples of engineering drawings that can be used for:
- manufacture
- installation
- maintenance

in each of the following fields of engineering:
- mechanical
- electrical
- electronic
- fluid power
- plant layout.

From each drawing extract the necessary information for use in manufacture, installation and maintenance.

Progress check

For each of the applications given in questions 1–4, describe the types of information required for manufacturing, installation and maintenance.
1. Mechanical assemblies, e.g. gearbox, internal combustion engine, machine tool, etc.
2. Electrical assemblies, e.g. car wiring circuit, electrics for consumer products such as vacuum cleaner, house wiring, etc.
3. Electronic assemblies such as audio/video circuits, motor control circuits, rectifying circuits.
4. Fluid power circuits such as hydraulic press, air-operated drill.

Assignment 6
Interpretation of information from engineering drawings

This assignment provides evidence for:
Element 2.3 Interpret information presented in engineering drawings
and the following key skills:
Communication 2.2: Produce written material
Communication 2.3: Use images
Communication 2.4: Read and respond to written materials
Information Technology 2.1: Prepare information
Information Technology 2.2: Process information
Information Technology 2.3: Present information

Graphical communication in engineering

Your work should be neatly presented as a set of structured notes and diagrams. A well-written set of word processed notes will give key skills credits in both Information Technology and Communication.

Your tasks

Using all sources available such as:
- textbooks
- service manuals
- British and International Standards
- manufacturers' literature such as catalogues, project kit drawings, etc.
- engineering drawings,

find and copy examples of drawings containing all kinds of components including:
- mechanical
- electrical
- electronic
- fluid power
- plant and equipment.

You should have at least one drawing from each of the areas listed above and they should be both single and multicomponent.

For the drawings for each of the above areas:
- recognise and describe the symbols for the individual components
- recognise and describe the physical situation, location, dimensions, functions and interrelationships of the components (this will mean describing the assembly, layout or circuit as applicable)
- detail the information required for:
    - making the product, assembly or circuit
    - installing the product, assembly, layout or circuit
    - maintaining the product, assembly or circuit.

Sample unit test for Unit 2

Note that this test provides only a representative sample of typical questions from Unit 2. As it is the standard test of only 30 questions, it cannot fully examine the subject. It is hoped that anyone involved in assessment will compile similar tests to fully cover the topics. The answers are given on page 460.

1. Identify the most suitable graphical method to provide detailed information for manufacturing purposes:
    a Photograph
    b Dimensioned scale drawing
    c Schematic sketch
    d Circuit diagram.

2. A block diagram is used for what purpose?
    a Representation of systems
    b Layout of machinery
    c Drawing of buildings
    d Computer programming.

3. Flow diagrams are used to:
    a Show the rate of movement of a fluid
    b Aanalyse changes in output
    c Provide information on costing
    d Show a sequence of events.

4. Manufacturing operations can be planned by using a:
    a Pictogram
    b Bar chart
    c Flow process chart
    d Circuit diagram.

5. (B2 + C2) * D3 = E4 is a spreadsheet model. If B2 = 5, C2 = 3 and D3 = 2 then E4 is equal to:
    a 16
    b 30
    c 64
    d 225.

6. A circuit diagram is essential for an engineer to be able to:
    a Test the functioning of a device
    b Take measurements for installation

c Position components correctly
d Plan the manufacturing process.

7 Which graphics would a firm selling washing machines send to prospective customers?
   a Circuit diagram
   b Three orthographic assembly views
   c Detailed drawings of all parts
   d Glossy brochure with photographs.

8 Select the largest paper size:
   a A0
   b A1
   c A2
   d A4.

9 If the outline of a drawing is drawn at a width of 0.7 mm, the correct line width for centre lines is:
   a 0.35 mm
   b 0.5 mm
   c 0.7 mm
   d 1.0 mm.

10 A scale of 1:5 means a drawing:
   a One-15th full size
   b One-fifth full size
   c Five times full size
   d Fifteen times full size.

11 Vector-based computer graphics are most suitable for:
   a Reproducing photographs
   b Painting in colour
   c Freehand sketching
   d Drawing straight lines.

12 In orthographic projection a depth in the plan view is equal to a:
   a Depth in end view
   b Depth in elevation
   c Width in end view
   d Width in elevation.

13 Isometric projection is used for:
   a Drawings dimensioned in metres
   b Three-dimensional representation of an object
   c Circuit and schematic diagrams
   d Two-dimensional representation of an object.

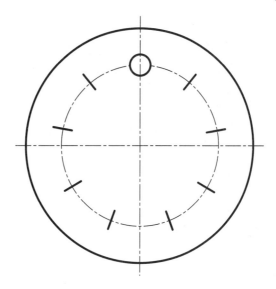

Figure 1

14 What is implied by the drawing of the part in Figure 1?
   a There is one hole in the part
   b There are no holes in the part
   c There are four holes in the part
   d There are nine holes in the part.

15 Hatching at 45° on a drawing usually indicates:
   a Sectioning
   b Knurling
   c Machining
   d Conductivity.

16 Identify the one correct statement about sectioning in drawings:
   a Cutting planes are always in one single plane
   b Holes are hatched
   c Screwed fasteners are not usually sectioned
   d Plain bearings are not sectioned.

17 What is the type of dimensioning shown in Figure 2?
   a Auxiliary dimensioning
   b Chain dimensioning

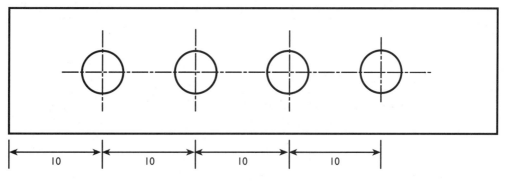

Figure 2

c Progressive dimensioning
d Non-functional dimensioning.

18 A feature on a part has limits of $^{12.500}_{12.490}$. The tolerance is:
a 0.010
b 0.990
c 12.490
d 12.500.

19 A thread is specified as M16-6H. 16 is the:
a Pitch
b Tolerance
c Minor diameter
d Major diameter.

20 The fastener shown in Figure 3 is a:
a Pan head screw
b Socket cap head screw
c Setscrew
d Fitted bolt.

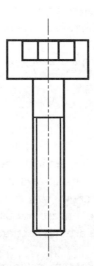

Figure 3

21 What is the part shown by the conventional representation in Figure 4?
a Plain bearing
b Spur gear
c Ball or roller bearing
d Bevel gear.

22 Figure 5 shows a draughting symbol. What does it represent?
a A cast surface
b A lubricated surface
c A machined surface
d A polished surface.

Sample unit test for Unit 2

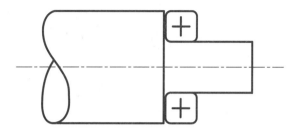

Figure 4

Figure 5

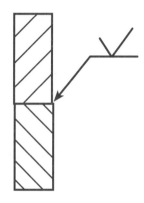

Figure 6

23 What is the type of weld shown in Figure 6?
   a   Double V-butt
   b   Single bevel butt
   c   Single-U butt
   d   Single-V butt.

24 Figure 7 shows the schematic symbol for a mechanical component that:
   a   Stores energy when compressed
   b   Is compressed radially
   c   Increases in length when compressed
   d   Releases energy when compressed.

25 Figure 8 shows the schematic symbol for a mechanical component. When this component is used with a spring it will:
   a   Increase the spring rating
   b   Reduce spring oscillations
   c   Reduce the spring rating
   d   Increase spring oscillations.

26 What is the function of the of the electrical component shown in Figure 9?
   a   Variation of voltage
   b   Circuit protection

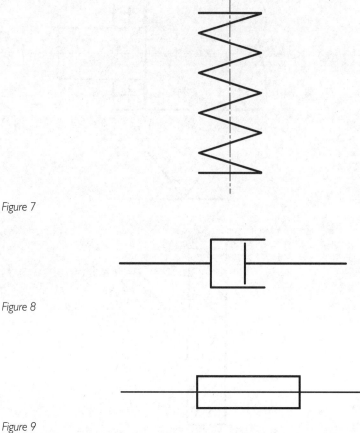

Figure 7

Figure 8

Figure 9

Figure 10

    c Variation of current
    d Control of circuit.

27 An electronic component is shown in Figure 10. The function of this component is to:
    a Emit light when a current is passed
    b Pass current when lit
    c Switch a circuit on
    d Emit sound when a current is passed.

28 Identify the fluid power motor shown in Figure 11:
    a Two-direction pneumatic
    b Single-direction hydraulic
    c Single-direction pneumatic
    d Two-direction hydraulic.

Sample unit test for Unit 2

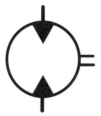

Figure 11

Figure 12

29 Figure 12 shows a fluid power valve. When put into a circuit this valve:
   a Opens if outlet pressure exceeds inlet pressure
   b Safeguards circuit against excessive pressure
   d Opens if inlet pressure exceeds outlet pressure
   d Maintains a constant outlet pressure.

30 What information is required for the installation of a fluid power system onto a machine tool?
   a Detail and assembly drawings of all parts
   b Circuit diagram and scale drawing of layout
   c Photographs of the different parts
   d A computer simulation of the system operation.

PART THREE: SCIENCE AND MATHEMATICS FOR ENGINEERING

Chapter 7   Scientific laws and principles applied to engineering
Chapter 8   The measurement of physical quantities
Chapter 9   Mathematical techniques

Sample unit test for Unit 3
Every trade has its tools, and engineering is no exception. Although the most obvious tools of the engineering trade are those to be found in the workshop, the most fundamental and most important are quite different. It is mathematics and science that allow the engineer to make progress, to develop new products and techniques, and to find uses for new materials.

Part three explores the basic scientific laws and principles and how they are applied to engineering. It also looks at how we make measurements, and how we use different mathematical techniques to help us understand engineering problems.

Chapter 7: Scientific laws and principles applied to engineering

> **This chapter covers**
> Element 3.1: Investigate engineering systems in terms of scientific laws and principles.
> **...and is divided into the following sections:**
> - Unit systems
> - Thermal systems
> - Static systems
> - Fluidic systems
> - Dynamic systems
> - Electrical systems
> - Investigate the science of engineering systems.

Engineering system

A system has at least one input and at least one output; there is also a relationship between the input(s) and output(s)

In each section you will learn about the variable and fixed (constant) quantities by which a system can be defined, and the formulae that relate these quantities, i.e. the scientific laws and principles by which the various systems operate.

When you switch on an electric light or apply the brakes of a bicycle you instinctively know what the result will be. The brightness of the light is always the same, provided that the same type of bulb is used, and the amount by which the bicycle slows will depend on the force applied to the lever on the handlebars. In other words, in both of these engineering systems there is a relationship between the input and output; by this we mean that the systems operate according to one or more scientific laws or principles.

Any device can be thought of as a system. A system always has at least one input and at least one output. There is also a relationship between the input(s) and output(s). Systems are often shown as block diagrams. Figure 7.1 shows the electric light and bicycle braking systems.

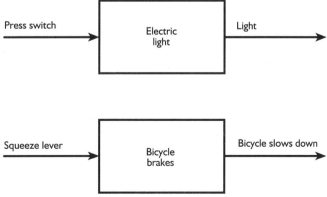

Figure 7.1 Electric light and bicycle braking systems

Science and mathematics for engineering

Because there is a relationship between the input and output of any system, it is possible to predict the behaviour of the system, i.e. what happens to the output when the input is changed in a certain way, or, conversely, to determine the input required to achieve a certain output. To enable the behaviour to be predicted, the scientific principles that govern the behaviour of a system are described by mathematical formulae, from which values can be calculated and the behaviour of the system predicted.

Unit systems

The SI system of units

The most widely used unit system in Europe is the SI (Système Internationale), which is based on the metric system. This system is built on a set of so-called base units.

SI base units

Table 7.1   SI base units

Quantity	Base unit	Abbreviation
Mass	Kilogram	kg
Length	Metre	m
Time	Second	s
Electrical charge	Coulomb	C
Temperature	Kelvin	K

There are a few other base units, which you may come across later in your engineering career. They have been left out to keep things simple.

From these base units all the other units used in engineering can be built up, for instance the units of density are kg/m^3 and of velocity, m/s. These are compound units. Some quantities, however, are not measured in compound units, but have been given their own special units. Those that you will find in this chapter are shown in Table 7.2.

Table 7.2   Special units

Quantity	Unit	Abbreviation
Force	Newtons	N (1 kg m/s^2)
Pressure	Pascals	Pa (1 N/m^2)
Electric current	Amperes	A (1 C/s)
Voltage	Volts	V
Electrical resistance	Ohms	Ω
Energy	Joules	J (1 Nm)
Power	Watts	W (1 J/s)

Scientific laws and principles applied to engineering

Multiples and submultiples of base units

In many situations the base units, or the units built up from them, are too large or too small. For instance, you would not normally measure the diameter of a small bolt as 0.006 m; it would be more appropriate to write 6 mm. Conversely, the output of a power station would be measured in megawatts not watts. Simply by adding a prefix to the base unit the order of magnitude can be changed. Table 7.3 gives these prefixes and their meanings.

There is just one hiccup in this system. The base unit of mass is the kilogram, i.e. the prefix 'kilo' has been used in the base unit. Hence, some adjustment to your thinking will be required when measuring mass, as shown in Table 7.4.

Methods of writing units

Consider the unit for velocity, the metre/second. Throughout this book this has been abbreviated to m/s. However, an alternative method of writing this unit makes use of indices, a mathematical technique with which you may not be familiar. Briefly, when using the 'index form' of a unit the following rule is used:

$$1/s = s^{-1}, \text{ hence, } m/s \text{ is } ms^{-1}$$

$$1/s^2 = s^{-2}, \text{ hence, } m/s^2 \text{ is } ms^{-2}$$

$$1/m^3 = m^{-3}, \text{ hence, } kg/m^3 \text{ is } kg\, m^{-3}$$

You will meet both forms of unit abbreviation during your engineering career, and you should be prepared to use either when you take the GNVQ exam. The index form is preferred by scientists, and will almost certainly be used if you go on to higher education.

Table 7.3   Base unit prefixes

Times multiplied	Prefix	Abbreviation
× 1 000 000 (× 10^6)	Mega	M
× 1000 (× 10^3)	Kilo	k
× 1	Base unit	
× 1/1000 (× 10^{-3})	Milli	m
× 1/1 000 000 (× 10^{-6})	Micro	μ
× 1/1 000 000 000 (× 10^{-9})	Nano	n

Table 7.4   Kilogram units

Times multiplied	Unit	Abbreviation
× 10^3	Megagram	Mg (better known as a tonne)
Base unit	Kilogram	kg
× 10^{-3}	Gram	g
× 10^{-6}	Milligram	mg

Science and mathematics for engineering

The imperial unit system

A description of unit systems would not be complete without mentioning the Imperial system, i.e. feet, pounds, etc. Although we in Britain claim to be metricated, old habits die hard and the Imperial system is still used widely. An important point is that one of the world's major manufacturing countries, the United States, uses the Imperial system universally. To make matters worse, the US system of capacity measurement varies from the standard Imperial units, i.e. the US gallon and pint are smaller than the British units of the same name. However, there is no need for you to worry about these units as far as the GNVQ Intermediate course is concerned. You should just be aware of the existence of the other systems.

Thermal systems

Temperature scales

The two temperature scales in everyday use are the Celsius and Fahrenheit scales. Of these, the Celsius (or centigrade) scale is compatible with the SI unit system and so is commonly used in engineering in Europe.

The Fahrenheit scale is still widely used in the US. Figure 7.2 shows the scales side by side.

At some stage you may need to convert from one scale to the other, in which case the following conversion formulae should be used:

$$\text{to convert °C to °F: °F} = (9/5 \times \text{°C}) + 32$$

$$\text{to convert °F to °C: °C} = 5/9 \, (\text{°F} - 32)$$

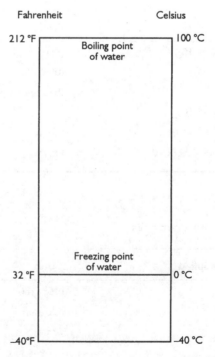

Figure 7.2 Fahrenheit and Celsius scales

Absolute temperature

When heat energy is supplied to a substance the energy is transferred to the molecules of the material as kinetic energy, i.e. the kinetic energy (motion) of the molecules increases. Conversely, when a substance is cooled, the motion of the molecules decreases. The theoretical temperature at which molecular motion ceases is known as absolute zero temperature.

In degrees Celsius the absolute zero temperature is slightly less than –273°C. However, a temperature scale, based on absolute zero temperature, is in common use; it is known as the Kelvin scale.

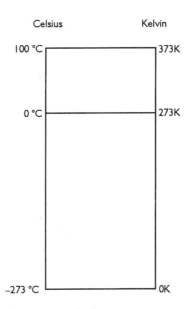

Figure 7.3 Celsius and Kelvin scales. Note: 273K not 273°K

The Kelvin and Celsius scales have 'degrees' of the same size, so that a temperature *change* of 1°C is the same as a *change* of 1 K.

There is also an absolute temperature scale consistent with the Fahrenheit scale, the Rankine scale. However, you are very unlikely to encounter this.

Latent heat

Latent heat of vaporisation

The heat energy required to convert 1 kg of liquid, at its boiling point, to a gas

When a substance is heated *one* of the following will occur:
- the temperature will increase
- the substance will undergo a change of state, i.e. from a solid to a liquid or a liquid to a gas.

The graph in Figure 7.4 shows what happens when a solid such as ice is heated until it first melts to become water and then vaporises to become steam.

An important point to note is that while the state is changing there is no temperature change. In this section we will look at the process of vaporisation only.

The latent heat of vaporisation is the heat required to convert 1 kg of liquid, at its boiling point, to a gas. Before examining the scientific principle behind the process of vaporisation, the fixed and variable quantities must be defined.

The fixed quantity, usually called a constant, is:
- the latent heat of vaporisation for the liquid under consideration. (symbol L, units J/kg).

Science and mathematics for engineering

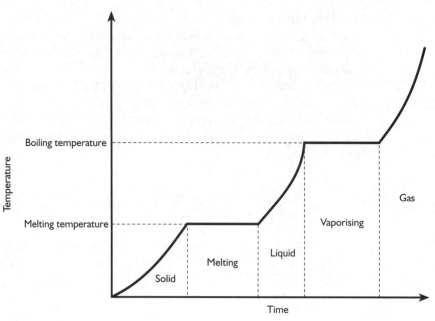

Figure 7.4 Changes of state from solid to liquid to gas

The variable quantities are
- the mass of liquid (symbol m, units kg)
- the heat energy supplied (symbol Q, units J).

The formula relating these quantities is
$$Q = Lm$$

Case study

The following worked examples will demonstrate the way in which this formula can be used.

1. Calculate the heat energy required to vaporise 0.75 kg of water at 100°C (i.e. its boiling point).

Data:
   mass of liquid = 0.75 kg
   latent heat of vaporisation = 2260 kJ/kg
   = 2 260 000 J/kg

Formula:
   $Q = Lm$

Heat required = 2 260 000 × 0.75 = 1 695 000 J = 1695 kJ.

(Note that it was unnecessary to convert kJ to J in this case. The joule is a very small unit of energy and it is quite common to use kilojoules.)

2. In an experiment to determine the latent heat of vaporisation of a fuel it is found that 280 kJ of heat energy is required to vaporise 250 g of the fuel. Calculate the latent heat of vaporisation.

Data:
   mass of liquid m = 250 g = 0.25 kg
   heat energy Q = 280 kJ

Formula:
   $Q = Lm$

The formula must first be transposed to make L the subject. If you are unsure about transposition refer to Chapter 9.

Scientific laws and principles applied to engineering

$L = Q/m = 280/0.25 = 1120$ kJ/kg
i.e. the latent heat of vaporisation is 1120 kJ/kg

In both of these examples the data given were written down first, then the relevant formula and then the calculation. It is suggested that you also use this format to make your work tidy and easily understood. Always state the answer in words at the end, and don't forget to put in the units!

Activity 7.1

1. Liquid ammonia (a refrigerant) has a latent heat of vaporisation of 1370 kJ/kg. Determine the heat energy required to vaporise 15 kg of ammonia at its boiling point.
2. Calculate the mass of water that can be vaporised at 100°C by supplying 1 MJ of heat energy.

We will now look in more detail at two engineering applications of the vaporisation process.

Case study

The steam boiler

Steam boilers are widely used in industry to generate steam for a variety of purposes, such as
- driving turbines for electricity generation or propulsion
- heating
- cleaning.

Boilers are made in many different sizes and to a wide variety of designs. However, one of the most important variable quantities to be considered when specifying a boiler is the steam generation rate, measured in kilograms per hour (kg/h). A typical small 'tube boiler' might generate 900 kg/h, so the data known are:

for water $L = 2260$ kJ/kg
steam rate $m = 900$ kg/h

(Note that the mass is given in kg/h. Don't worry; it means that the required energy will be calculated in kJ/h.)
Formula:
$Q = Lm$
$Q = 2260 \times 900 = 2\,034\,000$ kJ/h $= 2034$ MJ/h

This calculation gives the rate at which energy is required. This is also called the power required.

In the case of our steam boiler, the power required to provide 900 kg/h of steam can be calculated as follows:
2034 MJ = 2 034 000 000 J
1 h = 3600 s
So power required = 2 034 000 000/3600 = 565 000 W = 565 kW
i.e. the power required to provide 900 kg/h of steam is 565 kW.

Power

Power = energy/time
1 watt = 1 joule per second

Science and mathematics for engineering

Efficiency

Efficiency = output energy/ input energy or output power/ input power

Efficiency

Efficiency = output energy/ input energy or output power/input power. Efficiency is often given as a percentage, i.e.

% efficiency η = output energy/input energy × 100%

No engineering system can have an efficiency of 100% because there is always some energy lost to the surrounding environment. In the case of the boiler there will be heat losses no matter how good the insulation is.

A typical value for the efficiency of a small boiler would be 70%. Using this information the power supplied to the boiler can be calculated.

Data:
  power output (i.e. power supplied to heat water) = 565 kW
  efficiency = 70% or 0.7

Formula:
  η = power output/power input
  Transposing gives power input = power output/η
  = 565/0.7 = –807.1 kW,

i.e. the fuel supplied to the boiler must provide 807 kW of heat (i.e. 807 kJ/s)

Case study

Refrigeration

The second application we will look at is the refrigeration process. Ammonia is one of several substances used as a refrigerant, and is widely used in large-scale freezing plants.

Ammonia vaporises at about –33°C (240K) at normal atmospheric pressure. Liquid ammonia is fed into the freezer coil and the heat from the contents of the refrigerated space is 'extracted' by the ammonia as it vaporises. By the time the ammonia has reached the end of the freezer coil it has all vaporised, and is taken back to the rest of the system to be liquefied again. During the liquefying process heat is removed from the ammonia and transferred to the air or water outside the system.

Apart from its low vaporising temperature, the property that makes ammonia very suitable for use as a refrigerant is its high latent heat of vaporisation, i.e. as it vaporises it absorbs a lot of heat from its surroundings.

The next example considers an ammonia refrigeration system that can provide liquid ammonia to the freezer coil at a rate of 50 kg/h. From this information the rate of heat energy extraction can be calculated.

Data:
  ammonia L = 1370 kJ/kg
  m = 50 kg/h

Formula:
  $Q = Lm$
  $Q = 1370 \times 50 = 68\,500$ kJ/h,

i.e. heat can be extracted at a rate of 68 500 kJ/h.

Scientific laws and principles applied to engineering

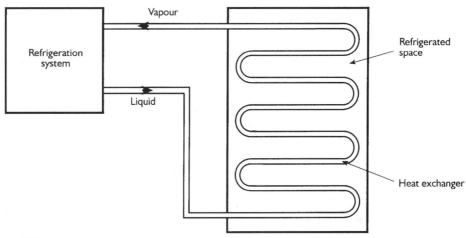

Figure 7.5 Refrigeration

Activity 7.2

A kettle is supplied with 2.2 kW of electrical energy and is 80% efficient. If the safety cut-out fails, determine how long it will take for 1 litre of water at 100°C to boil away completely? You will find further questions about the kettle in the section on electrical systems at the end of this chapter.

Linear expansion

Linear expansivity

The amount by which a 1-m length of a material expands if the temperature is increased by 1°C

When most materials are heated, they expand (increase in size), and when cooled they contract. Metals, in particular, expand and contract in a very predictable way, and it is important that allowance is made for 'thermal movement' when designing engineering products that are likely to undergo large temperature changes. Large structures such as bridges, railways and ships are often exposed to variations in temperature, which can lead to large forces within the structure as one part expands more than another to which it is attached. We will look at some examples in detail later in this section.

When an object expands, it does so in three dimensions (e.g. length, width, depth). Figure 7.6 shows a rectangular bar being heated.

Provided the temperature of the bar increases uniformly, each dimension will increase by the same percentage. The actual amount of expansion will depend on two variable quantities:
- the length of the bar before heating (or cooling)
- the temperature increase (or decrease).

The expansion will also depend on the material being heated. A property of any material is its linear expansivity (sometimes called the coefficient of linear expansion). This is a constant for the particular material.

The linear expansivity of a material is the amount by which a 1-m length expands if the temperature is increased by 1°C. To clarify this definition, we are talking about expansion in one direction only. However, as has already been described, the same percentage increase in size will occur in all directions. Expansion in one direction is called linear expansion. The actual amount of expansion that occurs is small, e.g. a 1-m length of aluminium would increase by 0.000023 m if the temperature was increased by 1°C. Table 7.5 gives values of linear expansivity for some engineering materials.

Science and mathematics for engineering

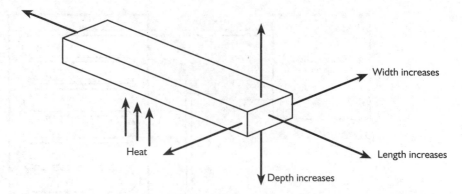

Figure 7.6 A rectangular bar being heated

Table 7.5 Linear expansivity for some engineering materials

Material	Linear expansivity (m/m°C or /°C)
Aluminium	0.000023 (23×10^{-6})
Bronze	0.000017 (17×10^{-6})
Mild steel	0.000015 (15×10^{-6})
Titanium	0.000009 (9×10^{-6})
Ceramic	0.000009 (9×10^{-6})
Glass	0.000009 (9×10^{-6})
Nylon	0.0001 (100×10^{-6})
Polythene	0.00025 (250×10^{-6})
Polypropylene	0.000062 (62×10^{-6})
Rubber	0.00022 (220×10^{-6})
Invar	0.0000009 (0.9×10^{-6})
Copper	0.000017 (17×10^{-6})

The scientific principle behind thermal expansion and contraction leads to the formula:

$$\Delta L = \alpha L_0 \Delta T$$

where ΔL is the change in length measured in metres (note that the symbol Δ (means 'the change in…')
α is the linear expansivity measured in metres/metre °C
L_0 is the initial length measured in metres
ΔT is the change in temperature measured in °C.

Case study

Steel bar

A steel bar 2.5 m long is used as part of the support structure of a furnace. During normal operating conditions the temperature increases from 25 to 150°C. If the bar is free to expand, calculate the increase in length in millimetres.
Data:
 initial length = 2.5 m

Scientific laws and principles applied to engineering

linear expansivity = 15 × 10⁻⁶ m/m°C
initial temperature = 25°C
final temperature = 150°C
Formula:
$\Delta L = \alpha L_0 \Delta T$
We must first determine the temperature change:
$\Delta T = T_2 - T_1 = 150 - 25 = 125°C$
Putting the figures into the formula:
$\Delta L = 15 \times 10^{-6} \times 2.5 \times 125$ m
   = 0.00469 m or 4.69 mm
The change in length = 4.69 mm
In reality, the final length would probably be of more importance.
Final length = initial length + increase
   = 2.5 + 0.00469 m
   = 2.50469 m

Case study

Nylon bearing

A nylon bearing has an internal diameter of 25 mm at 20°C. It is to be used in a piece of machinery in the Arctic, at a temperature of –40°C. Calculate the internal diameter of the bearing at this temperature.
Data:
   initial size = 25 mm = 0.025 m
   linear expansivity = 0.0001 m/m°C
   initial temperature = 20°C
   final temperature = –40°C
Change in temperature $\Delta T = T_2 - T_1$
Formula:
$\Delta L = \alpha L_0 \Delta T$
   = 0.0001 × 0.025 × –60 m
   = –0.00015 m
   or –0.15 mm
Note the minus sign. This tells you that there is a reduction in size or contraction, as you would expect when there is a temperature decrease.

Activity 7.3

1. A 12-m length of copper pipe is installed as part of a heating system. The room temperature when the pipe is fitted is 15°C. Calculate the length of the pipe when water at 80°C is flowing through it.
2. A steel pin of diameter 32 mm at 25°C is used in a freezer plant at a temperature of –30°C. Calculate the diameter at the operating temperature.

Science and mathematics for engineering

Case study

Shrink-fitted bearings

Ball bearings are often shrunk into housings to prevent the outer race turning (Figure 7.7). This is common practice in motorcycle engines where the high rotational speeds would cause considerable damage if the outer race did turn. In the fitting process the casing into which the bearing is to be fitted is carefully heated to avoid distortion; this is especially true if the casing is made from aluminium alloy. The bearing is then pushed into the expanded housing, which grips the outer race as contraction takes place during cooling. To make fitting even easier the bearing may be cooled, either by placing it in a refrigerator or by using a 'freezing' spray.

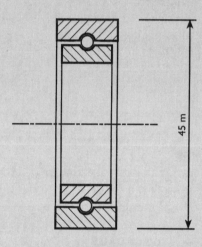

Figure 7.7 Ball bearing

Static systems

Springs

Work

Work done = force × distance moved in the direction of the force ($W = Fx$)

Everyone is familiar with springs and the type of 'bouncy' motion they can provide. However, in engineering this motion is often an unwanted aspect of spring behaviour, and springs are more commonly used as a means of storing energy in both static systems (i.e. those in which there is no motion) and dynamic systems (i.e. those in which components are in motion). The dynamic behaviour of springs is rather complex and is not included in the Intermediate GNVQ.

If a coil spring is subjected to a force F (N) (Figure 7.8) it will, in this case, compress by a distance x (m).

The amount of deflection will depend on the stiffness of the spring. As the force is applied, work is done in compressing the spring.

Work done = force × distance moved in the direction of the force ($W = Fx$). The units of work are joules (J), i.e. work is a form of energy. Hence, as long as the spring remains compressed, it is storing energy. To put this idea into context think about the following example. Springs are often used to return levers to a rest position. Figure 7.9 shows a lever to which a force F is applied.

This force causes the lever to move, and the return spring will be extended. When the force is removed the extended spring provides the energy required to return the lever to its original position. Return springs are used in braking sys-

Scientific laws and principles applied to engineering

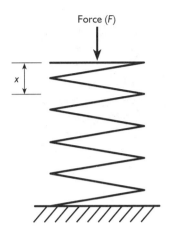

Figure 7.8 Coil spring subjected to force

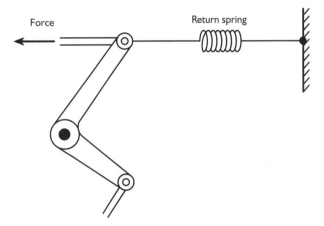

Figure 7.9 Force applied to a lever

tems to release the brake shoes or pads once the lever or pedal force is removed. Have a look at bicycle braking systems and you will see return springs in action. These springs also provide 'feel' to the system. Without them, the slightest force applied to the brake lever would result in an abrupt application of the brakes.

Types of springs
There are many different shapes and sizes of springs. Figure 7.10 shows some of the ones seen more frequently in engineering products.

The behaviour of springs under load
As long as springs are not deflected to the point at which their shape or size is changed permanently, their behaviour is easily predicted. To keep things simple we will consider coil springs in tension or compression, as shown in Figure 7.11,

The variable quantities that can be measured are:
- applied force F newtons
- spring deflection x metres

These quantities are related by a scientific law known as Hooke's law, which states that the extension or compression of a spring is proportional to the applied load. In mathematical terms:

$$F = kx$$

where k is the spring stiffness. The stiffness depends on the size and shape of the spring, and the material from which it is made, and is a constant for any particu-

Science and mathematics for engineering

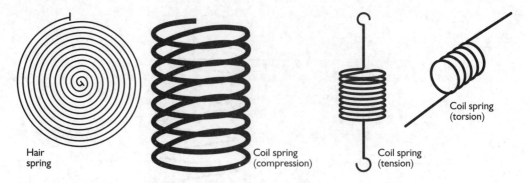

Figure 7.10 Types of spring

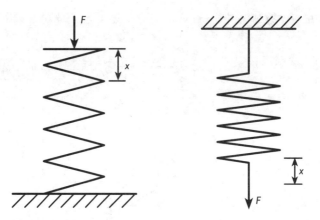

Figure 7.11 Springs in tension and compression

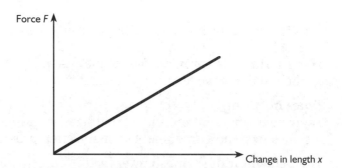

Figure 7.12 Graph of applied force against deflection

lar spring. Figure 7.12 shows a graph of applied force F(N) against deflection x (m) for a spring.

The relationship between force and deflection is linear, as you should have realised from the formula. If the formula is transposed to make stiffness k the subject, we get

$$k = F/x$$

The stiffness can also be obtained by determining the gradient of the graph. The technique for obtaining gradients is discussed in Chapter 9.

Scientific laws and principles applied to engineering

Case study

A compression spring of stiffness 50 kN/m is compressed by 15 mm. Calculate (a) the applied force and (b) the work done.
The applied force
Data:
  spring stiffness $k = 50$ kN/m $= 50\,000$ N/m
  deflection $x = 15$ mm $= 0.015$ m
Formula: $F = kx$
  Applied force $F = 50\,000 \times 0.015 = 750$ N
The work done
Data:
  applied force $F = 750$ N
  deflection $x = 0.015$ m
Formula: $W = Fx$
  Work done $= 750 \times 0.015 = 11.25$ J

A small tension spring is used to return the dutch pedal of a car to its rest position. When a force of 25 N is applied to the spring it extends 10 mm. Calculate the spring stiffness.
Data:
  applied force $F = 25$ N
  spring deflection $x = 10$ mm $= 0.01$ m
Formula: $F = kx$
This formula must be transposed to make stiffness k the subject:
  $k = F/x$
Hence, spring stiffness $= 25/0.01 = 2500$ N/m.

Activity 7.4

1. A tension spring of stiffness 7.5 kN/m is stretched by a force of 150 N. If the initial length of the spring is 800 mm, calculate the extended length.
2. Table 7.6 gives results obtained from a test carried out on a compression spring. Use these data to plot a graph of force against deflection and then determine the spring stiffness in N/m by finding the gradient.

Table 7.6 Tests on a compression spring

Load (N)	2.0	3.0	4.0	5.0	6.0	8.0	9.0	10.0
Deflection (mm)	6.0	9.2	12.1	14.9	18.0	21.1	24.0	26.8

Static equilibrium

Forces and moments

If a force is applied to an object one of two things will happen:
- if the object is free to move it will accelerate
- if the object is not free then some deformation of the object will occur.

It is also possible for a combination of these two to occur. For instance, when a golf ball is struck it initially deforms, but also, obviously, accelerates. Engineering systems in which motion, and especially accelerations, occur are known as dynamic systems. Those systems that are fixed or at rest are known as static systems. There are three scientific laws that neatly allow us to predict the behaviour of systems of forces. These laws are summarised below.

Newton's laws
1. An object will remain at rest or at constant velocity unless acted upon by an external force.
2. The acceleration of an object depends on the size of the force causing the acceleration.
3. Every force is opposed by an equal but opposite force.

In dynamic systems all three of these laws are used to predict behaviour. However, for static systems only the third law will be used.

Any object that is at rest, whether it is standing by itself on a surface or is joined to other components as part of a structure, is said to be in a condition of static equilibrium. For this to be true there must be no overall (resultant) force acting on the object. If there was, the object would accelerate in the direction of the force (first law). There must also be no turning moment acting. If there was the object would rotate. These conditions can be summarised by the following three statements:
- resultant horizontal force = 0
- resultant vertical force = 0
- resultant moment = 0.

In this section we will examine several systems, all of which are in a state of static equilibrium. To start with, Figure 7.13a shows a block resting on a horizontal surface.

Experience tells us that the block will not accelerate through the surface, so there must be a force acting that is equal but opposite to the weight. Figure 7.13b includes this reaction force, which is the force exerted by the surface, upwards, on the block (law 3). The conditions for static equilibrium can now be checked against the complete diagram:
- horizontal forces – there are none
- vertical forces – upward force = downward force
- moments – there are none.

We will now make the situation more complicated by attempting (but not succeeding) to push the block using a horizontal force (Figure 7.14a).

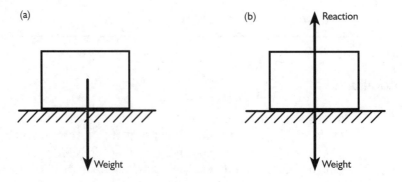

Figure 7.13 (a) Block resting on a horizontal surface; (b) reaction force

Scientific laws and principles applied to engineering

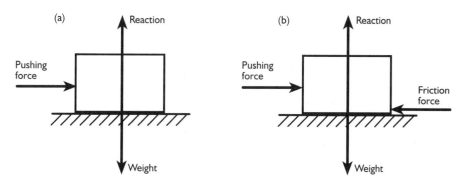

Figure 7.14 (a) Horizontal force; (b) friction force

The conditions for static equilibrium are not met because the resultant horizontal force is not zero. If the block is not moving there must be a force opposing the push; there is… friction. Figure 7.14b shows the friction force acting against the force that is attempting to move the block. But are the conditions for static equilibrium met now? Look carefully at the positions of the two horizontal forces: they are not directly opposite one another. Figure 7.15 shows this more clearly.

The result of the vertical displacement of these forces is a moment, i.e. the block will rotate. The only way to achieve static equilibrium is to push the block close to the surface on which it is resting (Figure 7.16).

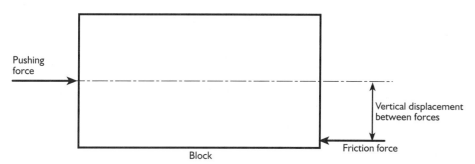

Figure 7.15 Displacement of horizontal forces

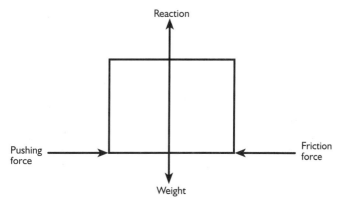

Figure 7.16 Static equilibrium

301

The principle of moments

Figure 7.17 shows a system with a pivot about which a bar is free to rotate. The system is at rest.

If the conditions for static equilibrium are checked:
- there are no horizontal forces
- the total downward force $F_1 + F_2$ must be equal to the reaction force that acts upward through the pivot, i.e. $R = F_1 + F_2$
- the turning moments (or just moments) due to the forces F_1 and F_2 must be equal but opposite.

This leads us to the principle of moments, which states that for a system to be in a condition of static equilibrium the total clockwise moment must equal the total anticlockwise moment. In fact, this is just a repetition of the third condition for static equilibrium, stated earlier, but in systems that have some form of pivot point this principle is useful, as is illustrated by the following examples.

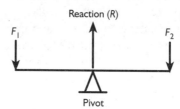

Figure 7.17 A force system

Case study

The system shown in Figure 7.18 is at rest. Calculate the magnitude (size) of the force F.

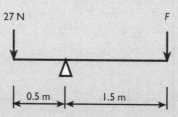

Figure 7.18

The first thing we should state here is the principle of moments, i.e. for static equilibrium the total clockwise moment = the total anticlockwise moment.
  Considering first the effect of the 25 N force:
  anticlockwise moment = $27 \times 0.5 = 13.5$ Nm
Next the effect of the unknown force F:
clockwise moment = $F \times 1.5$ Nm
As the two moments are equal if the system is in static equilibrium,
  $13.5 = F \times 1.5$ which gives $F = 13.5/1.5 = 9$ N
  i.e. force $F = 9$ N
The analysis of the forces on the system could now be completed by saying that the reaction force at the pivot = $27 + 9 = 36$ N.
  Figure 7.19 shows the completed force diagram for the system. You should get used to drawing this type of diagram to make your work on forces and moments clearer.

Scientific laws and principles applied to engineering

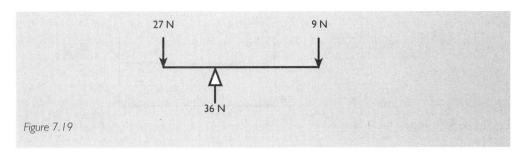

Figure 7.19

Case study

The system of forces in Figure 7.20 is at rest. Determine the position of the 10-N force, and the force on the pivot.

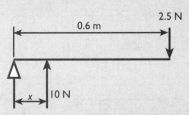

Figure 7.20

For static equilibrium the clockwise moment = the anticlockwise moment.
   Clockwise moment = 2.5 × 0.6 = 1.5 Nm
   Anticlockwise moment = 10x Nm
So, 1.5 = 10x
   x = 1.5/10 = 0.15 m
i.e. the 10-N force is 0.15 m (150 mm) from the pivot.
   Total upward force = total downward force
   R + 10 = 2.5 N
   R = 2.5 – 10 = –7.5 N
   This is an interesting result. The reaction force at the pivot is negative; why? If you look back a few lines you will see that the reaction force R was assumed to act upwards. The negative sign is telling us that this assumption was incorrect, and that the force at the pivot acts downwards. Hence, we can now state the answer as:
   the force at the pivot is 7.5 N downwards.

Activity 7.5

Make sure that you state the relevant condition for static equilibrium before each calculation.
1. Calculate the force F and the reaction force at the pivot shown in Figure 7.21.
2. Determine the distance between the 12 N force and the pivot shown in Figure 7.22.

303

Science and mathematics for engineering

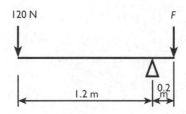

Figure 7.21

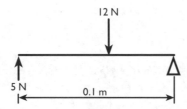

Figure 7.22

Case study

The principle of moments, and the conditions for static equilibrium in general, are two of the basic tools of any engineer designing static systems. The following case study will help you to see how these techniques are used when applied to real engineering systems.

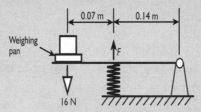

Figure 7.23 *Diagram of a simple weighing instrument*

Figure 7.23 shows a simplified diagram of a weighing instrument. When an object is placed on the weighing pan, the spring is compressed slightly and the pointer moves down the scale. In this case study we will look at the deflection of the spring that occurs when an object weighing 16 N is placed on the pan. The spring stiffness is 10 N/mm.

If the principle of moments is applied to this static system:
clockwise moment = $16 \times 0.21 = 3.36$ Nm
anticlockwise moment = force on spring $(F) \times 0.14$ m = $0.14 F$ Nm
For static equilibrium, the clockwise moment = the anticlockwise moment, so
$3.36 = 0.14F$
which gives $F = 3.36/0.14 = 24$ N
The spring stiffness is 10 N/mm, which means that the spring will compress by 1 mm for each 10 N of force acting on it. The spring deflection for a force of 24 N will therefore be
$24/10 = 2.4$ mm

Fluidic systems

Pressure

Pressure
Force/area over which the force acts

If a force is applied to the surface of a liquid, or to a gas, the molecules just move out of the way. However, if the molecules of the fluid (a liquid or gas) are contained they will be forced closer together. Figure 7.24 illustrates these ideas. These phenomena can be observed most clearly if a very viscous (thick) liquid is used.

As the molecules are forced closer together, the forces between molecules resist the applied force F and the fluid becomes pressurised. Pressure is defined as force divided by the area over which the force acts. The units of pressure are newtons/square metres or N/m^2.

However, another unit is commonly used, the pascal (Pa):

$$1\ Pa = 1\ N/m^2$$

As long as you work in newtons and metres, the pressure will be in pascals. It should be noted that the pascal is a very small unit. The pressure of the atmosphere around you is 100 000 Pa (10^5 Pa).

There are two points that should be remembered about fluid pressure systems:
- pressure always acts at 90° to a surface
- the pressure in a fluid pressure system caused by an applied force is the same throughout the system.

Engineering fluidic systems

There are also two types of fluid pressure systems used in engineering:
- hydraulic systems, which use liquid, usually oil, as the pressurised fluid
- pneumatic systems, which use gas, usually air, as the pressurised fluid.

These systems are used to transfer force and/or motion from one part of an engineering system to another, i.e. they transfer energy.

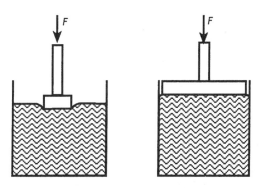

Figure 7.24 Force on surface of liquid or gas. (a) Liquid not contained. The liquid molecules move away and then close around the back of the piston. (b) Liquid contained. Liquid molecules are forced closer together

Case study

A hydraulic system is pressurised by a force of 50 N acting on a piston of diameter 20 mm. Calculate the pressure in the system.
Data:
  force = 50 N
  piston diameter = 20 mm = 0.02 m

hence, piston area = $P \times 0.02^2/4 = 0.00126$ m²
Formula:
$P = F/A$
$P = 50/0.00126 = 39\,789$ Pa
i.e. the pressure in the system is 39 789 Pa.

Case study

Calculate the diameter of piston required if a pneumatic cylinder pressurised at 200 000 Pa is to provide a force of 1 kN.
Data:
  pressure = 200 000 Pa
  force = 1 kN = 1000 N
Formula:
  $P = F/A$
In this case the diameter of the piston is required. This can be obtained from the piston area, so A must be made the subject of the formula:
  $A = F/P$
Putting in the data
  $A = 1000/200\,000 = 0.005$ m²
  Area of a circle $A = \pi d^2/4$
Transposing to make d the subject gives:
  $d = \sqrt{(4A/\pi)}$
  (if you did not understand the transposition, look ahead to Chapter 9)
  $d = \sqrt{(4 \times 0.005/\pi)} = 0.08$ m or 80 mm
i.e. the piston diameter is 80 mm.

Case study

Energy transfer using fluid pressure

Figure 7.25 shows a simplified fluid pressure system. The cylinder on the left, the piston of which is subject to an applied force, is known as the master cylinder. The cylinder on the right, the piston of which is being 'driven' by the fluid pressure, is known as the slave cylinder.

The pressure in the system is provided by the force of 500 N. This pressure will be transmitted to the slave cylinder, and apply a force to the larger piston. This is the principle of operation of hydraulic jacks and braking systems.
  pressure in fluid = $500/(\pi \times 0.01^2)$ Pa
  force on large piston = $P \times A = [500/(\pi \times 0.01^2)] \times (\pi \times 0.025^2) = 3125$ N
Hence, the system acts as a force amplifier. However, there will also be motion involved. If the master cylinder piston moves 30 mm, then a volume of fluid will be displaced. This volume can be thought of as a cylinder:
  volume of fluid displaced = $\pi r^2 h = 9.42 \times 10^{-6}$ m³
(Note that because we are working in cubic metres, the volume seems very small.)

The same volume of fluid will be passed to the slave cylinder, causing the piston to move up, but as the slave cylinder is larger, the motion will be less. Using the formula for the volume of a cylinder again:
  $V = \pi r^2 h$

Scientific laws and principles applied to engineering

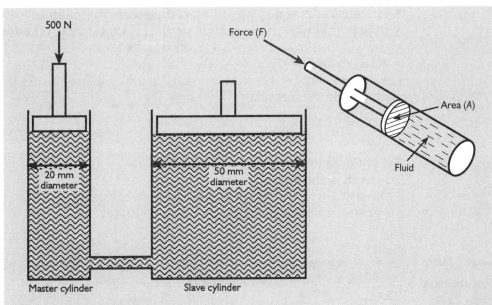

Figure 7.25 Simplified fluid pressure system

This time h becomes the subject:
$h = V/\pi r^2$
and the diameter is 50 mm or 0.05 m and the radius is 0.025 m.
$h = (9.42 \times 10^{-6})/(\pi \times 0.025^2)$
$h = 0.0048$ m
i.e. the slave piston moves up 0.0048 m or 4.8 mm.

We have said nothing about any energy losses in the system, so the work done by the 500-N force in pressurising the system should be the same as the work done on the slave piston.

Work done by 500-N force = force × distance
= 500 × 0.03 = 15 J
Work done on slave piston = 3122 × 0.0048 = 15 J

Our system is apparently 100% efficient!

Energy losses in fluid pressure systems

Earlier in this chapter it was stated that an efficiency of 100% was impossible. Unfortunately, that holds true for hydraulic and pneumatic systems, but where do the energy losses occur? The following list gives some of the major points:
- friction between pistons and cylinder walls
- deflection of piston seals, connecting pipes, etc.
- compression of the fluid, especially in pneumatic (gas) systems.

Having said this, fluid pressure systems are very efficient – hydraulic systems in particular as oil is virtually incompressible.

Hydrostatic pressure

If you dive into a swimming pool you will notice that as you go deeper the pressure on your ears increases (the ear is a particularly sensitive pressure sensor). The change of pressure in liquids with increasing depth has significance in engineering systems that:

- operate under water, e.g. oil rig components, submarines
- transport liquids through large vertical distances, e.g. oil refineries
- store liquids in deep tanks, e.g. oil storage tanks
- act as dams.

The relationship between the pressure in a liquid and the depth is linear and is modelled by the formula

$$P = \rho g h$$

where
P is pressure in pascals (a variable)
h is depth in metres (a variable)
ρ is density in kilograms per cubic metre (a constant)
g is gravitational acceleration (m/s²) (a constant).

Case study

A water storage tank has a depth of 3.5 m. Calculate the pressure at the bottom of the tank if it is filled completely.
Data:
$\rho = 1000$ kg/m³
$g = 9.81$ m/s²
$h = 3.5$ m
Formula:
$P = \rho g h$
$P = 1000 \times 9.81 \times 3.5 = 34340$ Pa
i.e. the pressure at the bottom of the tank = 34 340 Pa or 34.3 kPa.

Case study

A storage tank contains oil (density = 800 kg/m³). Calculate the depth at which there is a fluid pressure of 12 kPa.
Data:
$P = 12$ kPa $= 12\,000$ Pa
$\rho = 800$ kg/m³
$g = 9.81$ N/m²
Formula: $P = \rho g h$
which transposes to $h = P/\rho g$
$h = 12\,000/(800 \times 9.81) = 1.53$ m
The depth at which a pressure of 12 kPa exists = 1.53 m.

Activity 7.6

1. Figure 7.26 shows a dam that is part of a hydroelectric scheme. Calculate the pressure at the turbine inlet due to the depth of water.
2. Atmospheric pressure is approximately 100 kPa. How deep would you have to dive in the sea to feel this pressure due to the liquid above you? You can assume that the density of the sea water is about the same as that for pure water (it is actually slightly greater owing to the dissolved salts).

Scientific laws and principles applied to engineering

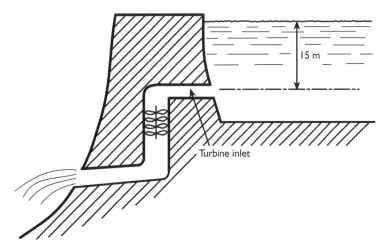

Figure 7.26 Hydroelectric dam

Total pressure

This last example will help to illustrate an important point. Figure 7.27 shows a liquid surface open to the atmosphere.

In such a situation the total pressure at the bottom of the tank equals atmospheric pressure + pressure due to the depth of liquid, i.e.

$$P = P_0 = \rho g h$$

Your ears are quite used to operating at atmospheric pressure, but if you look at your answer to the previous question, you will see that when you dive to a depth of 10 m the pressure on your ear drums has doubled. No wonder it feels uncomfortable.

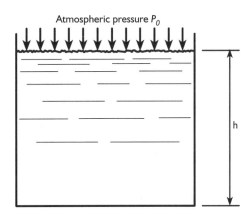

Figure 7.27 Liquid surface open to the atmosphere

Case study

Figure 7.28 shows a water supply system. The storage tank is situated on a hill, the top of which is 600 m above sea level. When full, the tank holds water to a depth of 25 m.

Near to the base of the tank is situated an inspection hatch to allow maintenance of the tank lining. The hatch is secured by 16 bolts. As part of the design process, the pressure on the hatch and the force on the bolts have to be checked.

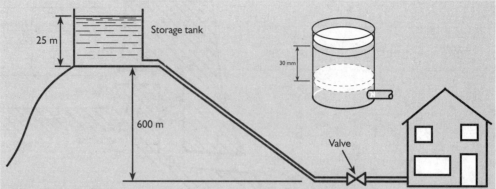

Figure 7.28 Water supply system

Pressure on the hatch

The centre of the hatch is 1 m above the tank base so the depth of water above the hatch is:

$h = 25 - 1 = 24$ m

Data:
$h = 24$ m
$\rho = 1000$ kg/m³
$g = 9.81$ m/s²

Formula:
$P = \rho g h$
$P = 1000 \times 9.81 \times 24 = 235\,400$ Pa $= 235$ kPa

You might wonder why no allowance was made for atmospheric pressure. Figure 7.29 shows that atmospheric pressure acts on both sides of the hatch, and is therefore cancelled out.

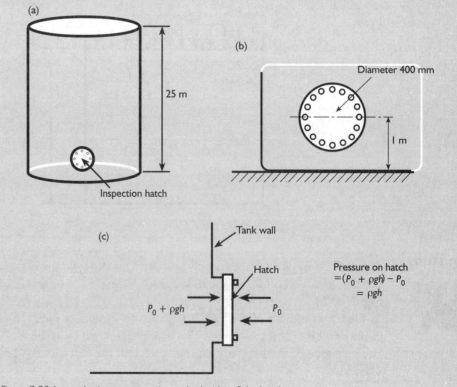

Figure 7.29 Atmospheric pressure acting on both sides of the hatch

Scientific laws and principles applied to engineering

Pressure on hatch $= (P_0 + \rho gh) - P_0$
$= \rho gh$

Force on the hatch

If you refer back to the first part of this section you will see the relationship between pressure (P), force (F) and area (A):
$P = F/A$
For the hatch:
Data:
  $P = 235\,400$ Pa
  diameter $= 400$ mm $= 0.4$ m
  hence, radius $= 0.2$ m
  $A = \pi r^2 = \pi 0.2^2 = 0.1257$ m^2
Transposing the formula to make F the subject:
  $F = P \times A$
  $F = 235\,400 \times 0.1257$ N
  $F = 29\,590$ N
  or 29.6 kN

Force on each bolt

The force on the hatch will be transmitted to the bolts, so that each bolt is subjected to a sixteenth of the force, i.e.
  force on each bolt $= 29\,590/16 = 1849$ N
  or 1.85 kN

Activity 7.7

Pressure in the water supplied to the house
The water is supplied to the house, the internal diameter of the supply pipe at the house being 18 mm. Your task is now to calculate the:
- pressure at the valve
- the force acting on the valve when it is closed.

Dynamic systems

Displacement

Distance travelled in a given direction

Velocity

Velocity is displacement divided by time = s/t

Engineering systems that involve motion are referred to as dynamic systems. Although the types of motion involved in some systems are very complex, all motion can be defined by five variable quantities:
- distance travelled or displacement s metres
- direction of travel
- time t seconds
- velocity v or u metres/second
- acceleration a metres/second^2.

An important point to note is that displacement, velocity and acceleration are vector quantities, i.e. they have both magnitude (size) and direction. Because this section will consider motion in a straight line only, this added complication will not affect your work much. The direction of motion can only be either forwards (positive) or backwards (negative).

Science and mathematics for engineering

Distance–time graphs

Acceleration
Acceleration is the rate of change of velocity

Graphs are a very effective way of illustrating motion. Figure 7.30 shows a simple distance–time graph for an object travelling at constant velocity.

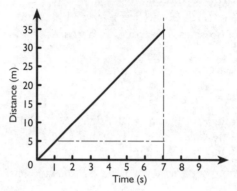

Figure 7.30 Distance-time graph

The gradient of the line represents the magnitude of the velocity, i.e. the rate of change of distance with time. In this case the gradient is given by

$$(33 - 5)/(7 - 1) = 28/6 = 4.6 \text{ m/s}$$

Figure 7.31 shows a more complicated journey, which can be summarised as follows:
- between 0 and 3 seconds the object travels with constant velocity $v = (10 - 0)/(3 - 0) = 3.33$ m/s
- between 3 and 5 seconds the object is stationary because there is no increase in distance
- between 5 and 7 seconds the object again travels at constant velocity; this time though the gradient of the graph is steeper: $v = (35 - 10)/(7 - 5) = 25/2 = 12.5$ m/s
- between 7 and 8 seconds the object is stationary.

This graph in Figure 7.32 shows a journey in which the velocity is changing continually. It would be difficult, using the mathematical techniques available in the GNVQ Intermediate course, to obtain values for the velocity at any particular moment. However, the average velocity for the journey is easily determined, as the average velocity is just the total journey distance divided by the total journey time:

$$v = 30/9 = 3.33 \text{ m/s}$$

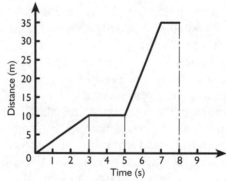

Figure 7.31 Distance–time graph for a more complicated journey

Scientific laws and principles applied to engineering

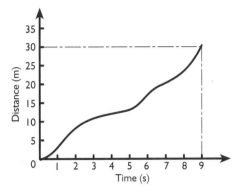

Figure 7.32 Distance–time graph for a journey in which velocity is changing continuously

Table 7.7 Distance and time values for a car journey

Distance (km)	0	15	30	55	85	115	145	145	165	185	205
Time (h)	0	0.25	0.5	0.75	1.0	1.25	1.5	1.75	2.0	2.25	2.5

Activity 7.8

Table 7.7 gives distance and time values for a car journey. Plot these values and draw a distance–time graph for the journey.
Now use the graph to determine:
- the average speed for the journey
- the speed at which the car travels for the first half-hour
- the speed at which the car travels between 55 km and 145 km
- the length of time during the journey that the car is stuck in a traffic jam.

Velocity–time graphs

In many ways velocity–time graphs are more useful than distance–time graphs as they allow both acceleration and distance travelled to be calculated, as well as showing the velocity of the object at any particular moment in time.

Figure 7.33 shows the velocity–time graph for an object travelling with constant acceleration. The acceleration is represented by the gradient, i.e. acceleration is change in velocity divided by time.

Putting in figures gives:

$$a = (35 - 5)/(7 - 1) = 30/6 = 5 \text{ m/s}^2$$

Figure 7.34 shows a velocity–time graph for an object accelerating, travelling at constant velocity and then decelerating:
- between 0 and 2 seconds acceleration = $(15 - 0)/(2 - 0) = 7.5 \text{ m/s}^2$
- between 2 and 4 seconds acceleration = 0; velocity = 15 m/s^2
- between 4 and 7 seconds acceleration = $0 - 15 = -15/3 = -5 \text{ m/s}^2$

Note that deceleration (slowing down) is indicated by a negative acceleration value.

Science and mathematics for engineering

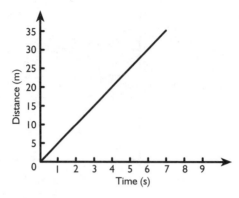

Figure 7.33 Velocity–time graph for an object travelling with constant acceleration

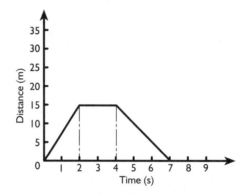

Figure 7.34 Velocity–time graph for an object accelerating, travelling at constant velocity and then decelerating

Distance travelled

The distance travelled during a journey can be determined by obtaining the area under the velocity–time graph. Referring back to Figure 7.33, the area under the graph is simply a triangle, the area of which is

$$(7 \times 35)/2 = 122.5 \text{ m, i.e. the distance travelled is } 122.5 \text{ m}$$

For Figure 7.34 the distance travelled during the journey is given by the areas of two triangles and a rectangle. So,

$$\text{distance travelled} = (2 \times 15)/2 + (2 \times 15) + (3 \times 15)/2$$
$$= 15 + 30 + 22.5 = 67.5 \text{ m}$$

Having obtained the distance travelled, the average velocity can also be obtained. For Figure 7.33 this is 67.5/7 = 9.6 m/s.

Activity 7.9

Figure 7.35 shows the velocity–time graph for a machine component that is in motion. Determine
- maximum acceleration
- the deceleration
- average velocity between 0 and 8 s.

Scientific laws and principles applied to engineering

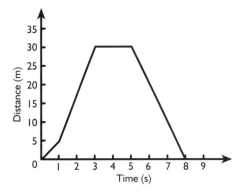

Figure 7.35 Velocity–time graph for a moving machine component

The equations of motion for constant acceleration

Provided the motion of an engineering system is fairly simple, it can be described by a set of straightforward equations or formulae. You may have already realised this having seen that linear graphs are produced when the acceleration of an object is constant. The equations that are used in this section all describe constant acceleration motion.

The equations of motion can be worked out or derived from the graphs already studied, i.e. the distance–time and velocity–time graphs, and the definition of acceleration, $(v - u)t$.

Equation 1
The definition of acceleration is usually transposed to make final velocity v the subject, i.e.:
$$a = (v - u)/t \text{ becomes } v = u + at$$

Equation 2
You should by now be familiar with velocity–time graphs, and know that the area under such a graph gives the distance travelled s. Using Figure 7.36, the area under the line is made up from:

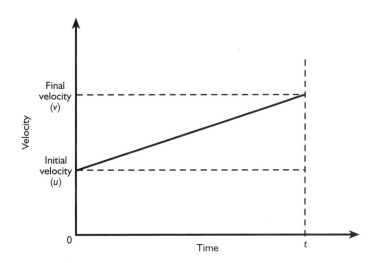

Figure 7.36

315

- a rectangle, area = ut
- a triangle, area = $(v-u)\,t/2$

So distance travelled $s = ut + (v-u)\,t/2$

This formula does not include acceleration, but we know that $a = (v-u)/t$ so, rewriting we get:

$$s = ut + (v-u)\,/t^2$$

$$s = ut + at^2/2$$

Equation 3

The third equation is more difficult to prove and so will just be stated:

$$v^2 = u^2 + 2as$$

The GNVQ syllabus does *not* demand that you should be able to derive these formulae yourself. As long as you can use them correctly in appropriate engineering situations you will have satisfied the requirements of Element 3.1. The following examples show how the equations of motion can be used. Remember, however, that they can only be used in systems undergoing constant acceleration.

Case study

The acceleration of a racing car

This case study looks at the acceleration of a racing car, and the subsequent deceleration when it brakes. The car in question is capable of reaching 100 mph (46 m/s) in 3 s, and then stopping again in another 2 s.

Using Equation 1 we can work out the acceleration and deceleration of the car during these two periods.

To calculate the acceleration:
Data:
  $v = 46$ m/s
  $u = 0$
  $t = 3$ s
Formula: $v = u + at$
  Transposing gives $a = (v-u)/t = (46-0)/3 = 15.33$ m/s²

To calculate the deceleration:
Data:
  $v = 0$
  $u = 46$ m/s
  $t = 2$ s
  $a = (v-u)/t = (0-46)/2 = -23$ m/s²

The negative sign of course indicates deceleration, as discussed earlier in this chapter.

We will now calculate the distances required for the acceleration and deceleration phases using Equation 2.

Acceleration phase:
Data:
  $u = 0$
  $t = 3$ s
  $a = 15.33$ m/s²
Formula:

$s = ut + at^2/2$
$s = (0 \times 3) + (15.33 \times 3^2)/2 = 69$ m

Deceleration phase:
Data:
 $u = 46$ m/s
 $t = 2$ s
 $a = -23$ m/s
Formula:
 $s = ut + 1/2at^2$
 $s = (46 \times 2) + (-23 \times 2^2)/2$
 $s = 92 - 46$
 $s = 46$ m

In other words, the distance required for the car to accelerate to 100 mph is 69 m, and the stopping distance from this speed is 46 m. In reality, neither the acceleration nor the deceleration of the car would be constant. Nevertheless, the equations of motion allow engineers to obtain a good approximation in such situations.

So far Equation 3 has not been used. In this case study it could have been used as an alternative to Equation 2.

Activity 7.10

Use Equation 3 ($v^2 = u^2 + 2as$) to calculate the distance for the acceleration and deceleration phases, and check that you get the same answers.

Force, weight, mass and acceleration

When a force is applied to an object that is free to move, the object will accelerate.

Newton's second law gives us a relationship between
- force F measured in newtons (N)
- mass m measured in kilograms (kg)
- acceleration a measured in metres per second^2 (m/s^2).

The formula relating to these quantities is $F = ma$.

Case study

A mass of 0.35 kg accelerates at 5 m/s^2. Calculate the force that causes the acceleration.

Data:
 $m = 0.35$ kg
 $a = 5$ m/s^2
Formula:
 $F = ma$
 $F = 0.35 \times 5 = 1.75$ N

i.e. the force causing the acceleration is 1.75 N.

Science and mathematics for engineering

Case study

A force of 300 N is applied to mass of 25 kg, which is free to move. Determine the resulting acceleration.
Data:
$m = 25$ kg
$F = 300$ N
Formula:
$F = ma$
Transposing to make acceleration a the subject:
$a = F/m$
$a = 300/25 = 12$ m/s^2
i.e. the acceleration of the mass is 12 m/s^2.

Activity 7.11

1. A mass of 85 kg accelerates at 0.6 m/s^2. Calculate the size of the accelerating force.
2. An object is observed to accelerate at 3.0 m/s^2 when a force of 30 N is applied to it. Calculate the mass of the object.

Case study

Free fall

When an object falls under the influence of gravity it accelerates towards the ground. Figure 7.37 shows the forces acting.

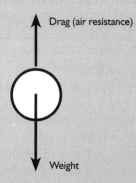

Figure 7.37 Forces acting on a falling object

If there was no air resistance then the only force acting would be the weight. This is almost the case when dense objects fall.
Using Newton's second law, force = mass × acceleration:
- force = weight, in this case W
- mass = mass of object m
- acceleration = acceleration due to gravity g.

So $W = mg$

This gives us a relationship between weight and mass. The acceleration caused by gravity, g, is taken as 9.81 m/s^2, and the fact that it is a constant means that all objects fall with the same acceleration, provided air resistance is negligible.

Electrical systems

Electricity or electrical energy is transmitted through wires by a flow of subatomic particles called electrons. Electrons carry a negative charge and so are attracted towards positive charges. Electric current is the rate at which charge flows through a conductor when electrons are attracted towards a positive charge. The strength of the attraction or 'electromotive force' (EMF) is the voltage. As the electrons pass through the conductor they interact with other particles and are slowed down. The amount of slowing down is called the resistance. In summary:

- voltage is a measure of the force pushing the electrons, known as the electromotive force or EMF
- current is a measure of the flow rate of electrons
- resistance is the slowing-down effect on the electrons.

Electrical units

The best place to start a discussion on electrical units is by thinking of each electron as a tiny packet of negative electricity or 'charge'. Just for the record (you don't have to remember this) the charge carried by each electron is 0.00000000000000000016 coulombs, i.e. 1.6×10^{-19} C. The coulomb (C) is the SI unit of charge. To bring you back to the real world the unit of electric current, the ampere or amp (A) is defined as a flow rate of 1 C/s; you can work out for yourself how many electrons per seconds that represents!

Activity 7.12

Use your calculator to work out the number of electrons that make up 1 coulomb of charge. You will have to use the exponent button, which is usually labelled EXP.

The volt is the SI unit that indicates the amount of force available to move the charge. When 1 joule of energy is required to move 1 coulomb of charge between two points in a circuit, the voltage between the two points is 1 volt.

Finally, resistance: when a current of 1 ampere flows with a voltage of 1 volt there is a resistance of 1 ohm (Ω or R).

A useful way to picture electricity is as a flow of water through a pipe going from a high level to a lower level (Figure 7.38).

Electrical circuits

If a complete circuit is set up with a source of EMF such as a battery, then current will flow around the circuit until the source is removed. The battery in this case

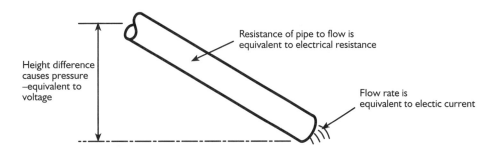

Figure 7.38 Water model for electricity

can simply be regarded as an energy provider for the electrons, which then carry the energy to other parts of the circuit. For a current to flow then:
- the circuit must be complete
- there must be a source of EMF (voltage).

Electrical energy

As current flows around a circuit, through the wires and circuit components such as resistors, energy is lost in the form of heat. You will have noticed that many electrical systems warm up after they have been left switched on for a while. This dissipated heat can be a source of danger if there is no way of removing it; many house fires are started by overheated wires. On the other hand, this heat energy can be used; electric kettles, irons and cookers are just three examples.

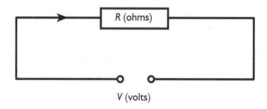

Figure 7.39 Simple circuit

Figure 7.39 shows a simple circuit in which a current (I ampere) is flowing through a resistance (R Ω). The current is driven by an EMF of V volts. As the current flows, energy is 'used' by the circuit. The rate at which energy is used (i.e. power) in joules per second (i.e. watts) is given by the formula

$$\text{electrical power} = \text{voltage} \times \text{current}$$

$$\text{or as a formula, } P = VI$$

Having established a relationship between power, voltage and current, it should be possible to write down a formula that will give the energy used.

$$\text{Power} = \text{energy}/\text{time}$$

$$\text{so energy} = \text{power} \times \text{time}$$

$$\text{electrical energy} = VIt$$

where t is the time (in seconds) for which the current is flowing.

Case study

A 9-V electric motor driving a model boat draws a current of 0.5 A. Calculate the rate at which the motor uses electrical energy. (The question is really asking, 'what is the electrical power used by the motor?'.)
Data:
  voltage = 9 V
  current = 0.5 A
Formula:
  $P = VI$
  power $P = 9 \times 0.5 = 4.5$ W
i.e. the electrical power used by the motor is 4.5 W.

Scientific laws and principles applied to engineering

A computer monitor uses electrical energy at a rate of 0.15 kW. Calculate the current drawn if the device is connected to a 230-V supply.

Data:
  power = 0.15 kW = 150 W
  voltage = 230 V

Formula:
  $P = VI$

Transpose to make I the subject
  $I = P/V = 150/230 = 0.65$ A

The current drawn by the monitor is 0.65 A.

If the monitor in the above question is left on for 2 hours how much energy is used?

Data:
  power = 0.15 kW = 150 W
  time = 2 h = 7200 s

Formula:
  power = energy divided by time

Transpose to make energy the subject:
  power × time = energy
  energy = 150 × 7200 = 1 080 000 J
  or 1.08 MJ

The energy used by the monitor in 2 is 1.08 MJ. (This may seem like a lot of energy; the joule, however, is a very small unit.)

Activity 7.13

1. The motor of a model car is powered by a 9-V battery and draws a current of 1.2 A. Calculate the electrical power supplied to the motor.
2. The headlights of a car are rated at 55 W each. If the battery voltage is 12 V determine the current drawn when both lights are working.

Case study

The electric drill

Good-quality electric drills (mains powered) are fitted with motors that use about 700 W of electrical power on full load, i.e. when the motor is delivering maximum torque (a torque is the same thing as a moment – refer back to the section on static equilibrium). Electric motors are very efficient electro-mechanical systems, and efficiencies of over 80% are common. Under full-load conditions a large current is drawn; if you have used an electric drill you will have noticed that it beomes quite warm because of the heating effect of the current. This heat is lost energy and is one reason why the motor is not more efficient.

To calculate the power available to turn the drill bit we use the relationship,
  efficiency = output power divided by input power

Data:
  η = 90% or 0.9
  P_{in} = 700 W

Formula:

Science and mathematics for engineering

$\eta = P_{out}/P_{in}$
Transposing to make P_{out} the subject:
$P_{out} = \eta \times P_{in}$
$P_{out} = 0.9 \times 700 = 630$ W
i.e. 630 W is available to turn the drill bit.

It is important that mains-powered electrical systems are fitted with cables and a fuse that can carry the maximum design current without overheating. The current at full load can be determined using the formula:
$P = VI$
Data:
 power $P = 700$ W
 voltage $V = 230$ V
Formula:
 $P = VI$
Transposing the formula to make current, I, the subject gives
$I = P/V$
$I = 700/230 = 3.0$ A
i.e. the current at full load is 3 A.

In fact, there will be situations in which this current will be exceeded in normal use. For instance, if the drill bit jams the current will rise rapidly as the motor attempts to keep turning. It would be inconvenient if the fuse blew every time this happened so a higher current rating is used. Likewise, the cable is fairly heavy to allow for higher currents.

Activity 7.14

In the section on thermal systems you carried out calculations on the output energy of an electric kettle. The kettle could be represented by a system block diagram (Figure 7.40).
Using the data given earlier, carry out the following tasks.
1. Complete the block diagram by stating the forms of energy at input and output.
2. Calculate the current drawn by the kettle when operating at full power.

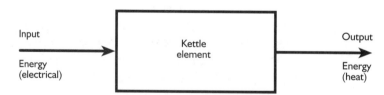

Figure 7.40 System block diagram of a kettle

The relationship between voltage, current and resistance in a circuit (Ohm's law)

This relationship was investigated by George Ohm in 1828. He concluded from his experiments that there was a simple relationship between:
- voltage or EMF (V) measured in volts
- current (I) measured in amps
- resistance (R) measured in ohms.

Scientific laws and principles applied to engineering

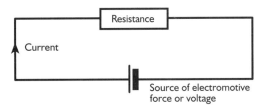

Figure 7.41

The formula that relates these quantities, Ohm's law, is

$$V = IR$$

Case study

In a circuit similar to that in Figure 7.41, a current of 120 mA flows through a resistance of 500 W. Calculate the supply voltage.
Data:
 current I = 120 mA = 0.12 A
Formula:
 $V = IR$
 $V = 0.12 \times 500 = 60$ V
i.e. the supply voltage is 60 V.

Case study

The resistance of a car light bulb filament under normal operating conditions is 4 Ω. Calculate the current that will be drawn.
Data:
 resistance R = 2.4 Ω
 supply voltage V = 12 V
(Note that car electrical systems operate at 12 V.)
Formula:
 $V = IR$
Transpose to make current I the subject:
 $I = V/R$
 $I = 12/2.4 = 5$ A
i.e. current drawn by the bulb is 5 A.

Activity 7.15

1. Calculate the EMF required to drive a current of 0.5 A through a resistance of 120 Ω.
2. An unknown resistance is connected to a 24-V supply. The measured current is 20 mA. Calculate the resistance.

Activity 7.16

You are now in a position to be able to calculate the resistance of the electric kettle when operating under the conditions described earlier. Use Ohm's law.

The effect of temperature on resistance

Although the heating effect of electric current has been mentioned, the effect of heat on resistance has so far been ignored by saying that electrical systems are at their operating conditions. However, as you will have observed, electrical systems warm up from the temperature of the environment (e.g. room temperature) to operating temperature. This increase in temperature causes the resistance of metallic conductors to increase also. There are a few exceptions. Two alloys, constantan and manganin, show almost no change in resistance when subjected to temperature changes, a property that makes them useful for some applications.

The two graphs in Figure 7.42 show how resistance varies with temperature for constantan and copper.

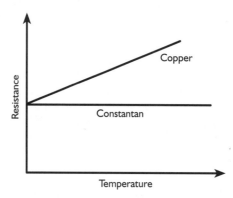

Figure 7.42 Resistance–temperature graphs for constantan and copper

The amount by which the resistance of a component varies with temperature depends on the:
- initial resistance of the component (a variable) R_0
- temperature change (a variable) ΔT
- temperature coefficient of resistance (a constant for any particular material) α.

The formula that relates these quantities is:

$$\Delta R = \alpha R_0 \Delta T$$

(Δ is just the Greek capital letter delta and is used to mean 'the change in').

Two points to note here are:
- the symbol α is used for the coefficient of resistance; the same symbol is also used for coefficient of expansion
- the formula is very similar to the one you use to calculate thermal expansion – have a look back at that section.

Scientific laws and principles applied to engineering

Table 7.8 Values of temperature coefficient of resistance for common metals

Metal	Temperature coefficient of resistance (°C)
Aluminium	$40 \times 10^{-4}/°C$
Aluminium alloy	$16 \times 10^{-4}/°C$
Brass (a copper alloy)	$15 \times 10^{-4}/°C$
Constantan	$0.4 \times 10^{-4}/°C$
Copper	$39 \times 10^{-4}/°C$
Manganin	$0.1 \times 10^{-4}/°C$
Silver	$40 \times 10^{-4}/°C$

Note the difference that alloying makes to the coefficients for aluminium and copper.

Case study

The temperature of an aluminium alloy conductor rises from 0°C to 27°C. If the resistance of the conductor at 0°C is 30 Ω determine the resistance at 27°C.
Data:
$\Delta T = 27 - 0 = 27°C$
$R_0 = 30\ \Omega$
$\alpha = 16 \times 10^{-4}/°C$
Formula:
$\Delta R = \alpha R_0 \Delta T$
$\quad = 16 \times 10^{-4} \times 30 \times 27$
$\quad = 1.30\ \Omega$
The resistance at 27°C would therefore be:
$30 + 1.3 = 31.3\ \Omega$

Case study

The electric kettle

You have already calculated the resistance of the heating element in the kettle at operating temperature. Using the formula $\Delta R = \alpha R_0 \Delta T$ you are now able to calculate:
- the resistance of the heating element
- the current drawn
- the power used

when the kettle is first switched on. To do so you can make the following assumptions:
- the element is made of copper
- the temperature of the element at first is 20°C
- the temperature of the element when the kettle is boiling is 120°C.

Now think back to Ohm's law, $V = IR$. In many applications current is supplied at a constant voltage, e.g.
- mains electricity is supplied at 230
- vehicle electrical components are supplied at 12.

If the resistance R increases owing to an increase in temperature then the current I will decrease if IR is to remain constant, which it must if the voltage V is constant. This means that as domestic appliances or vehicle components warm up, they draw less current, i.e. at operating conditions they use less current than at the moment they are switched on. Any fuse in the system must be capable of withstanding the initial high current.

Science and mathematics for engineering

Alternating current

You have already seen that electric current in a wire is a flow of negatively charged particles called electrons. When a battery is connected to a conductor, the electrons flow towards the positive terminal, and the direction of flow will remain constant. This type of electron flow is known as direct current (DC).

Mains electricity is produced by large generators. These are rotating machines which, owing to the nature of their construction, produce an electric current that, instead of flowing in one direction, 'shuffles' backwards and forwards. Figure 7.43 shows how the current varies with time. You will notice that for half the time the current is negative, i.e. it flows in opposite directions.

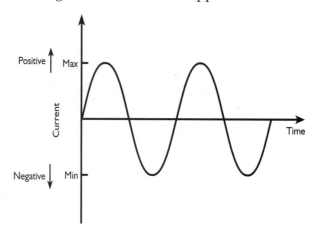

Figure 7.43 Current–time graph

In many electrical systems the effect of alternating current is similar to that of direct current, particularly if the system contains only resistance. However, there are important differences that are beyond the scope of the Intermediate GNVQ.

Alternating current is defined or described by several quantities:
- the frequency f
- the current I or voltage V
- the shape of the graph or 'waveform'.

Frequency

The frequency of an AC wave (Figure 7.44) is simply the number of complete waves or cycles per second. Alternatively, you could think of frequency as the number of times per second that the current reverses.

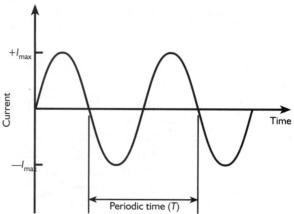

Figure 7.44 Defining an AC wave

Scientific laws and principles applied to engineering

The time for one cycle is the periodic time T seconds.

$$\text{Frequency } f = 1/T \text{ cycles/second}$$

Although it would seem logical to measure frequency in cycles/second, there is an SI unit, the hertz (Hz). So,

$$f = 1/T \text{ Hz}$$

Activity 7.17

Find out the frequency of:
- the mains 230-V supply
- the clock in one of the computers that you use.

Alternating current waveforms can be drawn for either voltage or current as both quantities alternate similarly. In some systems they do not alternate together; there is a time or 'phase' difference between them (Figure 7.45).

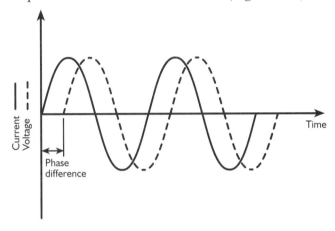

Figure 7.45 Phase difference

Amplitude

The height of the wave or 'amplitude' is probably the easiest way to describe the current or voltage. The amplitude corresponds to either the maximum current I_{max} or maximum voltage V_{max} (Figure 7.46).

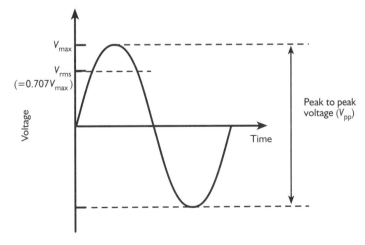

Figure 7.46 Peak to peak and rms current

RMS (root mean square) current

The alternating current that would have the same heating effect, when passed through a conductor, as a DC current of the same value

Another way of defining the current or voltage is as a 'peak to peak' value $I_{p.p}$ or $V_{p.p}$ (Figure 7.46). This method of describing the height of a waveform is frequently used by manufacturers of inexpensive hi-fi equipment to make the power output seem large.

Root mean square (rms) current and voltage

The rms (root mean square) current is the alternating current that would have the same heating effect, when passed through a conductor, as a DC current of the same value. For a sine wave (the most common AC waveform):

$$I_{rms} = 0.707\, I_{max}$$
$$V_{rms} = 0.707\, V_{max}$$

Activity 7.18

1. The mains electricity in your house is supplied at 230 V AC. This is the rms voltage. The maximum voltage can be calculated using the formula: $V_{rms} = 0.707\, V_{max}$ You will first have to transpose the formula to make V the subject.
2. In the last activity you found out the mains supply frequency and you have just calculated the peak value of the voltage. Now draw neatly the waveform for the mains supply showing the following:
   - V_{max}
   - V_{rms}
   - time period T – this should be calculated from the frequency.

Waveform shape

The way in which the voltage or current changes with time dictates the shape of the waveform. The most common shape is the sinewave or sinusoidal waveform. There are several other waveforms that you may meet later, especially when you learn how to use an oscilloscope. Figure 7.47 shows some common waveforms.

It is important to remember that the rms value of current or voltage for these waveforms cannot be worked out using the formula for a sine wave. The rms current/voltage depends on the shape of the wave as well as the maximum current/voltage.

Applications of alternating voltage and current

Apart from mains electricity there are many examples of AC in engineering. Most of these are electronic signals, i.e. the information that passes through systems such as
- radio
- amplifiers
- control equipment.

In fact, any system that makes use of analogue signals is an application of AC.

Scientific laws and principles applied to engineering

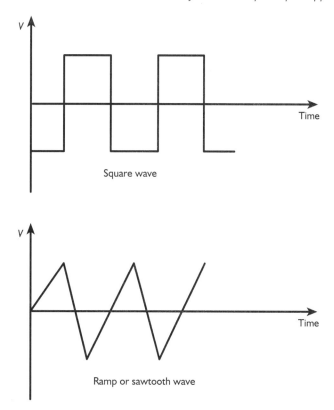

Figure 7.47 Common wave forms

Assignment 7
Investigate the science of engineering systems

This assignment provides evidence for:
Element 3.1: Investigate engineering systems in terms of scientific laws and principles
and the following key skills:
Communication 2.2: Produce written material
Application of number 2.3: Interpret and present data

This assignment is in three parts. Each part contains a description of an electromechanical engineering system which is fitted with instruments to carry out measurements of certain quantities.

Your tasks

For each part you should:
- state how the quantities given affect the way in which the system operates
- state which quantities are constants and which are variables
- give a description in words, graphs and formulae of the relationships between quantities

1. An electric winch consists of an electric motor which drives a pulley through a gearbox. The winch is used to raise a box.
   Quantities measured:
   - mass of box, m
   - height through which box is raised, h
   - time taken to lift box, t
   - supply voltage to motor, V
   - current drawn by the motor, I
   - resistance of the electric motor, R.

2. The output signal from a control amplifier is an AC waveform. This is checked using an oscilloscope.
   Quantities measured:
   - frequency of vibration, f
   - amplitude of vibration, A
   - time for one cycle of vibration, t.

3. The shaft of a steam turbine used to drive a generator undergoes tests to determine the effects of high temperatures.
   Quantities:
   - temperatures before and during operation, T_1 and T_2
   - length of the shaft before and during operation, L_0 and L
   - diameter of the shaft before and during operation, d_0 and d.

Chapter 8 The measurement of physical quantities

> **This chapter covers:**
> Element 3.2: Measure physical quantities used in engineering applications.
> **... and is divided into the following sections:**
> - Choosing the correct device for measuring a particular quantity
> - Common measuring devices and their applications
> - Recording measurements correctly
> - Select measuring devices for engineering situations.

Choosing the correct device for measuring a particular quantity

Whenever an engineering product is developed and manufactured, a considerable number, sometimes many thousands, of measurements have to be taken. These will range from simple linear measurements using rules, e.g. in the drawing office, to measurements of force, frequency, speed, etc. taken when the product is tested. The quantities measured will clearly depend on the nature of the engineering product itself. When choosing the best device for taking a particular measurement there are several factors to be considered:
- the quantity to be measured, e.g. force, speed, current, etc.
- the range of measurements likely to be taken
- the accuracy required
- the resolution that will be necessary.

In this section these factors will be examined so that when you have to look through instrument catalogues, or just the laboratory cupboard, for a device to carry out a particular measurement you will be able to make an informed choice.

Measuring the right quantity

There are two ways of determining the value of a quantity when you carry out an experiment or test:
- directly, i.e. the device being used gives a direct read-out of the value of the quantity you want to measure
- indirectly, i.e. some quantities are difficult to measure and it is often easier, and sometimes more accurate, to measure a different, but related, quantity and then calculate the value required.

In a later section some methods of indirect measurement are described, and in some cases it is the only method possible. An extreme example of this would be the measurement of the speed of galaxies. As yet no one has been able to attach a speedometer to a galaxy and their velocity is determined by measuring the frequency of emissions (light, radio, X-rays, etc.) over a long period and looking for a phenomenon known as the Doppler effect. You have probably noticed this effect when an emergency vehicle sounding its siren has approached you, and then moved away. As the vehicle gets close the pitch of the siren seems to get higher, and then gets lower as it travels away. In fact, the siren emits the same set of notes all the time, and the change in pitch is due to the velocity of the vehicle. The Doppler effect is used in engineering speed measurements also, but on a smaller scale!

Whenever possible, measurements should be made directly. There are two main advantages of doing so:
- the quantity can be measured quickly, and the behaviour of the engineering system examined as it happens, i.e. in real time
- the possibility of calculation error is avoided.

A further point to mention here is that, whatever device you are using, it should give readings in the correct units; converting from, for instance, Imperial units to SI is not only tedious, but can be a source of error.

The range

The range is the difference between the highest and lowest value to be measured. In many situations engineers have a feel for the values expected. This may be as a result of design calculations made or experience gained from similar work. As you become more familiar with your particular field of engineering interest you will find that you develop a 'sixth sense', and will often know, almost instinctively, that a measurement is right or wrong.

Instruments have a limited range over which measurements can be taken. Well-known examples are thermometers, rules and mechanical measuring devices in general (speedometers, pressure gauges, etc.). One big advantage of devices with electronic displays is that they can be wide-ranging or multiranging.

A wide-ranging instrument would require no adjustment at all to provide readings, as the name suggests, over a wide range of values. In such a device accuracy may be poor over part of the range and this should be borne in mind. The manufacturer's specification should therefore be studied carefully.

Multiranging devices are more common, and it is only necessary to turn a dial or flick a switch to select the appropriate range. Some instruments take multiranging one step further and are able to measure several different quantities. You may be familiar with digital multimeters (DMMs), which can be used to measure voltage, current and resistance, and sometimes frequency, over many ranges. Devices are also available that provide read-outs of distance travelled, time elapsed, velocity and acceleration from one instrument.

Multi- and wide-ranging instruments have the following advantages. They:
- can be used for many different applications
- will not be damaged if a measurement goes beyond the chosen range (think what would happen if you seriously overloaded a spring balance)
- can often be used to take readings from different points on the same apparatus (in the case of multiquantity instruments).

Accuracy

The term accuracy is often misused or misunderstood. When a measurement is made the value recorded is the *indicated value*, as opposed to the *true value* of the quantity. It is possible that the indicated and true values are the same, but there would be an element of luck involved if that were the case as, even in the best-made instrument, there will be sources of error.

The measurement of physical quantities

Before any measuring device is sold it must be calibrated. In the case of mass-produced instruments this would be carried out by inspecting samples at random. With more sophisticated equipment, especially that used to make critical and/or very accurate measurements, regular calibration may be required to ensure that readings are reliable. This calibration might be carried out by the manufacturers of the instrument or by a calibration department within the company using the equipment.

Calibration of measurement equipment is carried out by comparing readings from a particular instrument with those from a far more accurate device or standard. From these readings a calibration graph can be drawn, as in Figure 8.1.

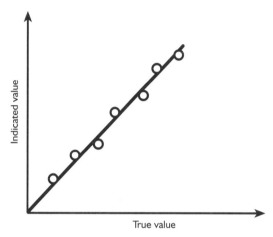

Figure 8.1 Calibration graph

If the indicated and true values were identical, all the points plotted would lie on the line, which would have a gradient of 1. However, because of inaccuracies, there will be some scatter either side of the line. The accuracy of any reading can therefore only be guaranteed to be within the tolerance shown in Figure 8.1.

Accuracy can be shown as a ± value either side of the indicated value, e.g. 3 V ± 0.002 V would mean that the voltage value measured lies between 2.998 V and 3.002 V.

An alternative method of stating the accuracy is as a percentage of the range of the instrument. This is usually shown as % f.s.d. (full-scale deflection). The accuracy of analogue meters is usually specified in this way. The worked examples in the following case study demonstrate the significance of this method.

Case study

The accuracy of analogue meters

An analogue meter is to be used to measure a DC voltage of about 9 V. The ranges available on the meter are:

0–2.5 V
0–10 V
0–50 V
0–250 V
0–1000 V.

Which range should be used?

Obviously the 0–2.5 V range would not be suitable, but all of the others would indicate 9 V. Most people would probably also dismiss the two higher ranges as

> the needle deflection would be too small. That leaves the 0–10 V and 0–50 V ranges.
>
> The accuracy of the meter is stated as ± 4% f.s.d. on the DC voltage ranges. For the 0–10 V range this means a possible error of ± 0.4 V as 4% of full-scale deflection is 0.4.
>
> For the 0–50 V range the possible error is ± 4% of 50 V, i.e. ± 2 V. Clearly, the 0–10 V range is the best one to use.
>
> This example highlights the point that for most analogue (needle and scale) meters, of any type, readings should, as far as is possible, be taken only in the top third of the scale. This would be achieved by changing the range when the reading falls outside the top third. For the meter in this example this would not be possible, and you might consider choosing another meter.

Types of error

In sophisticated measuring systems, there are many possible sources of error, which manufacturers strive to eliminate. On top of this there are errors that arise from the way in which the system is used. The following are just some of the more common errors.

Zero error

If the measurement device does not read zero when it is not measuring anything then there may well be any error in all the readings. For instance, if a spring balance indicates a force of 0.2 N before a load is placed on it then every reading will be 0.2 N too large.

This problem is usually easily overcome by making a simple zero adjustment before starting to take readings. If this cannot be done then the zero reading can be allowed for in all following measurements. However, if there is a large zero reading it may be that the instrument is faulty.

Reading error

Apart from simply misreading a measurement, error will exist because of the increments used on the scale or display. A ruler marked in mm would have a reading error of about ± a quarter of a millimetre in the case of a steel rule – higher in the case of a cheap plastic rule. All analogue meters suffer from the same problem and it is common practice to estimate the reading error as ± half the smallest difference between marks on the scale (increments).

One point of confusion arises when using digital measurement displays. The last digit on such devices is unreliable. A typical accuracy for a DMM on a voltage range would be 0.7% of the reading +2 digits. This means that the last figure might be out by as much as 2. On top of this the final digit is often subject to fluctuation, i.e. it keeps on changing.

Analogue meters are sometimes misread owing to parallax error. This occurs when the scale is viewed at an angle, e.g. from the side or above the meter. Try this for yourself; you might be surprised at the variation in readings. Some analogue meters are fitted with a mirror behind the needle. When taking a reading the needle is lined up with its mirror image so that parallax is removed.

Instrument errors

These are errors that occur within the measurement system due to friction, tolerances on components, etc. Although the accuracy of the instrument will be stated by the manufacturer, after a year or two of use (or misuse!) some components

The measurement of physical quantities

may deteriorate and accuracy may be lost. Accordingly, devices should be looked after and used and calibrated as recommended by the manufacturer.

Hysteresis errors

Either because of the measurement device or because of the method of reading, a situation may occur where measurements taken as a quantity increases will differ from those taken when the same quantity decreases, e.g. when a pressure gauge is used to measure increasing values of pressure it may indicate 29 kPa for

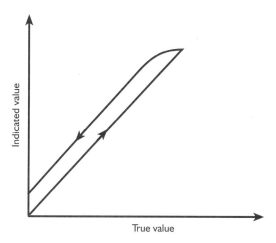

Figure 8.2 Hysteresis effect

true value of 30 kPa, but when the pressure is being decreased indicate 31 kPa, i.e. it is lagging behind. This effect is known as hysteresis. Figure 8.2 shows the hysteresis effect graphically.

Resolution

The resolution of a measurement device is the smallest change in the measured quantity to which it will respond. You should be careful when using digital instruments not to confuse the value of the '1' in the last digit display, with the resolution. As has already been described, the last digit in these displays is often unreliable.

Common measuring devices and their applications

Dimensional measurement instruments

Probably the first measurement you ever made was a dimensional measurement, and you most likely used a ruler to do it. In the engineering world there are several devices used for making dimensional measurements. The one chosen will depend on:
- the accuracy required
- the magnitude of the dimension to be measured.

The devices relevant to the Intermediate GNVQ are:
- the steel rule
- the micrometer
- the vernier calliper
- the dial test indicator (d.t.i.).

Steel rule

Steel rules are available in lengths from 75 mm up to 1 m and larger. For maximum accuracy they should be used near to the temperature at which they are calibrated, usually 20°C (this temperature is marked on the rule) because of thermal expansion and contraction. This is especially true when larger measurements are being made on objects made from materials that expand at a very different rate from steel. When used correctly, steel rules can provide dimensional measurements down to 0.25 mm.

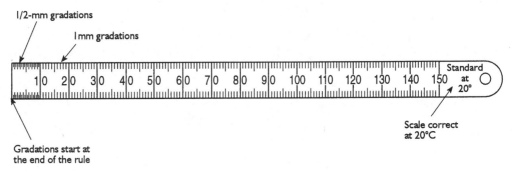

Figure 8.3 Steel rule

Micrometer

Micrometers are precision measuring instruments, i.e. they offer a high degree of accuracy. They are available as both internal and external measuring devices in sizes ranging from 0–2 mm (0–1 inches) up to 125–150 mm (5–6 inches) and bigger. Even large micrometers only have a range of 25 mm (1 inch), i.e. the barrel only moves a total of 25 mm between the ends of its travel.

Measurements are taken by turning the knurled ratchet knob until the ratchet itself can be heard clicking (Figure 8.4). At this point the barrel is locked in position using the lever, and the measurement taken.

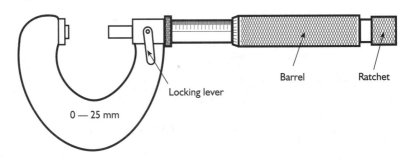

Figure 8.4 Micrometer

Care should be taken not to overtighten the barrel as permanent damage and loss of accuracy will result. Similarly, micrometers should be stored in such a way that the contact faces are not damaged. The best way to practise reading a micrometer is with a real instrument.

Vernier callipers

Vernier callipers are also precision instruments, although cheaper devices are available for use where a high degree of accuracy is not required. In common with micrometers, vernier callipers can be used for both internal and external measurements. However, unlike micrometers, one instrument will fulfil both roles. Vernier callipers are also made in a wide variety of sizes to suit different applications.

Reading a vernier calliper Figure 8.5 shows a vernier calliper being used to measure the diameter of a shaft. Note that the last digit of the measurement is obtained from the secondary or vernier scale. As with the micrometer, the best way to find out how to read a vernier calliper is to try making some measurements yourself.

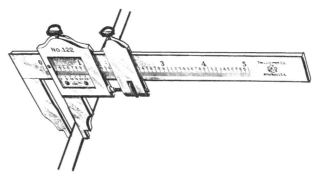

Figure 8.5 Vernier calliper

Temperature measurement devices

Mercury in glass thermometers

The mercury thermometer is probably the most common temperature measurement instrument, and can be used to good effect as long as a few points are borne in mind.

- Only the bulb should be immersed if the thermometer is being used to measure the temperature of a liquid, unless the manufacturer's instructions state otherwise.
- Care should be taken to read the temperature indicated at the top of the meniscus (Figure 8.6).
- Thermometers should be handled with care as they are easily broken; spilt mercury represents a hazard and should not be touched!

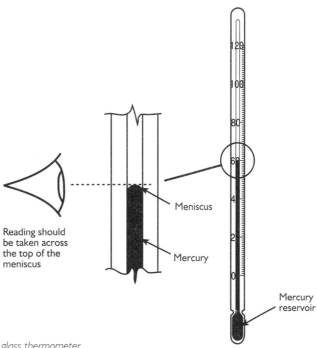

Figure 8.6 Mercury in glass thermometer

When choosing a thermometer for a particular application you should make sure that the instrument has a sufficiently large range and that the length will give the necessary resolution, i.e. the divisions or increments are small enough. The range of temperatures that can be measured depends partly on the thermometric liquid. Mercury can be used from about −30°C (it freezes at −35°C) to 350°C (it vaporises at 375°C). The range can be extended to 510°C by pressurising the space above the mercury with nitrogen. For temperature measurements below about −30°C alcohol coloured with a dye is often used as the thermometer fluid.

Digital thermometers

Many temperature measurements are now made using electronic devices with digital displays. These range from inexpensive devices available in high street stores to very accurate and expensive laboratory-standard instruments. Most of these operate on the thermocouple principle. Figure 8.7 shows the principal components of a thermocouple device.

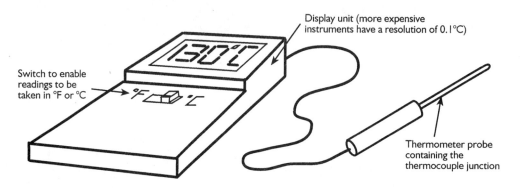

Figure 8.7 Digital thermometer

In general, digital thermometers are more robust than mercury in glass instruments and are therefore more portable. They are also wider ranging, typically from −50°C to +150°C. Thermometers are also available with analogue displays, although these are gradually being displaced by digital devices in manufacturers' catalogues. One significant advantage of an electronic thermometer is that the information can be fed directly to a data logger or control system.

Force measurement devices

Forces are often measured indirectly by observing the deflection of, for example, a spring or beam when subjected to a force. Devices for measuring force range from simple spring balances, of the type used by fishermen, to sophisticated electronic scales used in laboratories. The two, however, are remarkably similar.

Spring balance

The spring balance relies on the fact that the relationship between load and deflection is linear for a metal spring (see Chapter 7). When a load is hung on the balance the spring deflects, the amount of movement depending on the spring stiffness. A pointer attached to the spring indicates the load on a linear scale. The greatest source of error in a spring balance is friction; this is usually reduced to a minimum by using balances in a vertical position wherever possible.

The measurement of physical quantities

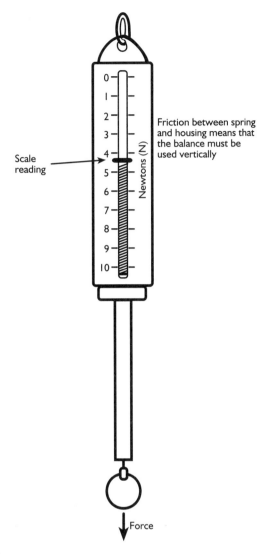

Figure 8.8 Spring balance

Electronic scales

These devices are rapidly becoming the preferred method of measuring weight and have the advantage of greater accuracy (when calibrated correctly) and a wider range. Most electronic scales (Figure 8.9) make use of strain gauges to respond to changes in load, just as in a spring balance the spring itself responds to changes in load.

The strain gauge is a small piece of metal foil shaped as shown in Figure 8.10.

The gauge is bonded to the surface of a beam so that as it bends the foil is stretched, increasing the length of the conductor, through which a small current is being passed. This results in an increase in electrical resistance, the effect of which can be measured and displayed.

In all force meters the display must be calibrated in units of force, as readings of deflection (on the spring balance) or change in resistance (on the beam balance) would be meaningless. In the case of spring balances, the range of force that can be measured with an acceptable degree of accuracy is small because large deflections are necessary. However, beam and strain gauge balances can be used to measure accurately a far wider range of forces.

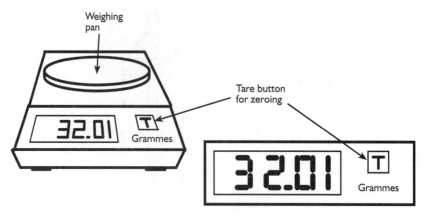

Figure 8.9 Electronic balance

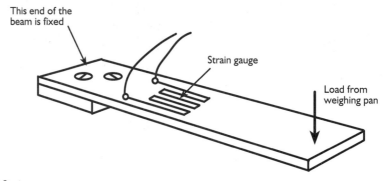

Figure 8.10 Strain gauge

Dead weight measurement of force

Devices that compare the force or weight to be measured with a known (calibrated) weight are known as dead weight scales. Balances of the type still used in some banks and post offices come into this category. The technique tends, however, to be reserved for calibration procedures as electronic devices have become the instrument of choice for everyday use.

Electrical measurement devices

The electrical quantities you are most likely to encounter in the Intermediate GNVQ are:
- voltage (AC and DC)
- current (AC and DC)
- resistance
- frequency.

The digital multimeter

Fortunately, modern technology has provided a means of reading all these quantities using just one instrument: the digital multimeter (DMM). These devices vary from inexpensive meters that have limited ranges, particularly for resistance, and that may not be capable of measuring frequency, to 'professional'-quality instruments that may also be capable of measuring capacitance.

Figure 8.11 shows a typical DMM with the main features labelled. Two important points to note are the maximum voltage and current that can be measured.

The measurement of physical quantities

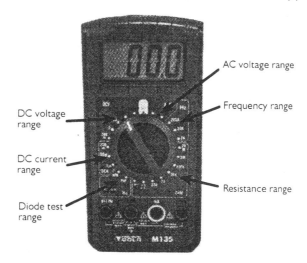

Figure 8.11 Digital multimeter

You are not likely to exceed the maximum voltage during practical work, but it would be possible to include the meter in a circuit with a higher current than the maximum allowable. However, DMMs are protected by internal fuses against damage, the maximum current for most DMMs being either 10 A or 20 A.

Using a DMM to measure current Figure 8.12 shows a meter included in a simple circuit to measure current.

As the current must pass through the meter as well as the resistor R, its resistance must be very low if the reading is to be of any value. DMMs are designed so that meter resistance is negligible when the current ranges are used, although accuracy may vary between ranges.

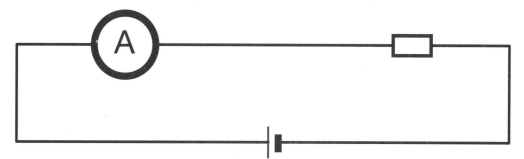

Figure 8.12 Measuring the current flowing through a resistor

Using a DMM to measure voltage Figure 8.13 shows a meter being used to measure the voltage across a resistor.

Unlike an ammeter, a voltmeter must have as high a resistance as possible to ensure that the circuit is disturbed as little as possible by its presence. Most DMMs have a resistance of 10 MΩ (10 million ohms), which means that the current flowing through the meter is very small.

Using a DMM to measure resistance The resistance measurement ability of DMMs is often used as a means of checking for continuity, e.g. checking that a connecting lead is not broken (Figure 8.14).

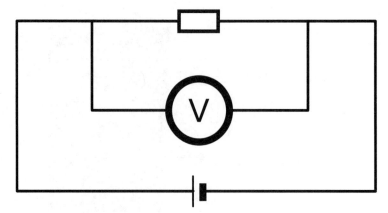

Figure 8.13 Measuring the voltage across a resistor

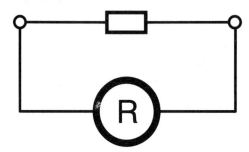

Figure 8.14 Measuring the resistance of a resistor

A resistance of an ohm or less indicates that the conductor and solder joints are intact. Many meters are fitted with a buzzer that indicates continuity. At the higher end of the range, resistances as high as 100 MW can be measured.

The oscilloscope

Oscilloscopes are widely used in engineering to observe and measure the dynamic behaviour of systems, i.e. the way in which quantities vary with time. Examples of the quantities that can be measured with an oscilloscope are:
- voltage and current signals
- vibrations
- pressure fluctuations
- speed variations.

If you look through an electronics catalogue it will become apparent that there is a wide variety of oscilloscopes available. They all, however, operate on the same principle, and display the results on a screen similar to that in a television. In a television a beam of electrons scans the screen in a series of horizontal lines, the intensity of the beam being controlled, indirectly, by the incoming aerial signal to give a picture. The oscilloscope picture or 'trace' is controlled by the quantity being measured as the beam of electrons scans the screen.

Any device or transducer that has an electrical output can be connected to an oscilloscope and its output signal displayed. Although there are several ways in which the trace can be controlled by the input, the most common method uses the measured quantity to vary the vertical motion of the trace (Figure 8.15), the horizontal motion being provided by the time-base system in the oscilloscope itself. The time-base causes the electron beam to pass across the screen at a constant speed.

The measurement of physical quantities

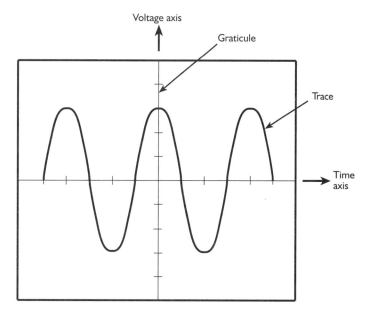

Figure 8.15 Oscilloscope trace

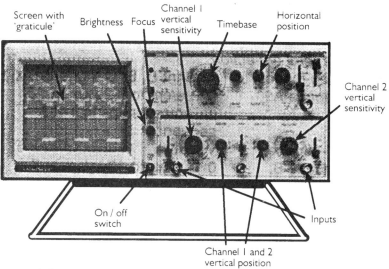

Figure 8.16 Front panel of an oscilloscope

Many oscilloscopes have two separate input channels so that two quantities can be observed at the same time. For example, the input and output of a system can be viewed simultaneously.

Using an oscilloscope At first sight the control panel of an oscilloscope is a daunting prospect. However, once you have used one a few times, it will become almost second nature. Figure 8.16 shows a typical front panel with some of the controls labelled.

343

Activity 8.1

You will need an oscilloscope to carry out the activities in this section. With the oscilloscope in front of you identify the following:
- channel inputs Y_1 and Y_2
- sensitivity controls for each channel, labelled V/cm
- time-base control
- focus control
- brightness control.

Recording measurements correctly

The ability to record measurements accurately and in a methodical way is as important as being able to use measuring equipment correctly. This section will examine ways in which measurements can be recorded. As you will find out, if you are fortunate enough to have access to a data recording system, connection to a personal computer (PC) can save a lot of laborious work.

Tabulating results

By far the most common and useful way of recording results is in a table, which allows you to set out figures in an orderly way. The headings for each column should contain the following information:
- the quantity being measured
- the units used
- an estimate of the accuracy of the reading.

Table 8.1 shows headings (not data) for readings of voltage and current taken during an investigation into Ohm's law.

Table 8.1 Headings for voltage and current readings (no data shown)

Voltage (volts) (± 0.4 V)	Current (amps) (± 6%)

Note that the accuracies have been quoted in two different ways, the voltage being measured with an analogue meter, so that a % f.s.d. accuracy applies, and the current with a digital meter, so that the possible error is a percentage of the reading taken.

It is normal to record the values of the quantity being adjusted between each reading in the left-hand column. This quantity is sometimes called the *independent variable*. The quantity(ies) that change in response to changes you make are then put in the following column(s). These quantities are sometimes called the *dependent variables*.

Readings should always be taken with the independent variable changing in one direction only. The correct and incorrect methods are shown in Table 8.2.

Increments between values of the independent variable (voltage in this case) do not have to be equal, but it can make graph plotting easier later if the test procedure can be organised in such a way that that they are.

The measurement of physical quantities

Table 8.2 Correct and incorrect methods of recording independent variables

Correct			Incorrect	
Voltage (volts) (± 0.4 V)	**Current** (amps) (± 6%)		**Voltage** (volts) (± 0.4 V)	**Current** (amps) (± 6%)
10	1.10		12	1.22
12	1.22		15	1.51
13	1.35		10	1.10
15	1.51		13	1.31
16	1.59		16	1.59

Repeat readings

To be sure that sets of readings are correct it is common practice to take several sets of readings and then to average the results. Table 8.3 shows force and deflection values taken during an experiment to determine the stiffness of a spring. As you can see, three sets of deflection readings have been taken for the same force values.

The average deflection in each case was found by adding the values of the three deflection columns together and dividing by 3, the number of sets of readings. So, for a force of 2 N, average deflection = (8 + 9 + 9)/3 = 26/3 = 8.667.

Now, bearing in mind that the estimated accuracy of each deflection reading is ± 0.5 mm, it would be unrealistic to record the average as 8.667. Hence, only one decimal place has been used and all average values are rounded off.

Very large and very small values

If very large or very small values of a quantity have to be recorded, the size of the results table, and its ease of use, can be prescribed by making use of either: alternative units, e.g. instead of writing 0.0036 m you could write 3.6 mm standard form, e.g. instead of recording a pressure of 813 000 Pa you could write 8.13×10^5 Pa.

Table 8.4 shows sample readings of pressure (in this case the independent variable) and deflection of the container being pressurised.

Table 8.3 Force and deflection values

Force (N) (± 0.05 N)	Deflection (mm) (± 0.5 mm)	Deflection (mm) (± 0.5 mm)	Deflection (mm) (± 0.5 mm)	Average deflection (mm)
0	0	0.5	0	0.2
2	8	9	9	8.7
4	17	18	17	17.3
6	26	26	26	26.0
8	35	34	34	34.3
10	42	43	44	43.0

Table 8.4 Pressure and deflection readings

Pressure (Pa $\times 10^5$) ($\pm$ 2%)	Deflection (mm) ($\pm$ 0.1 mm)
8.13	3.6

If you are not sure about the use of standard form refer to Chapter 9. The prefixes for unit multiples are listed at the beginning of Chapter 7.

Calculated results

Having taken results during the practical part of an investigation, you will probably have to carry out some calculations using the tabulated figures. You should not normally show all your calculations in a report. Usual practice is to show one sample calculation to demonstrate the methods used, and then to tabulate all the calculated values. Table 8.5 shows resistance and power values calculated from the results in Table 8.2.

Table 8.5 Resistance and power values

Voltage (V) ($\pm$ 0.4 V)	Recorded results Current (A) (6%)	Calculated results Resistance (Ω)	Power (W)
10	1.10	9.09	11.0
12	1.22	9.84	14.64
13	1.35	9.63	17.55
15	1.51	9.93	22.65
16	1.59	10.06	25.44

Sample calculations Using Ohm's law ($V = IR$)

$$R = V/I$$

At a supply voltage of 12 V

$$R = 12/9.84 = 1.22$$

$$\text{Power} = VI$$

Again for a voltage of 12 V

$$\text{Power} = 12 \times 1.22 = 14.64 \text{ W}$$

In some investigations there may be more columns of calculated values than can be fitted easily across the page. If this is the case, a separate table can be used.

When entering calculated values in your table avoid the temptation to put down every figure given by your calculator. Stick to the same kind of accuracy as you have used elsewhere in the results.

Graphs

The mathematics of graphs are dealt with in Chapter 9. In this section the correct method of presenting results graphically will be described. Having said that, there is no one correct way to present a graph – different people have different ideas. The following points should, however, always be borne in mind.

The measurement of physical quantities

- Unless there is a good reason for not doing so, put the independent variable on the horizontal (x) axis.
- Choose scales that are easy to use.
- Make the graph fill the paper, bearing in mind the previous point – accuracy of gradients and intercepts is lost if the graph is too small.
- Draw the graph using a sharp pencil; H or 2H grade is fine but do not use a B or 2B pencil.
- Label both axes and state the units for each.
- Title the graph.

The graph in Figure 8.19 shows the data in Table 8.2. Note that the best-fit straight line has to be drawn in. This technique is discussed in Chapter 9.

Plotting a graph while taking results

It is a good idea to plot results as they are taken. This will enable you to detect 'rogue' readings, or at least arouse your suspicion that something is going wrong. You will also have a good idea about the final outcome of the investigation by the time you have finished the practical work.

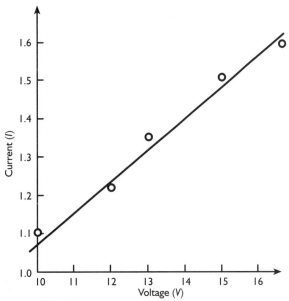

Figure 8.19 Graph of data from Table 8.2

Using information technology to record results

In extended tests, use is made of computers to record results as the investigation proceeds. Such systems are often referred to as data-logging systems. If you have such equipment in your school or college laboratories you may be able to make use of it to improve your IT key skills. Data loggers may be in the form of small hand-held devices that can be downloaded to a computer or may be complete systems connected through an interface to the computer.

Spreadsheets

A simpler, and possibly more accessible use of IT for results recording, is to make use of a spreadsheet package such as Microsoft Excel or Lotus 123. This can be used as a means of recording the results as an investigation is carried out and/

Science and mathematics for engineering

or later as a means of presenting your findings. Spreadsheets have the further advantage of being able to carry out calculations both across rows of figures and down columns. This means that, once you have fed in the recorded results, you just need to apply the formulae necessary to one set of figures and all other values will be calculated automatically. Finally, when you really become good at using spreadsheets, you can use the same package to plot graphs.

Keeping a record of the equipment used

It is important that a note is kept of the particular pieces of equipment used in case you find serious errors and have to repeat your practical work. Also, if it is necessary to leave an investigation part way through, by recording the serial numbers of the individual pieces of apparatus you can ensure that errors are not being introduced by a change in the measurement devices, which may be calibrated slightly differently. Once again the neatest way to write down this information is to tabulate it. Table 8.6 gives suggested headings.

Table 8.6 Headings for equipment used

Equipment	Manufacturer	Serial number
Bourdon pressure gauge	Hi-Press Gauges Ltd	AB 12345

Assignment 8
Select measuring devices for engineering situations

This assignment provides evidence for:
Element 3.2: Measure physical quantities used in engineering applications
and the following key skills:
Communication 2.4: Read and respond to written materials
Application of number 2.1: Collect and record data

In this assignment you will be selecting measurement devices for different engineering situations.

Your tasks

For each device selected you should:
- state which quantity the device measures
- the type of readout the device gives
- give two examples of situations in which *you* have used the device selected, using diagrams
- provide two sets of results which *you* have taken using the device

Select a device to:
1. measure voltage and current in a DC circuit
2. check the diameter of a steel bar
3. determine to a high degree of accuracy the mass of an engine component
4. obtain the boiling temperature of a liquid as it is heated in a beaker
5. time 20 swings of a pendulum
6. measure the peak to peak voltage of an ac waveform
7. measure the temperature inside a furnace while it is in use

Chapter 9   Mathematical techniques

> **This chapter covers**
> Element 3.3: Use mathematical techniques to investigate the behaviour of engineering systems
> **... and is divided into the following sections:**
> - Identifying relationships between physical quantities in engineering systems
> - Manipulating data using mathematical techniques
> - Obtaining values for physical quantities
> - Calculate the behaviour of engineering systems.

All engineering calculations involve relationships between physical quantities. In Chapter 7, some of the scientific principles and laws that govern the behaviour of engineering systems were examined in detail; in this chapter the mathematics involved in using these scientific laws and principles will be explained. The layout of this chapter may, at first, seem illogical, as the basic arithmetic section comes after the section that deals with equations and graphs, i.e. the relationships between physical quantities. The reasons for organising the chapter in this way are as follows.

- As you are an engineering student, the maths chapter is intended to be used as a reference source. This idea extends to the maths chapter itself as, once you have mastered the basic nuts and bolts of mathematics, the section likely to be of most use is the first one.
- The way in which the chapter is ordered closely follows the GNVQ specification, which will help you organise your evidence.

Identifying relationships between physical quantities in engineering systems

The relationship between two quantities can be described in three ways:
- as a statement in words, e.g. 'pressure increases as temperature rises
- as an algebraic equation or 'formula', e.g. $V = IR$
- as a graph, e.g. Figure 9.1.

All three methods should be part of an engineer's vocabulary, and being able to relate one method to another is of great importance when describing relationships. A graph quickly drawn on the back of an envelope can often be of great help when discussing engineering systems with other engineers!

Straight line relationships

Straight line relationships, so called because they can be described by straight line graphs, are the simplest form of relationship between two quantities. We have divided linear relationship into four categories:

Mathematical techniques

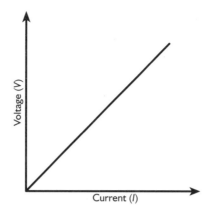

Figure 9.1

- proportional relationships
- inverse proportional relationships
- linear relationships
- inverse linear relationships.

Proportional relationships
Examples:
$$F = MA$$
$$V = IR$$
$$F = kx$$

All proportional relationships involve three quantities:
- two variable quantities
- one constant quantity.

Activity 9.1

For the three equations given as examples above (which are described in Chapter 7):
- describe or draw a diagram of the engineering situation to which each applies
- state which are the two variable quantities and which is the constant.

A typical way of describing, in words, a proportional relationship would be to say that *voltage is proportional to current in an ohmic conductor* or *voltage varies in the same way as current*.

This can be written algebraically as
$$V \propto I$$

or, introducing the constant, in this case R, $V = RI$ (or $V = IR$ in its more familiar form).

Figure 9.2 shows the graph of this proportional relationship between the two variable quantities, voltage and current. The feature of the graph that should be immediately apparent is that the relationship produces a straight line that passes through the origin, i.e. the point at which $V = 0$ and $I = 0$ in this case.

Because the line is straight it has a constant slope or gradient. The gradient of the line represents the constant involved in the proportional relationship, so that if the value of the gradient can be found the resistance of the conductor R can be determined.

Science and mathematics for engineering

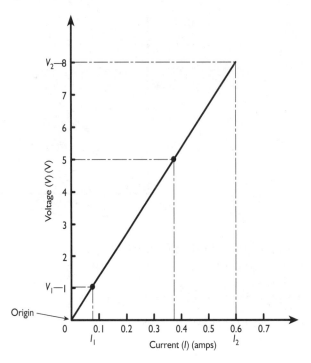

Figure 9.2 Proportional relationship between voltage and current

Figure 9.3 shows the procedure for determining the gradient of any straight line graph.

The method is as follows.
1. Mark two points on the graph line. These points should be as far apart as possible to ensure the highest degree of accuracy possible. Note that for a proportional relationship the line passes through the origin, so that, for this type of relationship only, the point ($x = 0$, $y = 0$) can be used.
2. Read off the values for x_1, y_1, x_2, y_2, i.e. the coordinates of the points.
3. Find the gradient using the formula

$$\text{gradient} = (y_2 - y_1)/(x_2 - x_1)$$

We will now apply this technique to the voltage–current graph in Figure 9.2. The first step is to write down the coordinates chosen. In order to demonstrate the technique, the origin has been purposely avoided.

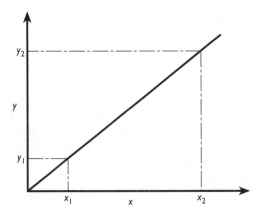

Figure 9.3 Procedure for determining the gradient of any straight line graph

Coordinate 2 $V = 8$ V $I = 0.6$ A

Coordinate 1 $V = 1$ V $I = 0.065$ A

Gradient = $(V_2 - V_1)/(I_2 - I_1) = (8 - 1)/(0.6 - 0.065) = 7/0.535 = 13.1$

From the graph we can now say that

$$V = 13.1\, I$$

or in words, the voltage (in volts) is always 13.1 times the current (in amps) However, as a budding engineer, you already know that the relationship between V and I is better known by the formula for Ohm's law

$$V = RI$$

so we can now say that resistance $R = 13.1\ \Omega$.

Activity 9.2

Now use the coordinate marked with a cross, and the origin, to find a value for R, explaining any difference between your value and that obtained using the earlier set of coordinates.

The next example is concerned with force and acceleration, a dynamics problem: the graph in Figure 9.4 shows the relationship between the acceleration of an aircraft and the thrust developed by the engine (ignoring air resistance!). From the graph determine the mass of the aircraft in kg.

We know that the relationship between force F (thrust in this case) and acceleration a is:

$$F = ma$$

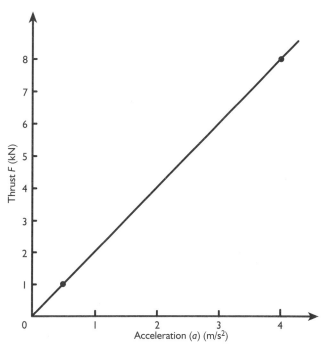

Figure 9.4 Relationship between the acceleration of an aircraft and the thrust developed by the engine

Science and mathematics for engineering

From the graph it is apparent that the variable quantities are F and a, the mass being the constant of proportionality. Hence, the mass will be given by the gradient.

Taking the marked coordinates:

$F_2 = 8$ kN $F_1 = 1$ kN

$a_2 = 4$ m/s^2 $a_1 = 0.5$ m/s^2

Before going further, look at the units in which thrust is measured – kilonewtons. The formula $F = ma$ holds true only if the units used are newtons, kilograms and metres/second^2. Accordingly, F_1 and F_2 must be converted to newtons:

$F_2 = 8000$ N $F_1 = 1000$ N

The mass can now be determined:

$$\text{mass } m = \text{gradient} = (F_2 - F_1)/(a_2 - a_1)$$

$$= (8000 - 1000)/(4 - 0.5) = 7000/3.5 = 2000 \text{ kg}$$

i.e. aircraft mass is 2000 kg.

Activity 9.3

Copy the graph in Figure 9.4 and then add lines that would show the performance of an aircraft of mass 1000 kg having the same engine, again ignoring air resistance. In the last activity you used the graph to predict the performance of an aircraft – graphs are powerful engineering tools! In fact you were really modelling the aircraft mathematically, a considerably cheaper way of testing aircraft performance than building a real prototype! Design engineers make use of formulae and graphs, often in conjunction with computers, to assess the performance of engineering systems before spending a lot of money to build the real thing.

Activity 9.4

This last example of a proportional relationship involves a spring. When a spring of stiffness k N/m is loaded with a force F N it deflects x m. Figure 9.5 shows the graph of load–deflection obtained during a test on a spring. Your job is to find the stiffness of the spring, but watch out for the units!

Linear relationships
Examples:

$v = u + at$ $L = L_0 + L\alpha \Delta T$

The proportional relationship between two quantities is a particular type of linear relationship for which the straight line graph passes through the origin. A more general type of linear graph is shown in Figure 9.6.

The equation that produces this type of graph is:

$$y = mx + c \text{ (or } y = c + mx\text{)}$$

where m is the gradient which, as with proportional relationships, is given by:

$$m = (y_2 - y_1)/(x_2 - x_1)$$

Mathematical techniques

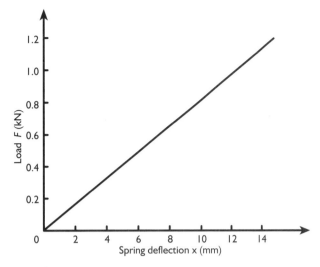

Figure 9.5 Load–deflection graph

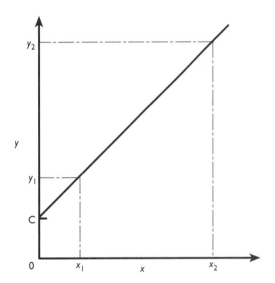

Figure 9.6 General type of linear graph

and c is the point at which the line crosses the y-axis, known as the y-intercept. It is the y-intercept that makes the relationship non-proportional.

The first example we will look at is the relationship between velocity and time for an object accelerating uniformly. Figure 9.7 shows this relationship graphically.

The velocity v m/s at time t s is given by the formula:

$$v = u + at$$

where u is the starting velocity at $t = 0$ and a is the acceleration in m/s. The variable quantities are velocity and time, but this time there are two constants, u and a.

Comparing the general straight line formula:

$$y = c + mx$$

355

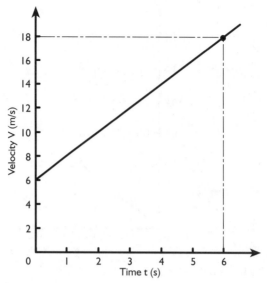

Figure 9.7 Relationship between velocity and time for an object accelerating uniformly

and the formula relating velocity and time:

$$v = u + at$$

we can see that u is the intercept on the y-axis and a is the gradient.

The initial velocity u can be read directly from the graph, so in this case:

$$u = 6 \text{ m/s}$$

a can be found by determining the gradient, so:

$$a = (v_2 - v_1)/(t_2 - t_1) = (16 - 6)/(6 - 0) = 10/6 = 1.67 \text{ m/s}^2$$

Hence, we can say that the starting velocity is 6 m/s and the acceleration is 1.67 m/s².

The next example concerns an object slowing down or decelerating. As you can see, the graph line (Figure 9.8) slopes in the opposite way to the previous example, i.e. as time increases the velocity decreases. However, the same rules apply when you come to find the gradient and intercept.

The two points used to determine the gradient are the intercepts with the velocity and time axes so:

$v_2 = 0$ $\qquad t_2 = 7$ s

$v_1 = 7.5$ m/s $\qquad t_1 = 0$

Therefore the gradient

$$= (v_2 - v_1)/(t_2 - t_1) = (0 - 7.5)/(7 - 0) = -7.5/7 = -1.07 \text{ m/s}^2$$

i.e. the acceleration = −1.07 m/s. The significance of the minus sign is that it indicates a deceleration. A better way of stating the answer would be to say that the deceleration = 1.07 m/s.

Now we come to the starting velocity. This is again given by the intercept with the velocity axis. So, starting velocity = 7.5 m/s.

Mathematical techniques

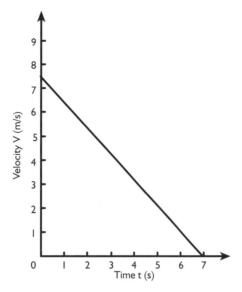

Figure 9.8 Velocity–time graph with constant deceleration

Activity 9.5

Figure 9.9 shows the velocity–time graph for an object speeding up and then slowing down again. Use the graph to find:
- the starting velocity
- the acceleration
- the maximum velocity
- the deceleration.

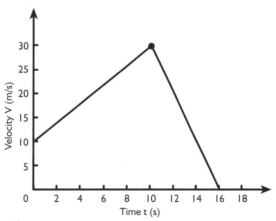

Figure 9.9 Velocity–time graph

Non-linear relationships

Unfortunately, from the point of view of ease of doing the mathematics, many relationships between physical quantities are non-linear, i.e. the graphs are curves. The mathematical equations that describe non-linear relationships are often very complex; similarly, a description in words can be difficult. However, a lot of information can be obtained from the graphs of such relationships.

One of the most important pieces of information that can be obtained from any graph is the gradient at a particular point. In the section covering linear relationships the gradient was found in order to determine the value of a constant, but, as Figure 9.10 shows, the gradient of a non-linear graph is not constant. So what is the significance of the gradient when discussing these types of relationships?

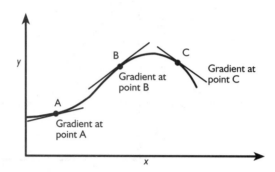

Figure 9.10 Gradient of a non-linear graph at different points

Rates of change

The importance of the gradients in Figure 9.10 is that they indicate the rate at which, in this case, the quantity y changes as x increases, or, in mathematical jargon, the rate of change of y with respect to x:
- at point A y is increasing slowly as x increases
- at point B y is increasing more quickly as x increases
- at point C y is decreasing as x increases.

Case study

Figure 9.11 is a 'cooling curve' for a steel component that is undergoing an annealing process. In such processes the rate of cooling, i.e. rate of change of temperature with respect to time, is often critical; the properties of the steel (hardness, toughness, etc.) after the process depend on this.

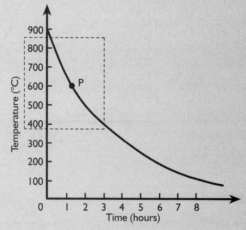

Figure 9.11 Cooling curve

The question now is, how can the gradient at a particular point on a curve be determined? If point P is the area of interest then the rate of change of temperature at this point can be found using the technique shown in Figure.9.12, which is an enlargement of the boxed area on Figure 9.11.

Mathematical techniques

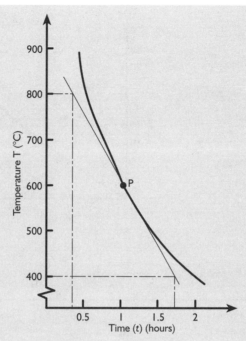

Figure 9.12 Enlargement of boxed area on Fig. 9.11

The gradient is found by drawing the tangent to the curve at point P, the aim being to get the angle between the tangent and the curve, either side of the point, approximately equal. This method gives a reasonable accuracy with practice. If greater accuracy is required then more sophisticated mathematical techniques must be employed.

Gradient at point P = change in temperature divided by change in time or, as a mathematical formula, gradient = $(T_2 - T_1)/(t_2 - t_1)$

$T_2 = 400°C$ $t_2 = 1.7$ h
$T_1 = 800°C$ $t_1 = 0.3$ h

Gradient at point P = $(400 - 800)/(1.7 - 0.3) = -400/1.4 = -286°C/h$ or, in terms of the engineering situation being investigated, the cooling rate at point *P* is 286°C/h.

Note that the negative sign indicates that the temperature is falling. Negative gradients were discussed in the section on linear relationships.

Activity 9.6

The next example involves an electrical system. The non-linear relationship between voltage and current in non-ohmic conductors was mentioned in Chapter 8. Figure 9.13 shows the voltage–current graph for such a conductor. Note that the units of current are milliamps (mA). The system is designed to operate between 20 V and 70 V, so the resistance values at these limits are of importance. Your task is to determine these values. To start with you should draw the graph on a larger scale. Table 9.1 gives voltage and current values to enable you to do this. These figures were obtained experimentally so expect some scatter of the points. Having drawn the graph you only need to remember the relationship between voltage, current and resistance in order to be able to find the resistances at the limiting voltages. If you cannot recall the formula look back at Chapter 7.

Science and mathematics for engineering

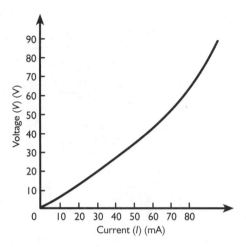

Figure 9.13 Voltage–current graph

Table 9.1 Voltage and current values

Current I (mA)	0	10	20	30	40	50	60	70	80
Voltage V (V)	0	8	25	23	32	43	57	70	88

The area under a graph

In some relationships between physical quantities the area under the graph is of as much importance as the gradient.

Case study

A box is moved along a horizontal surface through a distance of 3 m by a constant force of 25 N as in Figure 9.14. How much work is done in moving the box?

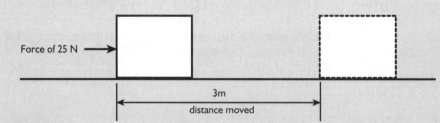

Figure 9.14

You may recall that work done, a form of energy, can be found by using the relationship:

work done = force × distance moved in the direction of the force, in this case

work done = 25 × 3 = 75 J

An alternative way of approaching the problem would be to draw a graph as shown in Figure 9.15.

Figure 9.15 shows the force–distance graph for the situation described. If you look at the shaded area under the line you will see that it has an area of 25 N × 3 m = 75 J, i.e. the area under the line represents the work done.

360

Mathematical techniques

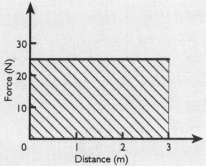

Figure 9.15 Force–distance graph

However, what happens if the force does not stay constant, i.e. the graph is no longer linear?

If the force starts at 25 N but then increases by 5 N every metre, the work done would be:

25 × 1 = 25 J for the first metre
30 × 1 = 30 J for the second metre
35 × 1 = 35 J for the third metre
total work = 90 J

The force–distance graph for this changing force is shown in Figure 9.16.

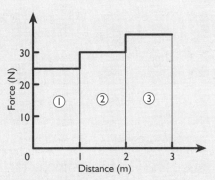

Figure 9.16 Force–distance graph for changing force

Calculating the areas under the graph gives
area (1) = 25 × 1 = 25 J
area (2) = 30 × 1 = 30 J
area (3) = 35 × 1 = 35 J
total area = 90 J

Again the area represents the work done.

The next refinement of this method is to increase the force by 1 every 100 mm (0.1 m). As you can see, the graph is now getting close to being a slope (Figure 9.17).

The final step is to increase the force by 0.1 N after the first 100 mm, 0.2 N after the next 100 mm, 0.3 N after the third 100 mm and so on. The resulting force–distance graph will be, effectively, a curve. Table 9.2 gives force and distance values for the first 1.5 m.

The area under this graph can be found using the method outlined in the next section.

Science and mathematics for engineering

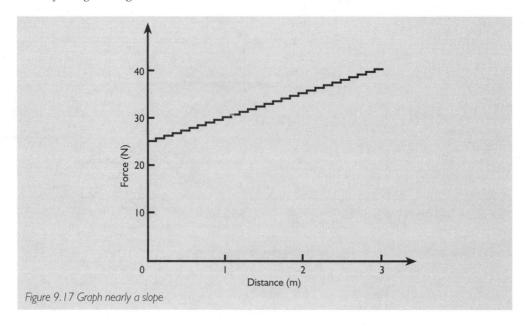

Figure 9.17 Graph nearly a slope

Table 9.2 Force and distance values

Distance (m)	0	0.1	0.2	0.3	0.4	0.5	0.6	0.7	0.8
Force (N)	25	25.2	25.3	25.6	30.0	30.5	31.1	31.8	32.6
Distance (m)	0.9	1.0	1.1	1.2	1.3	1.4	1.5		
Force (N)	33.8	34.9	36.0	37.2	38.5	39.9	41.5		

Activity 9.7

For the first metre of the distance moved by the box determine the work done by calculating the work done in each 100-m step.

Determination of the area under a curve

The simplest method of determining the area under a curve, from the mathematical point of view, is to count the squares and then apply the scales of the two axes to the figures obtained. This approach, however, is extremely tedious, and is not particularly accurate as part squares cause a problem, and it is all too easy to lose count.

There are several more mathematical ways of determining areas under curves. The following is one of the more easily understood. The method is as follows.

1. Divide the area to be determined into vertical strips of equal width. Note this width, remembering that the width is measured not with a rule, but by reading off the scale on the x-axis.
2. Mark the mid-point of each vertical strip and draw a line up at this point. These lines are shown in Figure 9.18 as h_1, h_2, h_3, etc.
3. Use the scale on the y-axis to measure the heights of h_1, h_2, h_3, etc.
4. The area of the first strip is approximately $h_1 \times W$. The area of the second strip is approximately $h_2 \times W$. The area of the third strip is approximately $h_3 \times W$, etc.

Mathematical techniques

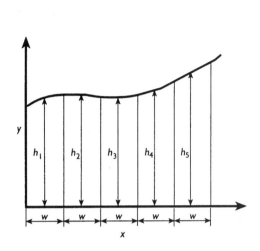
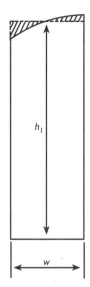

Figure 9.18 Finding the area under a curve

So the total area under the curve is
$(h_1 \times W) + (h_2 \times W) + (h_3 \times W) + (h_4 \times W) + (h_5 \times W) = W(h_1 + h_2 + h_3 + h_4 + h_5)$

Activity 9.8

Plot the force–distance graph from the coordinates given above and use the graph to determine the work done between 0 and 1.5 m.

Manipulating data using mathematical techniques

Fractions

For many students, dealing with fractions is something to be avoided at all costs, and it is true that when dealing with numerical fractions it is often possible to convert the fraction to a decimal fraction, e.g.

$$3/4 = 0.75$$

However, accuracy can be lost in rounding up or down, and the process is time consuming. Furthermore, those students going on to Advanced GNVQ and beyond will find it necessary to deal with algebraic fractions. Without a sound knowledge of numerical fractions they will be lost!

What is a fraction?
A fraction is simply one number divided by another. The number being divided is called the numerator and the other is the denominator, so:

$$5/16 \; means \; 5 \div 16$$

where 5 is the numerator and 16 is the denominator.

The numerator does not have to be smaller than the denominator; fractions can be 'top-heavy', e.g. 32/5 is a fraction.

363

Simplifying fractions

If both the numerator and denominator of a fraction can be divided by the same number then the fraction can be simplified as shown in the following examples.

1. 6/8 can be simplified by dividing the top and bottom of the fraction by 2.

$$6/8 = 3/4$$

2. Simplify 121/246. In this case the number that will divide into both numerator and denominator is not quite so obvious. The procedure in this case is as follows.

try dividing by 2:	126/252 = 63/126
try dividing by 2 again:	can't be done
try dividing by 3:	63/126 = 21/42
try dividing by 3 again:	21/42 = 7/14

Now simply keep going until you have either reduced the numerator (or denominator in the case of a top-heavy fraction) to 1 or tried dividing by all the numbers up to 9.

Multiplying fractions

This is the simplest mathematical process involving fractions, e.g.

$$\frac{3}{7} \times \frac{5}{4} = \frac{3 \times 5}{7 \times 4} = \frac{15}{28}$$

$$\frac{4}{9} \times \frac{3}{7} = \frac{4 \times 3}{9 \times 7} = \frac{12}{63}$$

Note that both the numerator and denominator can be divided by 3: 12/63 = 4/21.

Dividing fractions

This is almost the same as multiplying, but with a twist, e.g.

$$\frac{3}{7} \div \frac{5}{4} = \frac{3}{7} \times \frac{4}{5}$$

i.e. the dividing fraction is turned upside down and multiplied:

$$\frac{3}{7} \times \frac{4}{5} = \frac{3 \times 4}{7 \times 5} = \frac{12}{35}$$

Another way in which you might find a fraction divided is:

$$\frac{4/3}{3/7}$$

which is the same as:

$$\frac{4}{3} \div \frac{3}{7} = \frac{4}{3} \times \frac{7}{3} = \frac{28}{9}$$

Adding and subtracting fractions

These processes are slightly more tricky, but if you follow the method, with practice you should be able to master them.

1. 3/4 + 2/3

Mathematical techniques

The first thing to do is to find a common denominator, i.e. a number that both individual denominators will divide into. The easiest way to find a common denominator is to multiply the individual denominators together, e.g.:

$$4 \times 3 = 12$$

We will now convert both fractions to 'twelfths'.

$$\frac{3}{4} = \left(12 \times \frac{3}{4}\right) \div 12 = \frac{9}{12}$$

$$\frac{2}{3} = \left(12 \times \frac{2}{3}\right) \div 12 = \frac{8}{12}$$

Now that both fractions have the same denominator they can be added, so

$9/12 + 8/12 = (9 + 8)/12 = 17/12$

Remember, you can only add or subtract fractions if they have the *same* denominator.

2. $2/3 - 5/9$
   Common denominator = $3 \times 9 = 27$

   $2/3 = (27 \times 2/3) \div 27 = 18/27$

   $5/9 = (27 \times 5/9) \div 27 = 15/27$

   $2/3 - 5/9 = (18 - 15)/27 = 3/27$

   Note that both the numerator and denominator can be divided by 3 so the fraction can be simplified: $3/27 = 1/9$.

3. $5/8 + 3/5 - 1/4$
   Common denominator = $8 \times 5 \times 4 = 160$

   $5/8 = (160 \times 5/8) \div 160 = 100/160$

   $3/5 = (160 \times 3/5) \div 160 = 96/160$

   $1/4 = (160 \times 1/4) \div 160 = 40/160$

   $5/8 + 3/5 - 1/4 = (100 + 96 - 40)/160 = 156/160$

   Dividing top and bottom by 4 gives:

   $156/160 = 39/40$.

Positive and negative numbers

Vector and scalar quantities

A vector quantity has both size (magnitude) and direction. A scalar quantity has magnitude only; it does not have a specific direction.

It is a fairly simple matter to describe, in words, the motion of a spring bouncing after it has been stretched. However, as engineers, we are sooner or later going to have to describe the motion mathematically.

Figure 9.20a shows the spring in its central position, i.e. the position at which the spring would eventually come to rest; this is the datum position from which the deflection of the spring is measured. Figure 9.20b shows the spring compressed by a distance x, and Figure 9.20c shows an extension of distance x. By using the words 'compression' and 'extension', the direction of the deflection is indicated, but if the deflection is to be described mathematically we must have a way of defining direction using maths symbols.

This is done using positive or negative numbers. The first step is to define what we mean by a positive deflection. In this case deflection upwards, i.e. com-

Science and mathematics for engineering

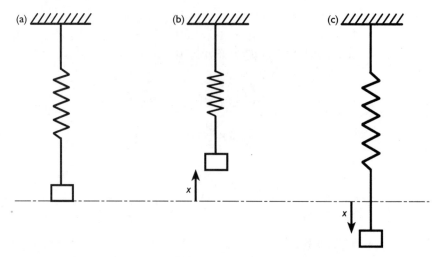

Figure 9.20 (a) Spring in central position; (b) Spring compressed by distance x; (c) extension of distance x

pression, will be defined as positive, and deflection downwards, i.e. extension, as negative. Hence, Figure 9.20b shows a deflection of $+x$ and Figure 9.20c shows a deflection of $-x$.

The use of positive and negative signs to indicate the direction of a quantity is of great importance when dealing with vectors.

Activity 9.9

The spring example demonstrates that distance moved, correctly called displacement, is a vector quantity. Now look at the list of familiar quantities listed below and decide which are vectors and which are scalars:
- mass
- velocity
- resistance
- frequency
- acceleration
- force, pressure
- temperature
- current
- time
- energy
- power.

Multiplication and division of positive and negative numbers

The effect of multiplying or dividing any number by a negative number is to change the sign of the answer, so:

$$+2 \times (-2) = -4$$
$$-3 \times (-4) = +12$$
$$12/(-2) = -6$$
$$-15/(-3) = 5$$

(Note the way in which brackets have been used to separate and indentify the negative numbers.)

This process can be summarised by four rules

$$(+) \times (+) = (+)$$
$$(-) \times (+) = (-)$$
$$(+) \times (-) = (+)$$
$$(-) \times (-) = (+)$$

Activity 9.10

Work out the answers to the following:
1. $(-2) \times 14 =$
2. $120/(-4) =$
3. $(-16) \times (-3) =$
4. $-36/(-9) =$

Addition and subtraction of positive and negative numbers

Our earlier example of positive and negative displacements will again provide a method of picturing the processes of addition and subtraction. When adding or subtracting positive and negative numbers think of
- positive numbers as being steps taken in a forward direction
- negative numbers as steps taken in a backward direction.

Also remember that a negative sign always changes the sign of the number that follows it.

Case study

1. $-3 - 4 = -7$ (three steps backward followed by four steps backward makes seven steps backward altogether)
2. $6 - (-8) = 14$ (the first thing to note is that $-(-8)$ is the same as $+8$ because the first negative sign changes the second to a plus), we now have $6 + 8 = 14$
3. $-5 + 9 = 4$ (five steps backward followed by nine steps forward gives a total of four steps in a forward direction).

Activity 9.11

Complete the following additions containing positive and negative numbers:
1. $-4 + 14 =$
2. $16 - (-14) =$
3. $-12 - 23 =$
4. $-24 - (-12) =$
5. $12 - 23 - (-8) + 14 =$

Standard form

Manipulating very large or very small numbers in calculations can be a source of error as mistakes are made, for example when entering such numbers into the calculator, e.g.:

$$\frac{0.00036 \times 2510 \times 2.64}{0.0085 \times 560100}$$

It is also difficult to estimate the value of expressions like this. The end result is that the answer shown on the calculator display cannot be checked easily to see if it is 'about right'.

When numbers are written in standard form they consist of two parts, e.g.:

$$3000 = 3 \times 1000 = 3 \times (10 \times 10 \times 10) = 3 \times 10^3$$
$$0.0008 = 8 \times 1/1000 = 8 \times 1/(10 \times 10 \times 10 \times 10) = 8 \times 10^{-4}$$

so in standard form $3000 = 3 \times 10^3$ and $0.0008 = 8 \times 10^4$.

When there are more integers (numbers greater than zero) the same procedure is used but you must remember to leave only one integer to the left of the decimal point. Hence:

$$58100 = 5.81 \times 10000 = 5.81 \times 10^4$$
$$386 = 3.86 \times 100 = 3.86 \times 10^2$$
$$0.105 = 1.05 \times 1/10 = 1.05 \times 10^{-1}$$
$$0.000091 = 9.1 \times 1/10000 = 9.1 \times 10^{-5}$$

Now try some for yourself.

Activity 9.12

Rewrite in standard form:
1. 291
2. 0.384
3. 76000
4. 0.0000265

Rewrite as ordinary numbers:
1. 4.05×10^{-4}
2. 1.62×10^5
3. 9.0×10^2
4. 8.361×10^{-1}

Obtaining values for physical quantities

Transposition of formulae

There will be many occasions when the formula used to calculate the value of an engineering quantity is 'round the wrong way'. For instance you may know the voltage across a known resistance, and want to determine the current. You know that

$$V = IR$$

but you want I to be the subject of the equation. The method by which the terms in an equation or formula are moved around is known as transposition, and may well be one of the most frequently used bits of maths that you learn!

Mathematical techniques

In this section we will deal with just three types of formulae:
- formulae of the form $V = IR$ (products)
- formulae of the form $a = b + c$ (sums)
- formulae of the form $L = L_0 + \alpha L_0 \Delta T$ (combined products and sums).

For each of these three types of formula worked examples are given to provide you with a general method. There are two important points to be borne in mind when transposing any formula:
- when a term has to be moved ask yourself, 'What is the term *doing* to the one next to it?', i.e. is it adding, subtracting, multiplying or dividing. Then do the opposite mathematical operation, e.g. addition and subtraction are opposites, multiplication and division are opposites
- whatever you do to one side of an equation or formula you *must* do to the other.

Now have a look at the worked examples and try Activity 9.13 for yourself.

Case study

Transposing products
1. Make a the subject of the formula $F = ma$
   Method: divide both sides by m, which gives $F/m = a$
2. Make A the subject of the formula $P = F/A$
   Method: multiply both sides by A, which gives $PA = F$
   then divide both sides by P, which gives $A = F/P$

Transposing sums
1. Make u the subject of the formula $v = u + at$
   Method: subtract at from both sides, giving $v - at = u$
   Notice that although at is in fact made up from two terms it can be treated as one in this case because we are not trying to mathematically separate a from t.
2. Make x the subject of the formula $y = x - a$
   Method: add a to both sides, giving $y + a = x$

Transposing combinations of products and sums
1. Make t the subject of the equation $v = u + at$
   Method: subtract u from both sides, giving $v - u = at$
   divide both sides by a, giving $(v - u)/a = t$
2. Make c the subject of the formula $a = b - c$
   Method: subtract b from both sides, giving $a - b = -c/d$
   multiply both sides by d, giving $d(a - b) = -c$
   to make the c positive multiply both sides by -1, giving $d(-a + b) = c$
   which is the same as $d(b - a) = c$

Activity 9.13

In each of the following examples, make the term marked * the subject of the formula given.
1. $V = IR$ $I*$
2. $\rho = m/V$ $V*$
3. $a = b/c$ $b*$
4. $\Delta T = T_2 - T_1$ T_2*
   (hint: remember that ΔT is just one term meaning 'the change in T')
5. $R = R_1 + R_2$ R_2*

6. $L = L_0 + L_0 \alpha \Delta T$ α^*
7. $I = I_1 + v/R$ v^*

Assignment 9
Calculate the behaviour of engineering systems

This assignment provides evidence for:
Element 3.3: Use mathematical techniques to investigate the behaviour of engineering systems
and the following key skills:
Communication 2.4: Read and respond to written materials
Application of number 2.1: Collect and record data

This assignment consists of six parts. Each part describes a different engineering system and the conditions under which it operates.

Your tasks

For each system you should:
- state the relationship which models the way the system operates using words, graphs and formulae (you should make sure you use each of these at least once in the assignment)
- calculate the required quantity
- where necessary research the value of constants.

1. The current which flows throught the filament of a 12-V car headlight is 5 A. Calculate the resistance of the filament.

2. The average speed of a train for a journey of 50 km is 85 km/h. What is the journey time?

3. A spring of stiffness 2 kN/m and unloaded length 50 mm is extended to a length of 65 mm. Determine the force which will cause this extension.

4. An electric current is passed through a coil of copper wire. At 20°C the resistance of the coil is 90 ohms. After a few minutes the temperature has risen to 70°C. What is the resistance of the coil at this temperature?

5. 10 kg of water at 50°C is heated to boiling point and then evaporated. The steam is then heated further to 140°C. Determine the amount of heat energy required to convert the water to steam at the boiling temperature.

6. A space vehicle of mass 800 kg is decelerated by firing a small rocket in the opposite direction to the course. The rocket thrust is 4 kN. Calculate the deceleration of the vehicle.

Sample unit test for Unit 3

The answers are given on page 460.

1. Which quantity would affect the acceleration of an object?
   a Temperature
   b Time
   c Mass
   d Voltage.

2. Which quantities would determine the power consumed by an electrical device?
   a Mass and time
   b Voltage and time
   c Mass and current
   d Voltage and current.

3. Which quantity remains constant when water vaporises to steam?
   a Mass of liquid
   b Temperature of liquid
   c Mass of steam
   d Volume of liquid.

4. Which quantity changes when a steel bar is heated?
   a Mass
   b Length
   c Coefficient of linear expansion
   d Coefficient of friction.

5. When using the formula $v = u + at$, which quantity is a constant?
   a a
   b t
   c u
   d v.

6. The law that describes the way in which a load spring deflects is:
   a Ohm's law
   b Hooke's law
   c Kirchoff's law
   d Newton's law.

7 The pressure acting on a piston can be found using:
   a $P = F/A$
   b $P = FA$
   c $F = ma$
   d $F = m/a$.

8 The efficiency of an electromechanical system can be found using the formula:
   a Efficiency = power input/power output
   b Efficiency = power output/power input
   c Efficiency = power input − power output
   d Efficiency = power output − power input.

9 Shaft diameters are measured with:
   a Micrometers
   b Multimeters
   c Spring balances
   d Thermocouples.

10 The weight of a component can be measured with:
    a An oscilloscope
    b A vernier calliper
    c A spring balance
    d A multimeter.

11 A thermometer measures:
    a Heat
    b Expansion
    c Liquid
    d Temperature.

12 A stopwatch measures:
    a Velocity
    b Time
    c Acceleration
    d Displacement.

13 Which measurement device would you use to measure the resistance on an electrical component?
    a An oscilloscope
    b A thermocouple
    c A multimeter
    d A thermometer.

14 Which measurement device would you use to measure the temperature inside a high-pressure steam pipe?
    a An oscilloscope
    b A thermocouple
    c A multimeter
    d A thermometer.

15 Which measurement device would you use to determine the frequency of an electrical signal?
    a An oscilloscope
    b A thermocouple
    c A multimeter
    d A thermometer.

16 The following graph shows the way in which the current through a component varies with time. Which of the following statements best describes this relationship?
    a Current remains constant
    b Current increases with time
    c Current is proportional to time
    d Current decreases with time.

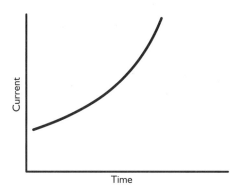

17 Which of the following readings would make you think about taking the results again?

No.	Force (N)	Velocity (m/s)
1	0	0
2	2	4
3	4	8
4	5	10
5	6	14
6	8	16

    a No. 1
    b No. 4
    c No. 5
    d No. 6.

18 The relationship between the load (F) on a spring and the extension (x) is:
    a $F = x$
    b $F \propto 1/x$
    c $F = 1/x$
    d $F \propto x$.

19 Which graph shows the relationship between the resistance of a copper conductor and its temperature?

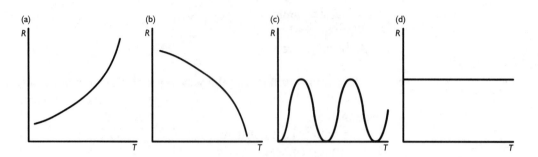

20 Make *I* the subject of the formula $V = IR$:
 a $I = VR$
 b $I = V^2/R$
 c $I = V/R$
 d $I = V^2R$.

21 Make *r* the subject of the formula $A = \pi r^2$:
 a $r = A/\pi$
 b $r = \sqrt{(A/\pi)}$
 c $r = A\pi$
 d $r = \sqrt{(A\pi)}$.

22 Make u the subject of the formula $v = u + at$:
 a $u = v - at$
 b $u = v - (a/t)$
 c $u = v + at$
 d $u = v + (a/t)$.

23 Make A the subject of the formula $P = F/A$:
 a $A = P - F$
 b $A = PF$
 c $A = P/F$
 d $A = F/P$.

24 Calculate the velocity of a vehicle that travels 10 km in 5 minutes:
 Formula: $V = s/t$
 a 0.5 km/h
 b 50 km/h
 c 120 km/h
 d 200 km/h.

25 A 2-m-long steel bar is heated from 10°C to 25°C. If the linear expansivity is $15 \times 10^{-6}/°C$, calculate the final length:
Formula: $L = L_0 + L_0 \alpha \Delta T$
  a   7.5 mm
  b   7.5 m
  c   2.0075 m
  d   2.75 m.

26 Calculate the cross-sectional area of a nylon bar 12 mm in diameter:
Formula: $A = \pi r^2$
  a   446 mm²
  b   37 mm²
  c   72 mm²
  d   144 mm².

27 The pressure on a hydraulic piston of area 600 mm² when acted on by a force of 3 kN is:
Formula: $P = F/A$
  a   5 GN/m²
  b   5 MN/m²
  c   5 kN/m²
  d   5 N/m².

PART FOUR: ENGINEERING IN SOCIETY AND THE ENVIRONMENT

Chapter 10:   The application of engineering technology in society
Chapter 11:   Careers in engineering

Sample unit test for Unit 4
Engineering applications are all around us. Whatever we do, wherever we go, we will be in an engineered environment. The contribution of the engineer cannot be underestimated, and as we enter the twenty-first century all the signs are that technology will impact more and more upon our lives.

   This chapter looks firstly at the applications of engineering technology in the world that we live in, and then at the role and work of the engineer. Qualifications and career options are all explored and advice given on how to present yourself for job selection and interview …and how to get that job!

Chapter 10: The application of engineering technology in society

> **This chapter covers:**
> Element 4.1: Describe the application of engineering technology in society.
> **... and is divided into the following sections:**
> - The range of engineering technologies
> - The application of engineering in home and leisure activities
> - The application of engineering in industry and commerce
> - The application of engineering in health and medicine
> - The environmental impact of engineering technology and activities
> - Managing the environmental impact of engineering activities
> - Engineering technology in society.

The aim of this chapter is to consider the application and environmental impact of engineering technology and engineering activity in relation to various sectors, including home and leisure, industry and commerce, and health and medicine.

Tracing the application and impact of engineering technology it is possible to identify key inventions and developments that underpin significant areas of change. Examples include electricity, the electric motor, the silicon microchip, and the development of materials such as plastics and stainless steel. A number of these are described in further detail below.

The range of engineering technologies

Engineering technology is defined within the GNVQ specifications as referring to information technology, automation and materials. The development of each of these areas, and the application of engineering capability, shapes the nature of the society in which we live, the way we live, our quality of life and our standard of living.

Engineering development and application is necessary for the productivity and wealth creation of our country. Without engineering we would not have roads, buildings or machines of any kind.

Most engineering developments have resulted from the identification of a problem or need, whether relating to an individual, community or society as a whole. Engineering skills and capabilities have then been used to address this situation, through the development of new products, processes or systems.

Engineering in society and the environment

Information technology

Information technology
The collection, processing, storage and use of data using electronic and telecommunication systems

Information technology (IT) describes the use of electronics and telecommunications to collect, store, process and use large amounts of data. It includes all uses of computer systems. Information technology is used as a means of communication and for the storage and retrieval of information.

The widespread adoption of information technology in all contexts – home, leisure, work, health – has affected the production, design, marketing, adoption and use of many products.

A major feature of information technology has been the incredible speed of change and development in recent years. During the last 20 years the use of multimedia, the Internet and the World Wide Web has become possible. Another major development has been the reduction in size of products made possible by the development of the silicon microchip.

The nature of modern information technology is shaped by the development of the silicon microchip.

Case study

The silicon chip

The silicon chip, or integrated circuit, was first produced in 1958. Essentially, it holds all the components necessary to make the central processing unit (CPU) of a computer. Basically, it consists of a number of tiny electronic components, connected to each other with microscopic strands of metal, placed on a tiny wafer-thin piece of silicon. At the same time as the size of the microchip has been reduced, the number of components that can fit on to it has increased dramatically. Now it is possible for literally millions of transistors to be contained on one chip.

There are different types of chip, for example a microprocessor chip, a memory chip, a central processing unit chip, a graphics chip.

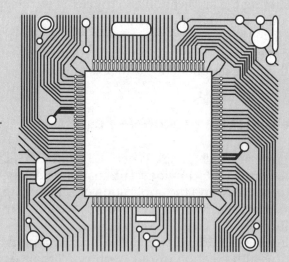

Figure 10.1 A silicon chip

A microchip has many uses and is to be found in many of today's products, in almost every context, including electronic calculators, cars, cameras, watches, televisions, washing machines and personal computers.

The application of engineering technology in society

Activity 10.1

Identify three products, other than those listed above, in which a microchip may be found. Briefly describe its purpose in each case.

Case study

Fibreoptics

Fibreoptics is the use of fine optical fibres made of very clear glass to transmit light signals over long distances. It works by translating information as electric pulses into pulses of light. Each fibre is made up of two layers of glass. The inner layer is the core along which the light travels. The outer layer acts as cladding and prevents the light from passing out of the side of the fibre. A single fibre can carry more than one beam of light. Each beam of light can carry more information than an electric current. Fibres do not need to be insulated as signals within neighbouring fibres do not affect each other. This technology has become possible as a result of the development of tiny lasers that are able to produce a narrow, but powerful, beam of light.

Case study

The laser

A laser is a source of light. It produces a strong and narrow beam of light, which can be focused accurately. The laser is an important element of fibreoptic technology. Lasers are used in the storage and retrieval of information. Compact disks are produced by using lasers to record the digital information on the disks. Lasers can also be used to record video and computer information. Laser is an acronym for light amplification by stimulated emission of radiation.

Figure 10.2 A laser beam

Engineering in society and the environment

Case study

Artificial intelligence
The use of computers to mimic human behaviour

Artificial intelligence

Artificial intelligence is the use of computer programs to mimic elements of human behaviour. Artificial intelligence usually combines aspects of electronic, mechanical and software engineering. Essentially, it knows how to react to given situations, and can also be used to aid decision making. This technology has been significantly advanced by the developments in neural networks. Neural networks are able to mimic the behaviour of groups of nerve cells. A significant recent advance is that they are able to learn from experience.

Automation

Automation
A machine-led activity or process, in which the human input is limited

Automation describes the use of one or more machines that can be programmed in order to achieve the work that was previously undertaken by human labour. These are usually simple, repetitive tasks. As long as the machinery is operating normally the human input will be limited, if not non-existent, for example setting up the machine for the process, undertaking tasks that would be too expensive or difficult for machines to perform and shutting down the process.

The use of automated systems in industry is usually related to mass production, for example car manufacture or canning processes. The type of work achieved within the industrial context includes machining, welding, transfer and assembly. Types of automatic machinery include computer numerical control (CNC) machine tools, robot arms and guided vehicles.

Case study

CNC machines

CNC machines are automated machines that incorporate control microcomputers and digital electronics. They can be programmed to perform tasks to high levels of precision.

The introduction of automated systems becomes viable when dealing with large numbers of production units. Automation of a process can help to reduce unit costs and waste because machines can be programmed to be precise, and to perform high numbers of repetitive actions over long periods of time.

Activity 10.2

Explain what is meant by automation. Give three examples of automation.

Materials

Materials
The matter from which items are made; materials can be defined as naturally occurring or manufactured

Materials exist either naturally, as a consequence of manufacturing processes or as a result of recycling. New types of materials and methods of processing materials are constantly being researched and developed.

Examples of materials that exist in nature include cotton and gold. Although they can be used in their natural state, they usually need the benefit of engineering processes in order for them to be suitable for use in commerce and industry. For example, clay is used in its natural state for lining canals and making pottery but needs to be heat processed before it is suitable for use in making clay bricks.

The application of engineering technology in society

Manufactured materials have been produced as a result of human activity. Examples include plastics, refined metals and composites such as composite boards (chipboard, plywood and hardboard).

Case study

Plastics

Plastics are a class of materials produced from petroleum. Methods used to process plastics include extrusion, injection moulding, vacuum forming and compression. Plastics benefit from being comparatively inexpensive and easy to use. They are also generally waterproof and lightweight, and can be produced in different colours. A disadvantage is the problem of waste disposal, as they are not usually easily degradable, and will only break down over many years. However, plastics can be recycled to produce other products or burnt as a source of energy. Examples of plastics include polyurethanes, acrylics, polycarbonates, nylons and epoxy resins. Attention is currently being paid to the development of plastics that will degrade more quickly than is possible at present.

As a result of new materials being developed, designers and manufacturers are able to consider new applications and product styles and variations.

The development of new materials can help further the development of engineering owing to the nature of the material and its properties.

Activity 10.3

Identify one product that is made from plastic. Find out what other materials have been used to produce this product in the past, and list the advantages of using plastic.

Recycled

Reclaimed for reuse; methods include product reuse, component reuse, energy recovery and reprocessed waste

Recycled materials

Materials can be recycled in a variety of ways. For example, recycling through material recovery (such as the recycling of waste paper and steel), product reuse (such as with milk bottles), component reuse (such as from cars and other motor vehicles) and energy recovery (such as nuclear fuel).

One of the reasons recycling is seen as important is because of its contribution to a reduction in the demand for non-renewable resources, for example the use of scrap steel reduces the need to mine iron ore. It also provides a solution as to what to do with waste that has to be disposed of in some way.

Recycling is therefore seen to be a way of saving energy and resources. However, individual recycling processes must be considered carefully as the use of recycled material can lead to a higher use of non-renewable resources, for example as a result of the collection and transportation of the waste material, and the conversion of the waste, which uses energy provided from fossil fuels.

Despite these disadvantages, recycling is an important issue. This has been recognised by the government, which in 1990 set a target of recycling 50% of all recyclable household waste by the year 2000.

Engineering in society and the environment

Table 10.1 Recycling levels by material (%), UK

	1990	1991	1992	1993	2000 (target)
Paper and board	32	34	34	32	–
Waste paper used in newsprint	26	26	31	31	40
Glass	21	21	26	29	50
Aluminium cans	6	11	16	21	50
Steel cans	9	10	12	13	37
Plastics	2	–	5	–	–

Sources: ACRA, BGMC, BP & BPIP, BS, SCRIB, PFGB, TTIC. Reproduced courtesy of the Office for National Statistics.

Case study

Recycling steel

A significant proportion of steel production involves the use of scrap steel, with roughly 25% of steel packaging made from recycled material.

There are two main methods of making steel:
- the electric arc process, which uses cold scrap steel
- the basic oxygen process, which uses molten iron and sometimes also scrap.

The scrap steel includes a mixture of waste from steel production processes and waste steel that has been recovered from products such as cars and steel cans.

After the steel has been made, it is sold to other steel companies to be processed into different shapes, such as sheet steel for car bodies and steel plate for building offshore rigs. In some cases, the steel will need further processing, for example adding tin for canning purposes and coating steel with plastic or zinc to provide protection or decoration.

Figure 10.3 Steel recycling process (reproduced courtesy of British Steel plc)

Activity 10.4

Identify three products or materials that are suitable for recycling. Research and describe the recycling process for each.

Progress check

1. What is meant by the term automation?
2. Give two examples of (a) natural materials and (b) manufactured materials.
3. Give a brief explanation of what is meant by 'information technology'.
4. What effect does automation have on human input?
5. Name three methods of recycling.

The application of engineering in home and leisure activities

Engineering technology, in terms of information technology, automation and the development of materials, can be found to have influenced almost every aspect of domestic and leisure activities. The following section highlights just a small selection of examples and offers many opportunities for home-based activities.

Information technology in home activities

Many examples of the use of information technology can be found in the home, for example within telephones, fax machines, music centres, televisions, video recorders and home computers.

Information technology has not only revolutionised the processes involved in producing many of the items designed for home use, it has also affected how many items are used, for example the ability to programme items such as washing machines and video recorders to operate at a given time.

The home computer is used for many purposes: working from home, as a means of communication via electronic mail (email) and the World Wide Web; and for leisure purposes such as playing games or surfing the Net. Individuals can also use the home computer for home management, for example paying bills and booking hotels and travel tickets.

Automation in home activities

Remote control
The control of a machine, from a distance, by means of signals transmitted from an electronic or radio device

Automation has an impact on home activities, both by being part of products used in the home and as a result of being used to produce products for the home. Whenever we talk about things working automatically, for example the automatic washing machine, we are describing something that involves automation. Many of the household items described as labour-saving products are examples of automation.

Products within the home that typify the automated process include the washing machine and dishwasher, both of which are machines that do work that was previously done by people. The human input on each is now limited to the preparation of the machine (filling the machine with the items to be cleaned, putting in the required amount of detergent and setting the programme) and the removal of items after the process is complete.

Automation, together with information technology, enables electrical appliances to be operated from a distance using 'remote control'. Although this has

Engineering in society and the environment

been available for some time with televisions and videos it is becoming more widely available for lighting and heating systems, and other household appliances, and also a growing range of appliances for the disabled.

Figure 10.4 Dishwashers

Activity 10.5

Identify two examples of automation in the home. Describe the social and environmental benefits.

Materials in home activities

Many new materials are being introduced into the home environment. Two key examples are plastics and stainless steel, which have replaced many traditional natural materials in the manufacture of products for the home. In many homes, both materials will have been used in the manufacture of kitchen furniture and utensils, such as washing up bowls or shelves in the dishwasher and refrigerator.

Both stainless steel and plastic offer the benefit of being easy to clean and rustproof; however, some types of plastic have the disadvantage of melting if they become too hot.

Home activities

Microwaves
Electromagnetic radiation of wavelength 0.001–0.3 m; used in microwave ovens to cook food quickly

Food preparation
Developments in engineering technology have contributed to the preparation and sale of convenience, preprepared, dehydrated, microwavable, freeze-dried, frozen, canned and processed food.

The development of kitchen furniture, or 'white goods' such as the refrigerator, freezer and other kitchen items relating to food preparation (e.g. kettles, food processors and toasters), has also been influenced by developments in engineering technology. The following case studies provide a brief outline of the development of the associated technologies to the present day.

The application of engineering technology in society

Case study

The food processor
Ceramic mixing bowls and wooden spoons have long been replaced by electric mixers, food processors and liquidisers. Each of these offers many options, including different speeds and different implements for different tasks, different sizes and cordless versions.

Activity 10.6

Select one kitchen implement. Investigate the technology used in modern versions. Describe the benefits of the modern products over those previously available.

Cooking
Home cooking has also benefited from developments in information technology, automation and materials. It has led to the development of such products as microwave ovens, self-cleaning ovens and convection cookers.

Case study

The toaster
The most basic method of toasting bread is using a toasting fork in front of a fire. The electric toaster first appeared in 1913, with the first toaster only able to toast one side at a time. It worked through a system involving a heating element that was essentially bare wires wound around strips of mica. The main problem with early toasters was that they did not have thermostats so tended to burn the toast. The UK did not see pop-up toasters until the 1950s.

The bimetal strip was introduced to work as a system of control. When a toaster is switched on, the heat produced by the heating element will cause the bimetal strip to bend. When it reaches a predetermined position the current supply to the heating element will be switched off. The time it takes for the bimetal strip to reach this position will affect how well the bread is toasted. The main problems with this system are the speed at which the bread is toasted and the inconsistency in the degree of toasting.

To prevent such problems many new toasters have a preprogrammed chip that is able to provide greater consistency and faster results.

Case study

Cookers
Cookers encompass engineering developments through information technology, automation and materials. In the late nineteenth and early twentieth centuries, a cooker was typically made of cast iron to a basic functional design and was heated by coal; examples included the Aga and Rayburn ovens. Today consumers face numerous possibilities when considering their choice of cookers: of fuel (gas, electricity, oil); of price range; of styles; of convenience factors. It is possible to preset the timer to cook at a preset temperature, automatic cut-off systems provide additional safety and other options include single or double ovens, separate or integrated grills and mixed fuel (e.g. gas oven and electric hob). Modern cookers will usually be made from a combination of stainless steel, ceramics, glass and enamel.

Microwave cookers work through the production of high-frequency electromagnetic waves, which heat up the food. They work differently from conventional cookers in that the inside of the food product is cooked first.

Electromagnetic waves

Electric and magnetic fields that travel at approximately 300 000 000 m/s

Engineering in society and the environment

Activity 10.7

Select one of the cooking products mentioned above. Identify the use of engineering technology in the production and use of a modern-day model.

Washing and cleaning

The nature of baths and showers has largely changed as a result of the development of materials and the availability of indoor plumbing and gas and electricity, and thus hot water in the home. In Victorian days personal hygiene involved washing in water that had been heated on a coal stove and carried to an iron bath. Nowadays, hot water is readily available, and bathroom furniture items are commonly made from plastics and ceramics.

Electricity and indoor plumbing have enabled improvement in personal hygiene and sanitation, which in turn has led to a reduction in the incidence of diseases such as cholera.

We take the availability of light, heat and water in the home for granted because the necessary technology has been with us for as long as we can remember. This is not the case in every country. For example, in countries such as Nepal and the Gambia there are communities that have no clean water supply or la-

Figure 10.5 Water collection in Nepal

trines. The charity WaterAid aims to support the introduction of safe water and waste-water facilities for such communities. In a project in the district of Piprapokharia, in the Terai region of Nepal, local communities are aiming to sink 49 tubewells and fit them with handpumps, as well as repair an existing hand-dug well, to provide a water supply that is safe, clean and close to their homes. They also plan to construct simple pit latrines. This will have both health-related and social benefits for the community as the women and children will no longer have to spend hours every day collecting the water for the community.

The task of washing clothes is another area of our everyday lives in which the contribution from engineering is taken for granted. Most people have access to a washing machine, whether within their own homes or at a local laundrette.

Case study

Washing machines

Washing machines have gone through various stages of development and refinement. From the early washboard and iron tub, to the twin tub that had separate tubs for washing and spinning, to the automatic washing machine that we have today with a multitude of programmes and wash cycles and the ability to set the wash cycle on a timer to start at a particular time. Early washing machines involved the movement of water rather than the movement of the clothes; this changed in the 1920s. Now, attention is being given to how to reduce the quantity of water used, without affecting the quality of the wash (solutions include using a system of recycling the water using a pump and jet-spraying the water back on to the wash load). The technology is now available to enable the washing machine to adjust the water intake automatically in accordance with the nature of the wash load, in terms of size and fabric. Key to this is the use of microchips.

One area of development that we are constantly made aware of through television advertisements is that of the detergent used in the washing machine and the method used to dispense it. Chemical engineers are involved in researching and developing new detergents: biological or non-biological, powder or liquid, standard or concentrated.

Activity 10.8

Select a washing machine and one other item used for cleaning purposes. Describe the main materials each is made from. Compare the application of information technology and automation in the operation of each item.

The application of engineering to the domestic situation has revolutionised the performance of household chores, from the development of appliances such as the dishwasher and vacuum cleaner to that of household cleaning fluids and polishes. Not only do these products reduce the time and effort needed in the performance of the chores, it can also be argued that a higher level of cleanliness is achieved. For example, the use of water of the correct temperature in washing dishes and laundry should achieve better results. In addition, vacuum cleaners that can be used to wash and dry carpets, or which include microporous filters to remove house dust and pollen (avoiding the need for disposable bags), should also improve results.

Engineering in society and the environment

Case study

The dishwasher
The first dishwasher was invented in 1855. This was a manually operated machine involving the need to turn a handle in order to move vigorously the cradle holding the dirty dishes through the water. A turning point in the development of the dishwasher was the development of the electric motor, although the dishwasher was not to gain general acceptance until able to benefit from greater reliability and improved detergents.

Activity 10.9

Investigate one of the household items described above, or one of your own choice. What are the most recent changes in this product's development. Identify the key areas of engineering technology involved in these developments.

Home entertainment

A major development in recent years has been the growth of home-based entertainment. Household entertainment products include televisions, music systems, video cameras and players, radios, camcorders, stereos, CD players, personal computers and games. Computer games are ever increasing in sophistication, in

Table 10.2 Participation in selected home-based leisure activities 1993–94

Great Britain	Hours per week (all persons)
Watching TV	17.1
Listening to the radio	10.3
Listening to CDs, tapes or records	4.0
Reading books	3.8
Reading newspapers	3.3
Caring for pets	3.1
Gardening	2.1
Cooking for pleasure	1.9
Watching videos of TV programmes	1.7
DIY or house repair	1.6
Sewing and knitting	1.3
Reading specialised magazines	1.0
Watching other videos	1.0
Reading other magazines	0.7
Exercising at home	0.5
Using games computer or console	0.5
Car maintenance	0.5

Source: The Henley Centre. Reproduced courtesy of the Office for National Statistics. Time spent in an average week in the three months before interview by persons aged 16 or over. Data relate to the 12-month period ending September 1994.

The application of engineering technology in society

terms of speed, detail, complexity and accuracy. Modern versions of these products rely on the use of programmed integrated circuits, enabling the ever decreasing size – and thus increasing portability – of these products, for example hand-held games machines.

Case study

Television

Television is still a relatively new technology, although few people nowadays would be able to imagine life without it. The last 50 years have seen a significant level of development from the original television, which was relatively big, bulky and heavy. Pictures were black and white and of relatively poor quality. Early significant developments included the introduction of colour and the improvement of sound and picture quality. The more recent introduction of satellite dishes enables programmes broadcast from outside the UK to be viewed. Other developments are the introduction of the remote control and the possibility of programming video machines to record a programme when it is actually broadcast, using the code number printed in the daily newspapers. Engineering developments have also led to the production of miniaturised televisions, even to the idea of a television small enough to be attached to the wrist in the same way as a watch.

Activity 10.10

Choose one leisure product used in the home. Trace the development of the product including developments in the materials used in its manufacture, and the application of information technology and automation.

Case study

Books

Even reading books has been affected by the development of engineering. Historically, the entire process of producing a book was first revolutionised with the introduction of printing presses. More recently, it has been revolutionised by computerisation, the materials used, including recycled paper, and the processes involved such as the development of desktop publishing, graphic packages, the use of electronic scanners and colour separators. Finding books has been made easier as a result of computerised systems and accessing books is now possible through the computer, with complete books now available on the Internet and on CD ROM.

Sport

Success in competitive sport can be as much about the equipment as the competitors themselves. In many sports it is necessary to pay attention to the equipment being used, whether it is a racing car, a tennis racquet, a bicycle or a training shoe. A prime example of this is Formula One racing, which is as much about the quality and reliability of the car and its engine as driver ability. The design and materials used in a driver's clothes are also essential, as they must provide the protection necessary in the case of crashes and fire, while not hampering performance during the race.

Engineering in society and the environment

Case study

> **Bicycles**
> Bicycles that were traditionally made of steel are now more likely to be constructed of much lighter composite materials. Magnesium alloy wheels add the benefits of being both strong and light. The quality, reliability and aerodynamic efficiency of the equipment and clothing can all be improved through engineering.

Home exercise

Home exercise is a growth area, with people generally aware of health issues and with a need to combat their increasingly sedentary way of life. Home-based weightlifting machinery, cycling and step machines are now available. Computerisation is often built in for the purpose of providing details about the amount of effort that has been used. Physical details such as weight and height can be programmed into exercise machines in order to help design a personalised exercise programme.

Leisure wear

Lycra
An elastic, polyurethane (plastic) fabric

Engineering is key to the development of new materials and the understanding of how they can be used. For example, the development of relatively new materials such as Lycra has revolutionised the manufacture of leisure wear, as has the use of plastics in sports shoes and protective clothing such as cycle helmets.

Activity 10.11

Choose one material, one example of automation and one example of information technology and describe their impact on leisure activities.

Personal transport

Catalytic converter
A device, attached to a car's exhaust, that converts poisonous gases into harmless ones

Personal transport includes a wide range of vehicles, such as cars, motor bikes, bicycles and wheelchairs. Engineering developments can be seen in all cases, from developments in the materials used, the production processes employed in their manufacture and the product specifications.

Car manufacture has seen a change in the materials used both in body construction and components within the car. Lighter materials are replacing heavier, traditional materials, giving an environmental benefit as the lighter the vehicle the lower the fuel consumption. Consideration is also being given to the fate of the product at the end of its useful life, for example the use of recyclable materials for components. Aluminium and plastic tend to be lighter than steel, but lighter steels are being developed, and steel has a good record for recyclability.

The development of motor vehicles is an example of an engineering technology that has had an impact on almost every aspect of our everyday lives. We use motor vehicles for domestic purposes, leisure, work and health. Cars have enabled much greater freedom in terms of where we live in relation to where we work, where we go on holidays, and the range of leisure activities available. Jobs have been created as a result of motor vehicles, not only in the production but also the use. Although it is true that accidents involving motor vehicles are significant in number, the use of motor vehicles for the purpose of reaching those in need of medical care has also improved the ability to save lives by reducing rescue time.

The application of engineering technology in society

Case study

Catalytic converters

One hundred years ago, when the motor vehicle was still at the experimental stage, different means of powering the vehicle were being investigated. The main systems included steam and petrol. At this time there was no indication that the use of petrol would cause any environmental concerns. As the levels of car ownership have increased over the decades, it has become evident that the petrol-powered motor vehicle has been responsible for a significant level of pollution of the atmosphere.

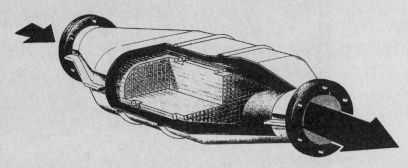

Figure 10.6 A catalytic converter

When this was realised, car manufacturers and petrol companies collaborated in research and development to address these problems. As a result of their efforts, modern cars can now travel one-third further on a gallon of fuel than cars produced in 1975 or before. Unleaded petrol has been on sale in the UK since 1986. Catalytic converters have been fitted to new cars since 1993, therefore reducing their pollution levels. A catalytic converter is a simple piece of equipment, with no moving parts, which is attached to a car's exhaust. It contains chemicals that can make most of the poisonous gases from the exhaust harmless. Because the chemicals are destroyed by lead they can only be run on cars using unleaded petrol. At the same time many car manufacturers continue to conduct research into alternative methods of powering.

Air bags

Many engineering developments have resulted from a perceived need or problem. The number of accidents involving motor vehicles, and the nature of the accidents, has led to such developments as the airbag, which is now becoming standard equipment on many new cars, along with more effective seat belts and child seats.

Case study

Aeroplanes

Air travel is today one of the safest means of transportation, although it had barely been thought of 100 years ago. The advances in air travel have been a result of the development of scientific and technological understanding and capabilities in product design and materials. The first aeroplanes constructed and flown by the Wright brothers were made of wooden frames over which canvas fabric was stretched. The canvas was made waterproof by being covered in a substance called dope, which was very much like wallpaper paste.

Engineering in society and the environment

Figure 10.7 Car testing at Ford

Figure 10.8 Development of engineering technology at British Aerospace

The application of engineering technology in society

Since this time aeroplanes have been constructed using metals such as steel, aluminium and titanium. Present-day aeroplanes are constructed using carbon-fibre composite materials, which are stronger per unit weight than the metals, easier to handle, can be cut with shears and able to stand much greater temperatures.

Activity 10.12

Choose one engineered product that contributes to your domestic life. Trace the development of that product over the last 50 years. Include reference to the impact of automation, information technology and materials on the production and use of the product.

Communications

Communications include the use of the telephone, the radio and television, fax machines, and computers via email and the Internet, to name but a few.

Early systems of communication relied on transmitting messages along wires, as used in telegraph messages and early telephone systems. As a result of the discovery of radio waves in 1901, it became possible to turn sound into electricity. The development of the diode and triode valves in the early twentieth century made it possible to increase the strength of amplified messages.

Case study

Telephones

The telephone was invented by Alexander Graham Bell, based on the transmission of sounds electrically along a wire. Subsequent developments included the invention of the automatic telephone exchange and boosting devices to enable long-distance calls. Previous telephone systems were analogue systems in which signals were sent as varying electrical currents.

Today's telephones rely on digital technology, and are often equipped with silicon-chip memories. It makes it possible to provide the wide range of telephone services such as the time and weather services, and to make long-distance calls that are routed along optical-fibre cables and microwave links.

Mobile phones combine telephone and radio engineering. Cellular phone systems operate on the basis that the country is split into a number of interlocking areas, each of which has its own radio transmitter/receiver. This system enables the same frequencies to be used in different cells. as long as the cells are not next to each other. This in turn enables a greater number of people to use mobile phones. The reduction in size and weight of mobile phones has been made possible as a result of digital technology and the reduction in size of individual components.

Modern telephones contain microchips for a variety of reasons, for example to translate analogue to digital signals and vice versa, to store telephone numbers, to provide automatic dialling facilities, and/or to provide a liquid-crystal display screen.

Engineering in society and the environment

Figure 10.9 Some telephone styles over the years

Case study

Radio technology

The transmission of sound signals by electromagnetic waves

Radio

In 1901 Marconi successfully transmitted a wireless signal from Britain to Canada using radio waves. Subsequent developments that aided the improvement of radio included the invention of such devices as the diode and triode, which enabled the development of a radio transmitter capable of sending speech and music. The superheterodyne receiver, which was invented in 1918, increased the sensitivity of radio sets to weak signals, so improving the tuning capabilities. This device is still used in modern radios.

Radio technology is used in many other products, such as telephones and televisions.

Progress check

1. Identify three applications of engineering technology to food in the home.
2. Describe one way in which engineering technology has affected hygiene.
3. List four examples of automation in the home.
4. Identify two developments relating to materials in sport.
5. Describe two examples of information technology applications in leisure activities.

The application of engineering in industry and commerce

There is no area of industry and commerce that has not been affected by engineering developments, whether as a result of developments in the automation of equipment and machinery, new information systems, communications or materials.

Manufacturing

Productivity
The rate of output per employee

Automated systems in manufacturing can help to increase productivity and business efficiency. With greater automation, businesses are able to achieve higher levels of output with fewer staff, therefore increasing productivity levels.

Manufacturing can involve many tasks that are simple, repetitive or even dangerous. Automated processes have tended to take over such tasks that were previously done by unskilled labour. These tasks are now done by machines that can be programmed to perform these tasks more quickly and more efficiently for longer periods of time. As machines can be programmed to levels of higher accuracy and precision, it is possible to reduce the wastage within the production process that might previously have resulted from human error.

Although automation has brought economic benefits to industry, and therefore the economy, it can also have negative social effects. Higher levels of automation can lead to the deskilling of a job, for example from semiskilled to nonskilled; this can lead to an employee feeling demotivated. A second disadvantage may be that people can become isolated at work, for example one person working with a machine, particularly a noisy machine, will have very little social contact during their working day. Feelings of isolation can lead to demotivation. Demotivation among the work force can bring its own problems, as it can lead to higher levels of error and absenteeism.

The introduction of automation into the workplace can affect the physical layout of the operational area in order to accommodate the machinery. It may lead to further increases in supporting mechanisation, to changes in the product design in order to accommodate the automatic process, and may also lead to increasing use of supporting information technology.

The use of automation in the industrial workplace saves on labour costs but increases capital costs, that is the cost of the machinery itself, as well as energy costs. As the machinery will generally be run on electricity this has implications concerning the burning of fossil fuels and the resulting pollution caused by the generation of electricity.

CNC
Computer numerical control – which refers to machinery that is digitally programmed

Computer-controlled equipment such as CNC cutters, lathes and grinders are controlled by the use of digital electronics and can be programmed to high levels of accuracy. This increases productivity levels and reduces wastage through error. The machines can operate continuously, not requiring the breaks to which human operators are legally entitled. They are most suitable for highly repetitive, unskilled tasks. However, effective though automated systems are, there are many jobs that machines cannot do.

Activity 10.13

Select a local engineering business. Research the ways in which automated systems could or do benefit the business.

Research and development into new materials is on-going in many industries.

Materials that have been developed in the past, and believed to be ideally suited to their purpose, such as lead for water pipes, have since been discovered to have unforseen disadvantages. This in turn has led to the development of materials that are more suitable for the task, and which do not have the associated hazards. With materials such as lead and asbestos the negative effects arose as a result of a lack of full understanding of their impact because the effect was cumulative, building up over time.

Engineering in society and the environment

Silicon chip
A microchip or integrated circuit

The introduction of new materials has led to changes in work processes. For example, people working with radioactive material have to wear protective clothing and work within procedures and systems introduced specifically relating to that material and the associated hazards. This might include wearing lead aprons, or working behind lead shields. The need to monitor exposure has led to the development of products such as the dosimeter and body monitor, both of which will register the amount of radiation to which the person has been exposed.

Manufacturing and production processes can lead to the pollution of air, water and land. All production processes produce some level of waste. It is important to understand the nature of the waste and how this waste should be disposed of. Different materials react in different ways, and this should influence how the material is used within manufacturing and as a waste.

A key material in modern industry is silicon. Modern computer technology is based on the silicon microchip. A significant amount of office and manufacturing equipment, and the products being manufactured and sold, rely on the inclusion of one or more microchips in order to operate. Silicon is used because it is a semiconductor, that is it will, under certain conditions, conduct a small amount of electric current. Silicon chips are also very cheap to produce.

People in all areas of industry and commerce will work with information technology, whether in terms of how they do their jobs or how they communicate

Figure 10.10 Using a computerised autorouter to establish connections on a printed circuit board

The application of engineering technology in society

with others. Most engineers will be expected to have some degree of computer literacy.

The design and manufacture of many consumer goods have been influenced by the use of information technology. This in turn affects the reliability and sophistication of product development and demand, all of which has a knock-on effect on industry, the products they make and how they are made.

Information processing

> **Information processing**
> *The storage, retrieval and use of information*

Information processing is a key area of any industrial or commercial organisation. Customer, financial, stock and supplier information are all important for the smooth and efficient running of the business operation.

A computer is now standard office equipment, even in small businesses. A computer can provide access to a range of information-processing services, not only the ability to word process documents, but also to access such communication facilities as the Internet and email. Other benefits have been to improve the efficiency of other services, for example telephone banking, computer conferencing and voice messaging.

Computers enable small businesses to deal with many areas of running a business that would previously have required a specialist, for example accounting and desktop publishing packages enable work to be done in house.

Computers also enable businesses of every size to establish efficient and easily retrievable systems for all aspects of business, such as employee records, financial accounts, stock control and reordering.

Another aid to information processing is the bar code. Bar codes are now commonly added to products to aid stock control and pricing. Each bar code represents a stock number that identifies the nature of the product and its price. When the bar code is passed over a scanner, it causes a beam of light from a laser to be sent out and reflected back to a sensor in the scanner. This converts the code into a digital signal that identifies the stock number.

Information technology makes it possible for industry to incorporate microelectronic systems such as sensors, control systems and robots in industrial equipment, so that they are interconnected. In this way, information can be collected, processed and communicated automatically.

Office work now tends to revolve around personal computers, which may be at the office, or based at the employee's home. Many workers also have portable computers, enabling greater flexibility in terms of the workplace.

> **PCB**
> *Printed circuit board – the board on which the microchips for each part of the computer are mounted*

With a greater range of computer programs available covering key office areas there has been a reduction in specialised areas, increased multitasking and demystification of many areas of business previously undertaken by specialists.

The computer industry is a large-scale employer. Careers in computing include the design and production of both software and hardware, and also support services including training and technical support.

Progress check

1. List three ways in which the application of engineering technology has reduced manufacturing costs
2. Name four benefits to industry of the application of engineering technology, other than reducing costs.
3. Describe two ways in which the manufacturing work environment has changed.

4. Name two modern materials. Briefly describe their impact on industry and commerce.
5. Describe two ways in which engineering technology has affected information processing in industry and commerce.

The application of engineering in health and medicine

Engineering technology has been applied to both health and medicine. The impact on health has been both direct and indirect. In an indirect sense engineering has helped to improve standards of health as a result of improvements in living conditions, for example with the introduction of indoor sanitation, hot and cold running water and heat. Improvements in health care and living standards have each contributed to an increased life expectancy, which now stands at over 77 for women and over 71 for men.

It can also be argued that engineering has had detrimental effects on health, for example the recent rise in incidence of asthma has been linked to pollution from cars and over 300 000 people are killed or injured every year in the UK as a result of car-related accidents.

However, the same engineering technology has helped serve the medical industry, with ambulances able to reach people and hospitals faster. Engineering developments in the medical industry have also led to being able to treat people at the scene of the accident and in the ambulance on the way to the hospital.

Engineering has also benefited the fields of health and medicine with the development of instruments that are capable of studying the human body and its behaviour under certain conditions. This has contributed to increased safety and improved treatment as a result of a greater understanding of the human body. A familiar example is the crash testing investigations carried out by all major car manufacturers, which have led to improved safety features including seat belts and air bags.

Information and data handling

Accurate and up-to-date information within the medical field is essential. Inaccurate information regarding a patient's medical history could lead to a misdiagnosis, with potentially fatal results. With medical records held on computer it is now quicker and easier to retrieve the appropriate details.

Case study

Touch-screen technology

It is increasingly possible for individuals to visit pharmacies and chemists and to investigate ailments, remedies and health care issues simply by touching the appropriate picture or words on a computer monitor. When the customer touches the screen, a combination of computer technology, communications technology and materials science work together to transmit the information via satellite to reach the computer monitor. The preprogrammed information is available almost instantly to view in full-colour graphics.

The application of engineering technology in society

Touch-screen technology

The use of computer and satellite technology to enable information to be obtained by touching a computer monitor

The combination of computers and modems enables access to an enormous range of medical information held on the Internet. With no need to wait for such information to be published, data is more accessible and immediate. This technology further facilitates ease of sharing of medical research data.

Use of the computer enables huge amounts of information to be stored and retrieved more easily. This can be a matter of life and death, for example when needing to access people with rare blood groups or who may be compatible for donation of organs or bone marrow.

In the field of medicine, precise diagnosis may be vital. Methods of providing precise information about an individual and their condition include X-rays, computerised axial tomography (CAT) scans, and magnetic resonance imaging (MRI).

X-ray

An image shown on a photographic plate, which is used to show the position of parts of the body. It is made using electromagnetic radiation

Computerised axial tomography is a technique of producing images of 'slices' of the body using a specialised X-ray machine. It is a medical procedure that produces a detailed picture of the soft tissues of the body using X-ray technology. An X-ray tube is rotated around the patient; this produces an X-ray that is intercepted by a bank of detectors, which then feed signals into a computer. A 'slice of the body' is then reconstructed on a television screen.

MRI involves the application of nuclear magnetic resonance techniques to obtain images of cross-sections through the human body. MRI produces scans containing different diagnostic information from CAT scans. The patient is placed in a strong magnetic field and is subjected to radio-frequency energy. When the radio waves are switched off molecules that have been energised give off signals that are specific to the type of atomic nucleus. These signals are then converted into an image by a computer. This development has benefited surgery by providing a far more precise view of the soft tissues.

CAT scan

A method of producing images of slices of the body to show detail of the soft tissues of the body

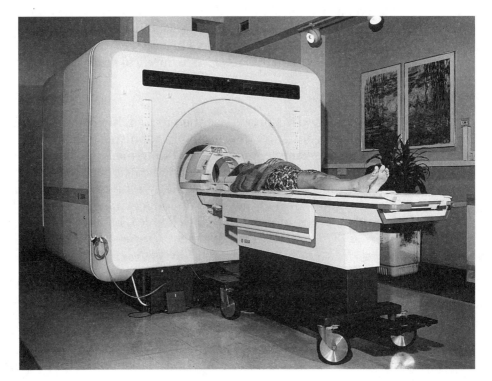

MRI

Magnetic resonance imaging – a method of producing pictures of cross-sections of the human body

Figure 10.11 Magnetic resonance imaging equipment

Engineering in society and the environment

Emergency services

Although the traditional idea of vehicles used by the emergency services is police cars, ambulances and fire engines, other vehicles have become more common as part of these services; for example the use of helicopters and paramedic motorcycles.

In the case of medical emergencies engineering technology has enabled an increasing amount of equipment to be carried within ambulances, enabling treatment of patients to begin immediately rather than waiting until they reach hospital.

In each of the above cases the essential feature is the speed with which the patient can be reached and their treatment can begin. This is not only down to the speed that the vehicle can travel, but also the use of telecommunications and computer technology to ensure that the vehicle travels the quickest route, avoiding any potential traffic delays.

The development of materials technology is important for emergency services such as the fire service. They need to be sure that the material on their clothes and helmets is able to withstand the heat of the fires that they are going to be tackling. For example, AB Chance(UK) Ltd has developed Nomex material, which is a synthetic fibre used in the manufacture of protective clothing for fire fighters providing a far greater protection from fire than was previously possible.

Surgery

Engineering and medical science have combined to make possible developments in prosthetics, replacement joints and replacement organs. Recent developments in prosthetics have included the use of materials and the development of electrically powered controlled movement. The use of electrodes on the prosthetic limb, placed against the skin, can register the tiny levels of voltages generated by the muscles. These signals are then translated, via an electronic system, into action.

Replacement heart valves have been developed through a combination of engineering design, materials science, medical science and surgery.

Intensive care

The human body is often described as a highly complex machine. Life support systems are constantly developing in sophistication and their capability to sustain life. For example, the heart and lung machine, which keeps the circulation and breathing of patients undergoing heart operations going, includes pumps, monitors and fail-safe systems, while a kidney dialysis machine is used by people when their own kidneys are no longer capable of removing waste products from their blood or regulating the chemical balance of their body fluids.

Health monitoring

Information technology has enabled the development of whole-body scanners that can detect tiny tumours. Early detection and diagnosis can help ensure that the right treatment is introduced, which in turn helps recovery.

Information technology has also enabled the development and use of 'ambulatory biomonitors', which are monitors that can be worn by patients while they move around, regularly sending body function details to the nurses' work station.

The application of engineering technology in society

Progress check

1. List three ways in which engineering has assisted medical information and data handling.
2. Describe two ways in which engineering has assisted the emergency services.
3. Outline two ways in which engineering has aided the understanding of the human body.
4. List three ways in which engineering has contributed to developments in surgery.
5. Describe one way in which engineering has improved monitoring of health.

The environmental impact of engineering technology and activities

Engineering technology is defined as the development and use of information technology, automated systems and the use of materials. Engineering activities are described in terms of tapping natural resources, production, servicing and the disposal of waste. In each case, it is possible to identify many positive and negative effects on the environment on both a general and specific level. The environment includes our physical surroundings, the impact on which can be identified in terms of resource use, pollution (of air, water and land), subsidence and spoil.

The impact of engineering technology and activities on the environment is of social, political and economic importance. It is an international issue as environmental impact is not necessarily limited to the country in which it originated and can have a worldwide effect, as with global warming.

The scale of the problem is linked to the size of the world population, the level of the consumption of resources such as energy and the production of waste.

Current concerns include the effects on the ozone layer, global warming, disappearing species of plants and wildlife, growing levels of waste, control of hazardous and toxic waste, the reduction in the quality of natural resources such as water and the depletion of natural resources such as coal and gas.

Awareness of environmental issues has grown among the general public, government and industry. Many industries now have a policy statement concerning their awareness of the environmental impact of their activities and their efforts to minimise harmful effects.

Case study

The Ford Motor Company recognises the implications of its industry on the environment in the following statement taken from their *Annual Report and Accounts (1994)*.

Low environmental impact is not just a function of vehicle operation and emissions; it includes the raw material the vehicle is made from, the processes involved in its manufacture and the way it can be recycled at the end of its useful life.

Ford has an environmental strategy that identifies objectives relating to materials, manufacturing processes, and waste. It plans to:

- '*minimise or eliminate materials of concern, such as heavy metals – chrome, nickel and lead*
- *develop new processes to reduce or eliminate environmental impact, for example the reduction of chlorofluorocarbons (CFCs) and organic solvents*

Engineering in society and the environment

> - *reduce the amounts of waste produced by manufacturing processes*
> - *reduce, re-use and recycle packaging, shipping materials and other industrial consumables*
> - *provide protection for wildlife habitats near Ford locations.'*
>
> As a result of this strategy Ford has developed chrome-free paint pretreatment systems, lead-free and water-based paints and nickel-free phosphates. All air-conditioning systems within Ford vehicles are CFC-free. CFCs have also been eliminated from other manufacturing processes. Substantial amounts of disposable packaging have been replaced with semidurable packaging. The development of new methods and processes has also led to significant reductions in chlorinated hydrocarbons (CHCs), which are often found in cleaning agents. For example, seat foaming moulds, previously cleaned with fluids containing CHCs, are now cleaned by a process that uses granulated walnut shells.

Activity 10.14

Choose a local engineering company. Find out what their environmental policy is.

Tapping natural resources

The term natural resources refers to those materials that occur in nature, for example stone, ores, minerals, wood and crops. The means of tapping these natural resources include mining and quarrying, drilling and pumping, forestry and agriculture.

Mining and quarrying

Natural resources include metal ores, such as iron ore, which is used to make iron and steel, rocks such as limestone, which are also used in the steel-making process, oil and coal.

Mining and quarrying for natural resources such as coal, iron ore and limestone involve the use of large expanses of land, and can leave pits, derelict or barren land, cutaway hillsides and slag and spoil heaps.

Figure 10.12 A quarry site

The application of engineering technology in society

Pollution of the land can be caused by the material that is left in the spoil and slag heaps, for example metal compounds left over from the extraction of heavy metals can seep into and contaminate the groundwater and soil.

Mining can also lead to problems of subsidence as a result of derelict mines being inadequately propped up and, if the slag or spoil is not correctly stored, slippage can occur, with serious consequences.

Oil and gas extraction

The extraction of oil and gas is currently necessary for the purpose of providing energy, such as used in the home or for motor vehicles and in industry.

One of the key issues with such natural resources is the fact that they are non-renewable, so once they have been used they are no longer available for further or future use. Although there is enough to meet our needs for many years, there is still the awareness that alternative sources of energy are needed for the future. Possible alternatives being investigated include wind, water and solar energy.

A second key issue is the creation of pollution, whether from drilling for gas and oil, from processing or from transportation. Environmental concern has arisen about the pollution caused by the decommissioning of oil and gas platforms. The example of the Brent Spar platform illustrated what a significant issue this was from an environmental and political viewpoint. Further concern relates to potential oil spills from tankers, as with the 1996 *Sea Empress* incident.

The third key area of concern relates to the pollution created from burning fossil fuels, such as coal and oil. This is necessary to create electricity, but it also leads to the production of carbon dioxide, a harmful gas that contributes to global warming.

Figure 10.13 Offshore rig

Forestry

Public concern about the effect of forestry on the environment has focused on the large-scale clearance of forest areas and the negative effects on the local environment. This is particularly emotive when considered in relation to the rain forests, and the devastation of local communities.

In many areas, trees are taken from managed forest areas, where stock is constantly replaced through the scheduled planting of young trees. A managed forest area will include trees of many different ages, which are spread throughout the area, so that trees can be cut down with no detriment to the local wildlife or wider environment.

Engineering in society and the environment

Forest areas are important to the environment because of their contribution to the recycling of carbon dioxide. Large-scale forestry clearance could lead to carbon dioxide build-up.

Large-scale forestry clearance can also result in areas of barren land and subsidence. The clearance of forest areas, particularly on hills and mountainsides, can lead to soil slippage or even avalanches, as well as land erosion, caused by the removal of the trees and other plant growth, leaving nothing to hold the soil.

Agriculture

Crop and animal farming have been significantly affected by engineering developments through the ages. Chemical engineering has influenced the development of fertilisers and pesticides used to improve crop growth and quality. The use of fertilisers and chemicals can also alter soil quality, which does not necessarily lead to long-term benefits.

Agriculture involves significant land use within the UK. The nature of the land use is changing, with agricultural emphasis being changed; for example, government subsidies have encouraged the prevalence of larger fields growing a limited number of crops.

Activity 10.15

Choose one example of tapping natural resources. Describe two potential effects of this on the environment.

Production, manufacture and processing

Production includes assembly of component parts, manufacture of parts and products from raw materials and the processing required to produce materials in the form appropriate for further production. Production processes are known to contribute to pollution levels, global warming, ozone layer depletion and acid rain.

As production processes become more automated so more energy is required to power the systems. The production of electricity through the burning of coal, gas and oil is one of the largest polluters of the atmosphere. A coal power station will emit carbon dioxide, particulates, sulphur dioxide and nitrogen oxide.

Many manufacturing businesses produce pollution from their factories. In the 1970s, in an attempt to reduce the pollution at ground level, taller chimney stacks were built. Unfortunately, this had the result of spreading the emissions over a larger area.

Global warming (the greenhouse effect) is understood to have been caused by the quantity of carbon dioxide, methane and CFCs produced by industrial production and other processes. Carbon dioxide is produced as a result of burning fossil fuels, such as coal, oil and gas; methane is produced from degradable waste; and CFCs result from a number of industrial processes and the use of aerosols. As a result of increased understanding of the environmental impact of CFCs, industry has reduced the amount of CFCs produced by developing and changing their production processes, and alternative sources of energy production, such as wind, water and solar energy, are under development and evaluation.

Ozone layer depletion is believed to be related to the emission of CFCs from the manufacture of polyurethane foam, refrigerants and aerosols. CFCs were originally used because they were thought to be much safer than other gases

The application of engineering technology in society

because they are non-flammable, and because they are effective as propellants for use in aerosols. As already mentioned, growing awareness of the impact of CFCs on the environment is resulting in alternative methods of manufacture and processing being developed.

Acid rain contains sulphur dioxide and nitrogen oxides, which are produced by industrial processes that involve burning coal and oil, such as at power stations, and from natural processes such as the eruption of volcanoes. Acid rain has a harmful effect on plant and animal life and soil and water quality. It can also have a destructive effect on the exterior of buildings. To combat this problem, alternative sources of power are being investigated, as are alternative processes.

Activity 10.16

Select one method of production. Investigate and describe the main areas in which this could have an environmental impact.

Waste

Toxic waste

Any substance that is poisonous if breathed in, eaten or absorbed by the skin

Waste is produced in three different states: solid, liquid and gas. Waste can also be categorised in terms of its potential harm. The correct method of treatment and disposal of waste depends on the nature of the waste and potential harm that can be caused by that waste.

Solid waste includes anything from packaging, left-over production material, broken tools or equipment and decommissioned machinery; liquid waste includes sewerage, oil and liquid chemicals; and gas waste includes CFCs from aerosols and methane from waste tips.

Toxic waste describes substances and objects that are poisonous if they are inhaled, digested or absorbed by the skin. Some will be more poisonous than others. There are different types of toxicity, for example ecotoxic waste is that which will have poisonous effects on the environment, including land, sea, air and biological systems.

Figure 10.14 Three forms of waste

Engineering in society and the environment

Special waste
All waste that has the potential to be so dangerous, difficult to treat, keep or dispose of, that special provision is required.

Non-toxic waste describes any waste that is not poisonous, although it can still be dangerous and so must be stored appropriately. Failure to do so can be fatal, for example in 1966 a slag heap at Aberfan, a Welsh mining village, slipped and buried part of the village, including the school. This caused the death of 144 people, of whom 116 were children.

Controlled waste describes all waste, other than special waste, that is subject to controls. The deposit, treatment, retention or disposal of controlled waste requires a licence.

Special waste includes all waste that has the potential to risk danger to human life, or has the potential to ignite or explode at a given temperature. It includes substances such as acids and alkalis and metals such as lead and mercury.

Radioactive waste is mainly produced by the nuclear power, defence and manufacturing industries and the health sector. The disposal of radioactive waste is governed by strict legal controls.

Degradable waste describes organic waste that will decay. Biodegradable means that the decay is caused by bacteria, whereas photodegradable means that the decay is caused by exposure to light. The speed at which the degradable waste will decay or decompose may depend on its treatment and storage. For example, waste that is deposited in a landfill site will decay more slowly after it has been covered with a layer of soil.

Non-degradable waste is inorganic waste, which means that it is produced from non-living things that will not degrade.

Organic waste
Waste produced from living things; organic waste will decay

Heat is a waste product that can be produced from many engineering activities. For example water is used by many manufacturing businesses as a coolant. As a result of the production process the water itself becomes heated. When this hot coolant water is then dumped into rivers or the sea it can raise the temperatures of the water near the outlet enough to destroy the natural aquatic animal and plant life.

Activity 10.17

Choose three products that you use on a regular basis. Identify the materials they are made from and their method of disposal.

Disposal of waste

Waste can be disposed of in a number of ways. It can be stored on a waste tip, slag or spoil heap, incinerated, buried underground in a landfill site, deposited in the sea or recycled. In each case there are implications for the environment.

The increasing automation of processes has helped to reduce waste as machines can be designed to be more precise, accurate and reliable. Such precision is further aided by the use of information technology in the programming of machines such as CNC lathes, cutters and grinders.

The collection, treatment and disposal of waste involves the consumption of resources, for example human energy, fuel for incinerators, land for spoil heaps or landfill sites, energy for recycling process and fuel required for collection vehicles.

Until the waste is safely disposed of, businesses are responsible for the safe storage of waste produced on their premises.

The application of engineering technology in society

Activity 10.18

Select one local small to medium-size business. Identify three types of waste produced as a result of their engineering activity. Describe the method of disposal of each type of waste.

By-products

A by-product is not the same as waste. Although it is produced as a result of the production of something else, it has a usefulness, and therefore value, of its own. For example, gas can be said to be a by-product of oil, as they co-exist underground.

Pollution

Pollution can affect the land, water and air. It can exist in solid, liquid or gas states. It may be seen, smelt or heard.

Land pollution

Pollution of the land can be caused by something as small as a sweet wrapper dropped on the ground, or as large as the slag heaps produced by the coal industry.

Industry must be very careful how it disposes of its waste, particularly if its production processes involve materials such as metals or chemicals, which could seep into and contaminate the ground.

The presence of metals in sewerage sludge that is going to be spread on agricultural land is subject to limits, and treatments are specified for particular metals. This is essential, as any contamination could be spread via the drinking water system. For example, the waste from electroplating plants may typically include residues of metals such as chrome and nickel, each of which are subject to special treatment in order to avoid harmful effects.

Landfill sites, such as the Welbeck site in Wakefield, West Yorkshire, have been designed to recover derelict land for local use, and to dispose of both industrial waste and domestic waste. The result is the use of waste as in-fill to build up the structure of the land, while landscape designers work with site engineers to create a safe and pleasant area of land appropriate for leisure activities. It is essential that safety criteria are adhered to as there is potential for future problems such as methane leakage, if not properly managed.

Air pollution

Air pollution is caused by smoke, gases and fumes. These can be produced by a variety of manufacturing and waste disposal processes, for example when waste on a waste tip or in a landfill site degrades it produces methane, which is a harmful gas and which creates air pollution and contributes to global warming. Damage to the ozone layer is connected to the use of aerosols and the production of carbon dioxide, which is a major pollutant caused by, among other things, burning of fossil fuels.

Water pollution

As water is essential to sustain life it is important to ensure that the quality is maintained. Therefore, effective water treatment and waste water treatment is an essential area of engineering.

Pollution of the sea and rivers ranges from the disposal of consumer products, such as plastic bottles and bags, to the dumping of chemical waste. In each case

Engineering in society and the environment

they cause danger to human and wildlife, as plastic containers can cause suffocation while chemical waste can lead to contamination of our drinking water.

The largest user of water is industry. For example the automotive industry uses approximately 50 000 litres of water in the production of one car. It is used for many purposes, including cutting, cooling, cleaning and transportation. Inevitably, during the course of its use it can become contaminated by the materials and substances with which it comes into contact, for example oil. Industry is obliged by law to ensure that the water has been appropriately treated before being returned to the water system.

Case study

Oil spills and recovery

The *Sea Empress* is a large oil tanker that ran aground off the Welsh coast in 1996, leading to the spillage of 70 000 tonnes of crude oil into the sea off the Pembrokeshire coast. The resulting oil slick and the amount of oil swept on to the beaches by the waves meant that wildlife was seriously affected. The impact on marine life and long-term effects are not yet known (Figure 10.15).

Special anti-pollution vessels have been developed in an attempt to control the effects of such accidents and minimise the impacts. Examples include units that are able to separate the mix of oil from the water before it is returned to the sea, oil-skimming vessels and dispersant application aircraft. Booms, pumps and skimmers are also essential pieces of equipment. Different designs deal with different conditions and situations, for example the belt skimmer works well with thick oil – it can recover about 90 tonnes of oil a day and can decant the water from the system back into the sea.

Progress check

1. Describe three categories of waste.
2. List three types of pollution.
3. Explain what is meant by global warming.
4. Outline the environmental impact of engineering and forestry.
5. Explain in what way heat can be described as waste.

Managing the impact of engineering activities on the environment

Managing the impact of engineering activities on the environment is not the responsibility of any one sector, nor is it limited to one geographic area. Action and cooperation on the part of industry, local authorities, government, environmental groups, the local community and individual householders are all necessary in order to ensure that the careful monitoring and control of the impact of engineering activities on the environment takes place. At its broadest level, international agreement and cooperation between governments is necessary.

Engineering activities are addressed by a number of legal statutes (acts of parliament), regulations (statutory instruments) and codes of practice that seek to manage and control the impact of engineering activities on the environment. Areas addressed include testing, classification, packaging, labelling, manufacturing processes, treatment, use and disposal of specified products and substances, including waste. As engineering technology progresses, and new processes and products are developed, so new methods of management and control are required. These laws may be initiated within an individual country, or may

The application of engineering technology in society

Figure 10.15 Engineers cleaning up after an oil spill

be developed in response to international agreements and European Community directives.

The government is further able to support legal methods of control by offering incentives, such as grants and tax incentives or, adversely, imposing fines or the cost of redressing the environmental impact on the business.

Another significant contribution to the management of the environmental impact of engineering activities is the provision of information to educate individuals and organisations about the impact of their actions.

Agencies that enforce legislation and provide information relating to environmental issues include local authorities and the Environment Agency. The Environment Agency, formed in 1996, is responsible for the environmental protection of air, land and water in England and Wales. Other parties involved in introducing and implementing controls include trade associations, the medical profession and interest groups.

Issues at the forefront of managing the impact of engineering activities on the environment include waste management, pollution control, energy conservation, accident prevention and the protection of endangered species.

It is important to realise that pollution is an unavoidable aspect of technological advances. The first caveman to light a fire could be said to be guilty of polluting the atmosphere, but we could not honestly say we would have wanted to stop his progress, and although much attention is currently being paid to the environmental impact of building roads, few people would argue for a return to the days before motor vehicles and roads were developed. This does not mean that unnecessary levels of pollution or other detrimental effects must be accepted. It is, therefore, important to establish and monitor acceptable and realistic levels at which minimal damage to the environment is caused by economic and technological progress.

Many industries recognise the benefits of considering the environmental impact of their activities, and of taking measures to minimise this impact; however,

Knock-on effects

The effects of one action that effect another action, that is the wider implications of any action

Engineering in society and the environment

it is also necessary for them to consider the potential impact on operating costs, and the knock-on effects on competitiveness and profits.

Controls relate to the testing, classification, packaging, labelling, manufacturing processes, treatment, use and disposal of specified products and (waste) substances.

Statutes, regulations and supporting codes of practice outline specific responsibilities on the part of business and industry.

Activity 10.19

Select one local engineering business. Identify the main areas in which their activities might be said to have potential environmental impact. Describe the means by which this impact is managed.

Disposal of waste

The disposal of waste is addressed in statutes, regulations and codes of practice. The legislation identifies different categories of waste and identifies the required level of treatment and disposal for each category. Failure to adhere to this could lead to fines. Those wishing to deposit, treat, keep or dispose of controlled waste must have a licence.

One highly controlled type of waste is that of nuclear waste, which can be divided into low-, intermediate- and high-level waste. How it is categorised will affect how it is treated and disposed of. Low and intermediate level wastes can be incinerated, or stored in shallow land, landfill and coastal site repositories. Radioactive waste used to be dumped in the sea. However, this has not been allowed since 1983.

Pollution offences can be punished by fines, imprisonment, or both. Fines may be be set at up to £20 000. Additional costs may also be charged, for example the costs of cleaning up after a pollution incident.

Discharges into rivers and the sea must have a licence, which will only be granted where no harm to human, river or marine life is foreseen. It is illegal to put untreated chemicals, oil or sewage into the water system.

Case study

Monitoring waste disposal and emissions

The Esso refinery at Fawley in Hampshire uses water as a coolant in its industrial processes. As it can become contaminated with oil during the cooling process it has to be cleaned in accordance with legal and internal regulations before it can be returned to the Solent. Esso environmental engineers are responsible for monitoring and advising on waste disposal and emissions from the refinery to the atmosphere and water. Specific areas of responsibility include reporting to the Environment Agency, keeping up to date with legislation, providing technical environmental advice and ensuring the proper treatment and disposal of effluent water and waste. Refinery processes are subject to authorisations issued by the Environment Agency and the environmental engineers have the prime responsibility for ensuring that the operations comply with the requirements. Effluent water is subject to stringent discharge standards set by previous legislation and tightened under the current authorisation.

The application of engineering technology in society

It is possible for manufacturers to begin their management of the impact of their waste at product design stage. At this point choices are made concerning the materials that are going to be used, the manufacturing processes that are going to be employed and the packaging for the product. Therefore, at this stage it is possible for choices to be made that will minimise the environmental impact, for example by identifying potential for recycling, or choosing a material that can be disposed of more efficiently than another.

On a small, but visible, level a common form of inefficient waste disposal is the dropping of litter on our streets. The same level of careless waste disposal by industry would cause a much larger scale problem, which would not only be unsightly but also a health hazard. In the same way that individuals can be fined for dropping litter, so businesses can be fined for careless or inappropriate disposal of waste.

One of the key problems with the disposal of waste is the sheer volume that is being produced. It is ,therefore, not only important to consider the means of disposal of waste, but also to address ways of reducing the amount of waste being produced.

Another form of waste disposal that is subject to a significant level of legal control is the decommissioning of such structures as oil rigs and production plants, which have to be decommissioned once their useful life is over. Decommissioning essentially describes the process of removal and disposal of the plant in question. In 1995 the decommissioning of the oil rig *Brent Spar* received substantial media and political attention owing to the public's concern about the potential environmental impact of the available options.

Control and monitoring of emissions

The term emissions usually refers to gases such as carbon dioxide, sulphur dioxide and nitrogen oxides, smoke, chemicals and particulates. In each case, emission levels are regulated by law. This is an area of significant interest to the EC, which has set standards and produced directives on areas such as CFCs, emission standards, etc.

Figure 10.16 An environmental engineer

Engineering in society and the environment

Businesses that produce noxious or offensive fumes are legally obliged to prevent or limit such fumes from entering the atmosphere. If this is not possible they must ensure that any fumes that are emitted must be harmless.

Different approaches can be taken. For example, in a number of countries there are tax incentives to be gained from installing catalytic converters on vehicles. In this country users of commercial vehicles that exceed emission levels can be prosecuted.

Conservation of resources

The conservation of resources is managed through a system of licences, inspections and penalties such as fines. For example, those wishing to take water for industrial purposes must have a licence. Businesses are required to be open to inspection of their premises or operations in order to ensure adherence to regulations and restrictions.

Health and safety

Health and safety issues relate to the health and safety of employees, the local community, the world at large and future generations.

Industry must protect the health and safety of employees, ensuring that they are not exposed to any unnecessary hazards. Inevitably, in some jobs, hazards cannot be avoided. In such cases, the employer is required by law to ensure that adequate protection is provided in terms of protective clothing, and protective procedures. Employees are protected by the legal system through such laws as the Health and Safety at Work Act 1974 and regulations such as the Personal Protective Equipment at Work Regulations 1992 and the Health and Safety at Work Regulations 1992.

Industry is also required by law to manage its operations in such a way that neither the health nor safety of the local community is adversely affected in any way. This covers issues such as safe disposal of waste, safe levels of emissions and noise control.

In the past, damage to health and safety has been caused by a lack of understanding of the long-term effects of contact with certain materials. For example, asbestos and lead are both materials that have caused health problems as a result of cumulative contact. Lead pipes were used by the water industry to carry clean water to households, asbestos was used in the building trade. Now that their effects are understood these materials have been replaced by alternatives. This illustrates the need to be aware of the potential for harm and the need to be prepared to act accordingly. As a consequence of the discovery of the impact on health it is now illegal to use these materials in the above-mentioned ways.

Activity 10.20

Identify the potential health and safety risks in one engineering-based working environment. Recommend precautions against those risks.

General coverage concerning the health and safety requirements of specific substances is provided by the Control of Substances Hazardous to Health Regulations 1988 (COSHHR), which were introduced following an EC Directive. There are also a number of approved codes of practice that relate to specific substances such as for the Control of Substances Hazardous to Health and the Control of Carcinogenic Substances. These codes of practice provide guidance for the safe control of hazardous substances generated by industrial processes and over 40 000 chemicals and materials.

Duty of care

Anyone, whether an individual or organisation, who is involved in the production, import, storage, treatment, processing, transportation, recycling or disposal of controlled waste has a duty of care in relation to that waste and its disposal. This includes a duty to store the waste safely and securely, and to make sure that no waste can escape while being stored. If they are transporting the waste in order to pass it on to anyone else they have a duty to secure it in an appropriate manner, for example in an appropriate container, or in a covered skip or vehicle. They have a duty to ensure that the person taking the waste has the legal authority to do so, and they are obliged to provide a written description of the waste and a transfer note.

People with authority to take waste include council waste collectors, registered waste carriers, exempt waste carriers such as charities and voluntary organisations, waste disposal or waste management licensees, and (in Scotland only) council waste disposers.

Environment Protection Act (EPA) 1990

The regulation and control of pollution and waste in relation to land, air and water is covered by this act. It outlines responsibilities for the condition of industrial premises, smoke, fumes and gas emissions, smoke and steam, also effluent from industrial or business premises that could be unhealthy, and potentially hazardous substances.

The Environmental Protection (Prescribed Processes and Substances) Regulations 1991

These regulations apply to those industrial processes considered to be most potentially polluting, and substances and materials considered to be particularly harmful or potentially polluting to the environment. The processes specified are categorised as fuel production processes, combustion processes, metal production and processing, mineral industries, chemical industries, waste disposal and recycling.

Clean Air Act 1993

This act relates to air pollution, including the emission of smoke fumes and gases, for example this act prohibits the emission of dark smoke from industrial and commercial premises and chimneys, and dust and grit from furnaces. It also regulates the height of chimneys. The type of smoke emitted from industrial premises is assessed by environmental health officers, who must consider whether all reasonable steps have been taken to prevent or minimise the smoke pollution. Reasonable steps are considered in terms of the design, installation, maintenance, and operation of plant and machinery, and in terms of the current level of technical knowledge and the financial implications. The Clean Air Act also addresses the composition and contents of motor fuel in relation to the pollution caused by motor vehicles.

Engineering in society and the environment

Progress check

1. Name three acts of parliament that seek to control and protect the environment
2. Explain the difference between monitoring and control
3. List three examples of penalties for pollution offences
4. Outline the main areas covered by the Environment Protection Act 1990
5. Provide two examples of industrial emissions that are regulated by law.

Assignment 10
Engineering technology in society

This assignment provides evidence for:
Element 4.1: The application of engineering technology in society
and the following key skills:
Communication 2.2: Produce written material
Communication 2.3: Use images
Communication 2.4: Read and respond to written materials
Information technology 2.1: Prepare information
Information technology 2.2: Process information
Information technology 2.3: Present information
Application of number 2.1: Collect and record data
Application of number 2.3: Interpret and present data

Your tasks

Task 1
Select one example of engineering technology.
Using examples, prepare a report in which you describe how this is applied in different contexts, such as home, leisure, industry and medical contexts.

Task 2
Identify the environmental impact of the applications described.
Support your work with examples of visual and numerical evidence.
Your report should include a bibliography in which you list all sources of information referred to as a basis of your work. You should also describe any software used in the preparation and presentation of your work.

Task 3
Describe methods by which the environmental impact you have described in task 2 can be managed.

Chapter 11: Careers in engineering

> **This chapter covers:**
> Element 4.2: Explore career options and pathways in engineering
> **... and is divided into the following sections:**
> - Employment in the main engineering sectors
> - Career options in engineering
> - Sources of information
> - Qualifications for a career in engineering
> - Presenting personal information
> - Investigate employment options in engineering.

Employment opportunities in engineering are many and varied. Entry into engineering work is possible at many levels.

This chapter aims to describe the main sectors within the engineering field and the different levels of job classifications within engineering and to outline possible career paths to be followed within this field. Although this cannot be an exhaustive list, it aims to provide some idea of the range of opportunities within an engineering career.

The second part of this chapter describes some of the sources of information for career opportunities and the various methods of presenting the personal information required by potential employers.

Employment in the main engineering sectors

Engineering can be subdivided into a number of categories, including such key areas as mechanical engineering, electrical and electronic engineering, chemical engineering and civil engineering.

Mechanical engineering

Mechanical engineering
Mechanical engineering focuses on working with machinery and mechanical products and systems

In broad terms, mechanical engineering can be described as being about working on mechanical products and equipment, such as vehicles, production machinery, medical equipment and other machinery.

Specific areas of work include the design, development, building, installation, operation, maintenance, testing and modification of machinery. The skills necessary to do the work include welding, turning, milling and design skills. Areas of required knowledge include mechanics, hydraulics and pneumatics, and thermodynamics.

Major employers of mechanical engineers include the manufacturing, aeronautical, power and medical industries. Many other sectors employ mechanical engineers, for example extraction, construction, transport and petrochemical industries.

Engineering in society and the environment

Mechanical engineers are represented by the Institution of Mechanical Engineers (IMechE). Further information about a career as a mechanical engineer can be obtained from its Schools Liaison Service (see page 456 for address).

Activity 11.1

Contact a local company. Find out whether it employs mechanical engineers and for what purpose.

Electrical and electronic engineering

Electrical engineering
Relates to power generation and supply

Electronic engineering
Relates to electronic circuitry necessary for telecommunications and computers

The work of the electrical engineer relates to power generation and supply, whereas an electronic engineer will focus more on the electronic circuitry related to telecommunications, computers and computer control.

Key employment areas include telecommunications, power engineering and the computer industry.

Electrical and electronic engineering support every sector, through the design and manufacture of everything from the circuits needed to provide electricity to buildings to computer-controlled temperature and lighting sensors for agriculture and the subassemblies required by the motor vehicle industry for modern cars.

Electrical and electronic engineers are represented by the Institution of Electrical Engineers (IEE), which will provide further information about careers (see page 456 for address).

Activity 11.2

Contact a local electrical engineering company. Interview an electrical engineer to find out about the key aspects of his or her work.

Chemical engineering

Chemical engineering
Relates to the chemical and physical nature of raw materials

The work of chemical engineers is concerned with the chemical and physical nature of raw materials such as food, oil, minerals and metal ores. Their work may involve the investigation of the behaviour of materials under specific conditions such as heat and stress, and within particular processes. Their work will support many other industries, such as agriculture in the development of pesticides and fertilisers, and the food industry in the development of processes for preparation and storage of food.

Key employers include the petrochemical, food and pharmaceutical industries.

Chemical engineers are represented by the Institution of Chemical Engineers, which can be contacted for further information about careers in this field (see page 456 for address).

Activity 11.3

Contact a petrochemical, food or pharmaceutical company. Find out about the nature of the work of a chemical engineer working within this company.

Civil engineering

Civil engineering
Relates to the design, building and maintenance of buildings and structures

The work of the civil engineer focuses on the design, building and maintenance of buildings and structures. Key employment areas include housing (houses, sewerage systems) and transport (roads, bridges, tunnels, airports and runways).

The specific industries within which an engineer may work are many and varied. It is important to realise that the type of engineer you are (mechanical, electrical, chemical or civil) does not necessarily determine the nature of the industry that you will work for. For example, a chemical engineer is as likely to work in the food or manufacturing industry as in the chemical industry. Specific fields of engineering include aeronautical, chemical, communications, health and biomedical, material, manufacturing, metallurgical, motor vehicle and power engineering. Outlined below is a brief overview of key industries, with just a few broad ideas as to how different types of engineers might be involved in each area. As will be evident from the following, in many cases one company may be defined as being part of more than one industry.

Aeronautical industry

This sector is concerned with the design and manufacture of commercial and defence aircraft and the production and provision of related facilities.

Electronic engineers are involved in the development of the computer-aided design technology used in building and testing engines, and for the instrumentation and automatic landing systems employed within the industry. Mechanical engineering contributes the design, testing and manufacture of the aircraft and engines, civil engineering is concerned with the design and construction of airports, runways and hangars, whereas chemical engineering is involved in the research and development of such areas as materials and fuels.

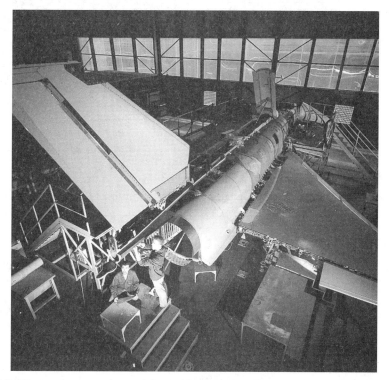

Figure 11.1 Design, construction and testing of aircraft at British Aerospace

Engineering in society and the environment

Activity 11.4

Research one organisation involved in the aeronautical industry. Research and describe the contribution of the main areas of engineering to this organisation's operations.

Chemical industry

The chemical industry is concerned with the design, development, manufacture, marketing and sale of chemical-based products, such as petrochemicals, plastics and paints.

Mechanical and chemical engineers are involved in the building, running and maintenance of plant, such as oil rigs and refineries. Electrical and electronic engineers are involved in the development and maintenance of communications and safety systems.

Figure 11.2 Engineers at an Esso refinery

Activity 11.5

Identify two national businesses involved in different fields of the chemical industry. Briefly describe the engineering contribution to each.

Communications industry

The communications industry includes personal communication systems, such as telephones and faxes, and one-way communication systems such as satellite TV and video. The development of the computer has constituted a major development in this field.

While electrical and electronic engineers are involved in the design of the systems and circuitry necessary for the product to work, the mechanical engineers are involved in areas such as the design and manufacture of the product in which the circuitry will be placed, for example telephones and televisions.

Careers in engineering

Activity 11.6

Research the range of communications-related engineering businesses in your local area. (Local telephone directories and a visit to the local library might help.)
Choose two different businesses and briefly describe the nature of their product or service.

Health industry

The health industry includes a wide range of services, from hospital treatment to health clubs. Mechanical engineering is involved in the development and production of replacement joints, such as hips, and prosthetic limbs, as well as wheelchairs and home aids such as chair lifts and adapters for controls on household items to make them easier for people with, for example, arthritis or in wheelchairs to use. Electronic engineering is involved in the development of health monitoring and computer analysis systems. Chemical engineering is very much involved in the pharmaceutical industry.

Figure 11.3 Special adaptors on gas cookers make it easier for the controls to be used

421

Engineering in society and the environment

Activity 11.7

Identify three different areas of the health industry, and areas of engineering related to each.

Manufacturing industry

The manufacturing industry covers any industry concerned with making products for sale. The products made may be designed for sale to the consumer or to businesses, possibly to aid the production of other products or the provision of services. The manufacturing industry, in particular, overlaps with many other industry classifications.

Activity 11.8

Identify three manufacturing firms in your local area: one small, one medium size and one large. Find out what product (range) each one produces and their main methods of production.

Metal products industry

The metal products industry involves any business involved in the production of the raw material, the processing of the material or the manufacture and sale of final products. Mechanical engineers will be concerned with areas such as the production and maintenance of production equipment and the design and production of products, chemical engineers will be concerned with the behaviour of the metal under certain conditions, whereas electrical and electronic engineers will be concerned with the electrical or electronic circuits necessary for the products to operate as required.

Activity 11.9

List three national or local businesses that operate within the metal products industry.

Motor vehicle industry

The motor vehicle industry is concerned with the design, manufacture and maintenance of motor vehicles. This includes cars, vans, lorries, buses and industrial vehicles such as fork-lift trucks. Mechanical engineering will be concerned with the design, testing and production of new engines, cars and other vehicles. Electrical and electronic engineering is concerned with the subassemblies, including the lights and air bags, whereas chemical engineering will be concerned with the development of new paint finishes, fuels and materials.

Activity 11.10

List three local engineering firms involved in the motor vehicle industry.

Careers in engineering

Figure 11.4 Fiesta production at Ford Motor Company

Power industry

The power, or energy, industry is concerned with the generation, distribution and use of energy. Mechanical engineers are involved in the design, production and maintenance of generation and distribution structures, chemical engineers are involved in the development of alternative sources of energy and investigations into the efficiency and environmental impact of different sources, whereas electrical and electronic engineers are concerned with designing, implementing and maintaining circuitry that will ensure the efficient transmission and use of electricity.

Activity 11.11

List three local engineering businesses linked to the power industry.

Progress check

1. What is the main focus of the work of an electrical engineer?
2. List five examples of skills and knowledge required by a mechanical engineer.
3. Describe the type of work with which a chemical engineer might be concerned.
4. Choose one industry. Briefly outline the nature of the industry and the type of work with which an electrical or electronic engineer might be involved.
5. Choose one type of engineer. Briefly outline the nature of their work within the motor vehicle industry.

Career options in engineering

The Engineering Council
The body that represents the engineering profession, as a whole, in the UK

The engineering profession offers an enormous range of variety in terms of potential specialist areas, levels of specialism, fields within which to work and specific job description. An engineer's career may change course numerous times as it progresses. A number of options exist in relation to the level at which an individual may enter the field of engineering.

A career in engineering is likely to involve working in teams. A key aspect of any engineering work is problem solving – making something work or improving the way it works through the application of engineering skills and understanding. A trained engineer may work on the practical aspects of engineering or may move into another sphere such as marketing, sales or finance. Other than these aspects of the work there is little else that could be described as typical of engineering jobs.

NVQ
A national vocational qualification that acknowledges the skills and knowledge acquired as a result of performing your job

An engineer may operate at any level of management within a business, for example as supervisor, project leader, project manager, department manager, director, partner or managing director

The Engineering Council has identified the following job classifications, which provide some indication of the anticipated level of input at the different levels of engineering.

Operators

SVQ
Scottish Vocational Qualification – the Scottish equivalent of the NVQ

Operators usually work on relatively simple, repetitive tasks in assembly work, machine operation, quality control or packaging, or they might possibly work on a more complicated task requiring a particular skill. They will be responsible for working within the appropriate health and safety regulations and for their own time-keeping. They will usually be responsible to a supervisor.

It is possible to begin as an operator without any qualifications. As an operator you can work towards NVQ/SVQ levels 1 and 2.

Craftsperson

A craftsperson will usually be involved with making products, working from engineering designs. Craftspersons will need to be able to set up and use the necessary equipment and produce the work accurately and to schedule. Craft skills include tool-making, maintenance work, welding, sheet-metal work and fitting.

As a craftsperson you can work towards an NVQ/SVQ level 3.

Engineering technicians

Engineering technicians are qualified to work as designers, estimators, quality assurance technicians, inspectors, planners or laboratory technicians. This can be at either operator or supervisory level.

To be eligible for registration as an engineering technician you must be at least 21 years old and have at least 2 years' experience in a responsible position, hold a BTEC National Diploma/Certificate/SCOTVEC or equivalent and have a minimum of 2 years' approved training. This means you can use the letters EngTech after your name.

Engineering technicians can work towards NVQ/SVQ level 3.

Incorporated engineers

Incorporated engineers are likely to be involved in detailed technical work, needing a detailed knowledge and understanding of current technology. They will be involved in the practical application of appropriate technology, specialist skills and knowledge. They may be independent or supervised practical engineers, and will often be responsible for projects.

In order to be eligible to be registered as an incorporated engineer you must be at least 23 years of age, hold either a BTEC Higher National Diploma or Certificate, a SCOTVEC Higher National Certificate or an engineering degree, and at least 4 years' approved training and 3 years' experience in a responsible position.

When qualified as an incorporated engineer you are able to add the letters IEng after your name.

While working as an incorporated engineer, or technician engineer it is possible to work towards NVQ/SVQ level 4.

Chartered engineers

An engineer at project team leader or management level is most likely to be a chartered engineer. Chartered engineers are expected to be receptive to new ideas and be able to develop and apply their own ideas, as well as new technologies, engineering concepts and management methods.

When qualified as a chartered engineer you can put the letters CEng after your name.

In order to be able to become a chartered engineer you must have an appropriate honours degree and a combination of at least 2 years' training or experience plus 2 years working at a professional level. Alternatively, you will have had to complete the Monitored Professional Development Scheme, which is a 4-year scheme consisting of 2 years' training followed by 2 years' approved career development. (Reproduced courtesy of the Engineering Council from *Engineering Job Classifications – Roles and Responsibilities*.)

In addition to defining the three levels of engineering qualification standards outlined above, the Engineering Council has also developed guidelines on the recommended roles and responsibilities at each level. These are reproduced in the Appendix.

Key areas for an engineer to become involved in include research and development, design, quality assurance, production and manufacturing.

Research and development

Modifications
Changes to an original concept or design

A patent
A legal recognition of an individual's or company's ownership of an idea or invention

Research and development is concerned with the development of new products and the refinement of existing products. In engineering businesses, research and development tend to involve engineers working together from product development, research and development and design teams. Specific aspects on which they might work include materials testing, design, modelling and building of prototypes, and the development of modifications of existing products and systems.

Within research and development, engineers will need to be aware of changes in the market, developments by the competition, scientific and technological discoveries, patents and on-going research projects.

Engineering in society and the environment

Design

Ergonomics
Describes the study of the relationship between people and machines

Engineering design is concerned with the interpretation of ideas into a product concept with workable specifications and a realisable form. Design has to take into account both function and aesthetic properties. Ergonomic issues in relation to how the product will be used and under what conditions will also need to be considered. Design work will often involve computer-aided design systems, although the traditional drawing boards also have a place. In addition to considering what the customer wants, the designer must also consider what the business is able to provide economically.

Figure 11.5 *Use of computer-aided design at British Aerospace*

Activity 11.12

Find out about the area of engineering design and the use of computer-aided design in this field.

Quality assurance

Quality assurance and quality control are concerned with ensuring that the quality standards of the end product meet those of the original specifications. Engineers working in this field will be concerned with establishing quality standards and

Careers in engineering

control systems. They will work closely with the production department to ensure that realistic quality standards are established and complied with.

Production and manufacturing

Production
Bringing together resources to make something more valuable than its component parts

Engineers working in production and manufacturing will be involved in the physical output of products. The term 'production' can be used to describe the primary function of any business, from the oil company extracting and processing oil, to the motor vehicle company assembling cars.

An engineer working in production may be involved in:
- producing one-off, made-to-order products
- working on batch production making relatively small numbers of products to the same specification
- mass-production systems involving the production of large numbers of products to the same product specification.

Mass production
The continuous production of large numbers of products to the same basic specification

In addition to the above fields, many engineers move into other fields within engineering industries, such as marketing, personnel and finance. In all fields, their understanding of the customer, product and systems can provide invaluable insights into the needs of the overall business. In addition to this, the skills of teamwork and project management can contribute to any business situation.

Progress check

1. List five job classifications as defined by the Engineering Council.
2. Explain the difference between the role of an operator and a craftsperson.
3. What qualifications and skills are required to become an engineering technician?
4. Explain the difference between an incorporated engineer and a chartered engineer.
5. List the NVQ/SVQ possibilities appropriate to each level of job classification.

Sources of information

There are a number of sources of information that can help you find out more about individual types of jobs, to identify the type of job you want and to help you find the organisations that might be able to offer you the right type of job.

Careers advisers

Careers advisers should be able to provide you with information about the wide range of engineering opportunities, particularly those in the local area. They will also be able to discuss the skills, knowledge and qualifications required for a given job. They will be able to suggest possible routes that you can follow in order to pursue the career you want and to enter employment at the level you prefer. They may also be able to advise you as to whether a job is suited to your abilities, helping you to assess your strengths and weaknesses in relation to career choices. Your careers adviser should be able to provide information about training opportunities in your local area, modern apprenticeship schemes and potential for work experience.

The Engineering Training Authority (EnTra) runs an Engineering Careers Information Service. It offers information about careers and training opportunities (see page 456 for address).

Engineering in society and the environment

Careers advisers

Careers advisers are trained to advise about the qualities needed for a particular job, and career routes and opportunities

Figure 11.6 *Careers advisers are able to tailor their advice to your needs*

Activity 11.13

Contact your local careers office or adviser. Find out what information they are able to provide about engineering careers.

Employment agencies

Employment agencies work as brokers between local businesses and people seeking employment, with most agencies specialising, to a certain extent, in a particular field. In order to get a job through such an agency, you will be required to register with it, providing details of your skills and abilities, the type of work you are looking for and any other requirements, such as the minimum payment you are seeking. The agency will interview you and, if you are considered to be suitable, will keep your details on its books. Each time a business approaches the agency concerning a vacancy, the agency will go through its files to find one or more people who may be suitable.

At the same time, the agency will be advertising the type of jobs and range of skills that it is able to offer to local businesses, through direct approaches, adver-

Figure 11.7 *An employment agency will promote job skills*

Careers in engineering

tising in local papers and advertising in its shop front. Although the posts are often short-term, temporary posts, it is always possible that the situation can become permanent, particularly if you are able to impress the employers with your job performance.

Activity 11.14

Look in your local telephone directories to find out how many employment agencies there are in your local area. Choose two and find out the type of jobs they specialise in.

Job centres

Job centres carry information about local job opportunities that have been registered by local businesses. Different firms will have different arrangements, but in most cases if you are able to show the job centre that you have the qualifications and skills requested you will be invited to an interview. The interview will give you the opportunity to find out more about the job and the prospective employer, and allow the firm to decide whether you are the right person for the job. There is no cost to the employer to place an advertisement at a job centre, which means that a wide range of jobs can be found there, although these may often be low-paid positions.

Activity 11.15

Visit your local job centre. Look at the range of jobs available. List the main categories under which the jobs are grouped.

Advertisements

Advertisement

A means of bringing a product or service to the attention of the general public using such media as television, radio, cinema or the press

Advertisements might be placed for job vacancies, open days or training schemes, or apprenticeship opportunities such as modern apprenticeships.

Figure 11.8 A job centre will hold information about local job vacancies

Businesses will advertise in a variety of media. Most common are newspapers, trade press and specialist magazines. Other media such as radio may also be used.

It is always useful to look at job advertisements in local papers as they provide a good idea of the type of vacancies being advertised in the area. They also provide an idea of the type of companies in the area, and their addresses, which can be used as the basis of research into a company.

Even if the job being advertised is not the one you want, you may like to write to the company on a speculative basis.

INSTALLATION/SERVICE ENGINEERS

Security Systems Plc operates in the electronic security industry. It is an expanding company. As a result of continued and steady growth we now have vacancies for experienced Installation/Service Engineers.

You must be fully conversant with BS4737 and NACOSS codes of practice. You must also be of smart appearance, hold a full driving licence and be able to work with the minimum amount of supervision.

In the first instance please forward your CV to:
D P Rodes, Personnel, Security Systems Plc, Fareham Industrial Park, Hampshire.

SUPERVISOR/ASSEMBLY ENGINEER

We require an experienced engineer to supervise our mechanical assembly unit. Knowledge and experience of stainless steel, mechanical/ electrical assembly and air handling systems is essential.

For further details and an application form contact:
J Jardin,
J J Engineering, Greenhays Park, Manchester.

Figure 11.9 Examples of engineering job vacancies

Activity 11.16

Collect local newspapers (including free papers) over a period of 2 weeks. Identify the engineering jobs advertised during this period.

Local directories

Directories such as *Yellow Pages* and *Thompsons* provide information on the names and addresses of companies that trade within specified business sectors. This information can be used as the basis of research into the type and size of a company before deciding which businesses to approach on a speculative basis. Although you may prefer to wait until they advertise vacancies, it can be a good idea to write on spec. While you may not get a job straight away, your letter and details will usually be kept on file, and this may stand you in good stead when a job is advertised. Make sure that your letter is addressed to the right person, and try to provide some indication that you know something about the company, and why you have applied to it.

Careers in engineering

Figure 11.10 Local directories can provide useful information

Activity 11.17

Look at your local directories. List the main directory sections that you think could provide you with information useful to your search for a career.

Engineers

People who are working as engineers or who are training to be engineers are, of course, a useful source of information. Organisations such as the Neighbourhood Engineers, run by the Engineering Council, can be of help in putting you in contact with engineers if you do not know anyone personally. They will be able to provide you with the practical, day-to-day picture of the work and the benefits of following a career in engineering.

Progress check

1. Explain how job centres and employment agencies can help you to find a job.
2. Describe how local directories could help you to find a job.
3. List the ways in which a careers adviser can help you.
4. Identify five potential sources of career information.
5. In what way can an engineer help you with career information?

Figure 11.11 Engineers can be a useful source of practical information

Engineering in society and the environment

Qualifications for a career in engineering

Different levels of qualifications are appropriate for entry at different levels of engineering and further qualifications can be gained by engineers while working to help career progression.

To become an engineer at craftsperson level you need to have at least three GCSEs, preferably in mathematics, English and science. Engineers working at this level can work towards NVQs as recognition of the skills and knowledge being learnt as part of their work. Alternatively, they can work towards part-time GNVQs.

An engineer wishing to achieve engineering technician level will require either a BTEC National Diploma or certificate, SCOTVEC National Certificates or City and Guilds Technician schemes.

An engineer wishing to achieve incorporated engineer status will require either BTEC Higher National Diploma or Certificate, SCOTVEC Higher National Certificate or Diploma, or City and Guilds Full Technological Certificate.

A chartered engineer must have an accredited honours degree in engineering.

Qualifications can be defined broadly as vocational, academic and professional.

Vocational qualifications

A vocational qualification

Relates directly to an occupation or type of employment; it will usually involve an element of practical application

Vocational qualifications include such schemes as GNVQs, NVQs and modern apprenticeships. Each of these is briefly described below.

GNVQs

If there is a need to provide information about GNVQs at this stage then perhaps you are on the wrong course!! However, a summary is provided below.

GNVQ courses are mainly available to students of age 16 and above. Each GNVQ, at whatever level and in whatever subject, includes a combination of mandatory and optional units. They also include key skills, including communication, application of number and information technology. All courses should involve some degree of links with industry.

The GNVQ at Foundation level is considered to be equal in status to four GCSEs at D–G level. This is usually studied as a 1-year full-time course.

The GNVQ Intermediate level is considered to be equal to four GCSEs at A–C level. This is usually studied as a 1-year full-time course.

The GNVQ Advanced level is considered to be equal to two GCE A levels at A–E level. This is usually studied as a 2-year full-time course.

It is possible to gain accreditation for individual units that have been completed satisfactorily even if the course as a whole is not completed.

GNVQs are suitable for vocational or academic students. They each contain a combination of studying and work experience. Although they are vocational in nature, this is on a broad basis and so are not specific to any single job, but rather a vocational field. Other than Engineering, subjects include Manufacturing, Science, Business, Construction and the Built Environment, Leisure and Tourism, Art and Design, and Health and Social Care.

The lead body for GNVQs is the National Council for Vocational Qualifications (NCVQ). The optional units are offered by the three awarding bodies: Business and Technology Education Council (BTEC), City and Guilds of London Institute (C & G) and the RSA Examination Board (RSA).

National Vocational Qualifications (NVQs)

NVQs (England and Wales) and SVQs (Scotland) are work-based qualifications that reflect your ability to do your job, recognising those skills and knowledge that relate to your job. You do not need to attend a course in order to be awarded NVQ levels, as assessment is based on whether you can show an assessor that you are able to do a particular job. This assessment can take place at your place of work. NVQs cover a wide range of occupational areas.

NVQs are awarded at five levels:
- operators can work towards levels 1 and 2
- craftspersons can work towards level 3
- engineering technicians can work towards levels 3 and 4
- incorporated engineers can work towards level 4
- level 5 is appropriate to professional and management levels, such as chartered engineers.

Levels (as quoted from NCVQ document)

Level 1: competence which involves the application of knowledge in the performance of a range of varied work activities, most of which may be routine and predictable.

Level 2: competence which involves the application of knowledge in a significant range of varied work activities, performed in a variety of contexts. Some of the activities are complex and non-routine, and there is some individual responsibility or autonomy. Collaboration with others, perhaps through membership of a work group or team, may often be a requirement.

Level 3: competence which involves the application of knowledge in a broad range of varied work activities performed in a wide variety of contexts, most of which are complex and non-routine. There is considerable responsibility and autonomy, and control or guidance of others is often required.

Level 4: competence which involves the application of knowledge in a broad range of complex, technical or professional work activities performed in a wide variety of contexts and with a substantial degree of personal responsibility and autonomy. Responsibility for the work of others and the allocation of resources is often present.

Level 5: competence which involves the application of a significant range of fundamental principles across a wide and often unpredictable variety of contexts. Very substantial personal autonomy and often significant responsibility for the work of others and for the allocation of substantial resources feature strongly, as do personal accountabilities for analysis and diagnostics, design planning, execution and evaluation.

Reproduced courtesy of National Council of Vocational Education

Modern apprenticeships

Modern apprenticeships are available to anyone who is 16–23 years of age, of GCSE standard (grade C+) or has NVQ level 2 and who can show that they are keen to be trained by a local employer.

The key aspect of the modern apprenticeship is that it provides training while in a full-time job, therefore enabling the individual to gain relevant job-related training while being paid a realistic wage.

It is an alternative to pursuing A-levels, GNVQ Advanced level studies or university.

The modern apprenticeship scheme is operated by the Training and Enterprise Councils (TECs). They have been approved by Industry Training Organisations (ITOs) for over 50 industry sectors.

Engineering in society and the environment

A modern apprenticeship will last approximately 3 years, depending on the rate of progress made by the individual. During this time the 'apprentice' will gain a mixture of on-the-job training, day release for study at a local college, and will possibly also attend a number of job-related training courses.

The nature of the modern apprenticeship and the relationship between the employer and employee will be determined by a training agreement made at the beginning of the scheme. This agreement involves the apprentice in making a commitment to being punctual, diligent and committed to his or her work and training, while the employer makes a commitment to provide employment for the duration of the apprenticeship and to provide a structured level of training and work experience. The local TEC is also party to the agreement and will evaluate the apprentices training. Each party will sign the agreement.

Modern apprenticeships have been designed to reflect the needs of industry so that the training being undertaken is relevant and appropriate to the requirements of future employers.

Figure 11.12 A modern apprentice

The aim of the modern apprenticeship scheme is to raise the number of skilled and adaptable technicians and junior managers.

The modern apprenticeship scheme can lead to NVQs levels 3 or 4.

Activity 11.18

Research into the modern apprenticeship scheme. Identify how it could help you.

Academic qualifications

Academic qualification
Qualification that focuses on theoretical study, rather than practical application

Academic qualifications include courses such as GCSEs and A-levels in England and Wales and SCE Standard grades and Highers in Scotland. Universities offer academic courses, including ordinary and honours degrees, masters degrees, diplomas and doctorates.

Useful academic subjects for engineers at GCSE level are English, mathematics and science-based subjects such as physics.

Most A-levels and theoretical degrees are best suited to the academically minded.

Careers in engineering

Engineering degrees tend to be less theoretical than strictly academic courses, but rather offer a balance of theory and practical experience. Degrees are offered in all the main areas of engineering, such as mechanical, electrical, electronic, civil and chemical engineering courses. Degree courses will generally last for 3 or 4 years. Many sandwich courses are available, which provides the opportunity to spend time working in industry, putting the skills and knowledge being learnt at university into practice. It is a valuable opportunity to see the reality of the industry before committing yourself to a particular career. It also provides valuable work experience, which can stand you in good stead when applying for jobs after you have achieved your degree. Sandwich courses are usually organised as a 'thick' sandwich, which means that you spend 12 months working with one company, or a 'thin' sandwich, which means you spend two 6- month periods out at work, usually with different companies.

Engineers are likely to follow Bachelor of Engineering (BEng) or Bachelor of Science (BSc) courses. Courses are available in each of the main engineering areas. Combinations and specialist areas are also available. Examples include: Air Transport Engineering, Chemical Engineering and Applied Chemistry, Civil Engineering with Surveying, Computer Science and Applied Mathematics, Electronic Engineering and Applied Physics, European Engineering, Innovation and Engineering Design, Mechanical and Electrical Engineering, Mechanical Engineering and Energy Management.

Professional qualifications

Professional qualifications are available in all fields of engineering. They are offered by organisations such as the Institution of Electrical Engineers, Institution of Mechanical Engineering and Institution of Chemical Engineering. These generally involve individuals in an accredited training course, undertaken while pursuing their career. For example, the Institution of Mechanical Engineers offers a 4-year monitored, professional development scheme, consisting of 2 years' training and 2 years' career development, into which annual assessments are scheduled.

Work experience

Work shadowing
The experience of accompanying and observing someone who is doing a particular job

The right work experience can be as important to securing a job and achieving the career progression you want as gaining the right qualifications. Engineering

Figure 11.13 Work shadowing

is a practical discipline by its nature. Evidence of being able to apply theory to practice is therefore important.

Work experience can be useful at every level. It can be gained while studying for GCSEs or GNVQs, or on a degree course.

Any work experience can be used to help secure employment. Employers will be interested, whatever the nature of the work, in qualities such as punctuality, reliability, honesty, application and effort. Work shadowing and work experience organised by schools and colleges can provide you with a useful insight into how various industries and individual businesses operate.

Sandwich course
A course that includes time working on an industrial placement

At university there are various options as to the nature of the duration and design of degree courses. It is possible to study on a 3-year full-time course, a 4-year 'thin' sandwich course or a 4-year 'thick' sandwich course. A 'thin' sandwich course indicates that you will have two 6-month periods of work experience during the duration of your course. A thick sandwich course indicates that you will have a 12-month work experience during your course. Although this means a longer course, it provides you with the opportunity to see how the theory relates to the practice, and to experience a work environment without long-term commitment.

Many people find long-term employment, at the end of their courses, with the firms with whom they undertook their work experience.

Progression routes

Different people will follow different paths to reach their target destination. Some people will choose to enter a profession at a lower level than others; some will prefer to follow an academic route, others will prefer a vocational route. Changes in direction may arise as a result of failed exams, lower or higher than expected grades, failure to be appointed to a particular position or simply changing targets.

There are different routes into engineering, and different levels at which engineering roles can be entered. Much will depend on qualifications, skills and experience. In engineering it is easier to provide examples of possible career progression rather than a typical career progression.

The three case studies that follow show the career progression of three engineers in very different fields. They have each followed a varied path to their present position.

Case study

Career progression

Dr Alan Rudge is the Deputy Chief Executive of BT. He is also the first chairman of the new Engineering Council, which represents the 300 000 chartered engineers and technicians in the UK.

Alan Rudge left school before A-levels to work in a City bank, and it was not until 1956, when he was conscripted into the Royal Air Force for national service, that his interest in radar and electronic engineering began. He restarted his education after demobilisation. This was initially part-time, while he worked in the telecommunications industry, and then full-time, when he studied electrical and electronic engineering at the London Polytechnic. In 1968 he received his PhD in Electrical Engineering from the University of Birmingham.

Dr Rudge emigrated to the USA after graduation, where he joined the Illinois Institute of Technology Research Institute (IITRI) in Chicago as a research engineer.

Careers in engineering

In 1971 he returned to the UK to teach in the electronic and electrical engineering department at the University of Birmingham, where he also pursued research in the field of antennas and electromagnetic theory.

In 1974 he resigned from the university and founded a joint Anglo–American Research Centre for Radio Frequency Technology at the Electrical Research Association (ERA) in the UK. The venture was commercially successful and ERA wholly acquired the Research Centre in 1979, when Dr Rudge joined the ERA Board, becoming managing director later in the same year. Dr Rudge took a leading role in the transition of ERA from a subscription-based collaborative research organisation to a fully independent research and technology organisation, and during this period was elected President of AIRTO (the Association of Independent Research and Technology Associations).

In January 1987, Dr Rudge was invited to take up the role of Director, Research and Technology, by British Telecom. He was appointed to the BT Management Board in 1988 and to the Main Board in 1989 as Group Technology Director. In 1991 he was appointed Managing Director, Development and Procurement, in which role he held line responsibility for one-quarter of all of BT's managerial and professional staff. In 1995, he became Deputy Group Managing Director of BT and Deputy Chief Executive the following year. He has been a member of the main board of MCI in the USA since 1995.

Dr Rudge serves, or has served, on a number of national advisory bodies in the UK including the MoD Defence Evaluation and Research Agency Council and the government's Advisory Committee for Science and Technology. He is a member of the DTI Multimedia Industry Advisory Group and serves as chairman of the Engineering and Physical Sciences Research Council, the largest of the six councils that manage government funding of university research. After playing a central role in the initiative to unify the UK engineering profession, he was elected as the first Chairman of the Senate of the New National Engineering Council at its formation in January 1996.

Dr Rudge was the 1993 President of the Institution of Electrical Engineers and was the Institution's Faraday Medallist in 1991. He was awarded the OBE in 1987 and the CBE in 1995. He was made a Fellow of the Royal Academy of Engineering in 1984 and was elected to Fellowship of the Royal Society in 1992. He has received honorary degrees of Doctor of Engineering from the Universities of Birmingham, Bradford, Portsmouth and Nottingham Trent; Doctor of Science from the universities of Strathclyde, Bath, Loughborough and Westminster; and an Honorary Doctorate at the University of Surrey.

Case study

Career progression

Pamela Wilson gained 10 O-level passes and three A-level passes, following which she studied for a Bachelor in Technology in Electrical and Electronic Engineering, which she was awarded by the University of Ulster, Northern Ireland.

Following university, Pamela joined British Aerospace Defence, Military Aircraft unit in 1988. As a graduate engineer she undertook 2 years' training with the Systems Engineering Department. This involved placements of approximately 9 months in equipment engineering, research and development and systems design.

On finishing her graduate training, Pamela remained within the systems design area on the Eurofighter Cockpit Group as a junior engineer. During her time on this group she has expanded both her technical and managerial responsibilities. As a senior engineer she is now responsible for a group of multidisciplined engineers and human factors specialists, including an Italian engineer seconded from a parent company, responsible for the design of information display, manipulation and control for the Eurofighter Cockpit Man–Machine Interface.

Pamela was recognised as woman engineer of the year in 1995.

Case study

Career progression

Colin Tompkins is Group Services Manager for Quantel. As such, he is responsible for managing the Drawing Office, Exhibition Services and Production Engineering and Property Maintenance Services. Within this he is responsible for running a multidisciplined drawing office of 50 staff and has responsibility for the complete organisation and logistics of company exhibitions in Europe and the US. He is also responsible for the general administration of the Research and Development Unit, Information Technology Department and Demonstration Centres. Other responsibilities include plant maintenance, security, canteen, switchboard and health and safety.

Colin's first job was apprentice instrument maker in a radio, television and radiograms manufacturer. This involved him in fitting, tool-making, sheet-metal work and plastics work, using injection-moulding machines, centre lathes, watch makers' lathes, a milling machine capstan, a surface grinder and drills. After 4 years, he moved to the drawing office of the same company as an apprentice, where he learned general drawing office systems and gained detailing and design experience in chassis design, plastic mouldings and printed circuits.

In 1967, he gained the position of detail draughtsman for a company making special-purpose machinery and transfer presses. He gained experience detailing and designing cam profiles, gear trains, castings, welded fabrications and mechanical assemblies associated with 40-ton presses. One particular machine was commissioned by Metal Box Ltd to produce fish paste tins, this machine having six rams with dual roller feeds and guillotines. In 1970, he was promoted to project draughtsman, which meant that he was responsible for the complete design and documentation of projects, for example production of a 12-ton cast transfer press with roller feed and guillotine scrap units (making pen caps) and a smaller steel fabricated press with vibration feed unit (making pen cap clip) of which about 12 were exported to the Soviet Union.

In 1973, Colin moved to his current employer as section leader for printed circuit boards. As such he was responsible for the design and production of all the company's printed circuit boards and procurement of reprographic services, and had four draughtsmen working for him. When promoted to chief draughtsman, his responsibilities included the design and documentation of printed circuit boards, the computer-aided design section, change section and general office administration. His further promotion to drawing office manager involved responsibility for running a multidisciplined drawing office of 50 staff.

Careers in engineering

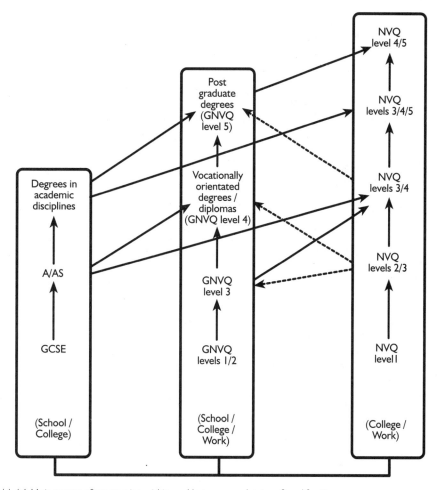

Figure 11.14 Main routes of progression within and between each type of qualification

Activity 11.19

Read the three case studies above. Trace the key points of career progression in each case.

Progress check

1. List three main categories of qualifications.
2. Briefly explain the modern apprenticeship scheme.
3. List three types of academic qualifications.
4. Explain what is meant by NVQ and what it entails.
5. List three different ways in which you can gain work experience.

Presenting personal information

The first thing that prospective employers are likely to know about you will be presented to them in the form of written communication. For this reason, it is essential that you make an effort to ensure that you provide the relevant and appropriate information in the best way possible and that there are absolutely no mistakes. It is always a good idea to ask someone else to check the information before you send it. It is unfortunately all too easy to miss your own mistakes.

You may decide to approach businesses on a speculative basis, that is you may write to a number of businesses that you think you would like to work for, or you may contact them in response to a job advertisement.

Most businesses will ask you to complete an application form, or to submit your curriculum vitae. In some cases, employers will ask you to support the application form with a letter of application; others will ask for a letter of application and a CV. Some will ask you to submit your national record of achievement at interview.

Think carefully about the nature of the personal information that will be required. What will an employer want to know and what do they need to know by law?

In addition to the basic information of your name and address and date of birth, the three main categories of information required are: qualifications, work experience and qualities that make you suitable for the job.

When describing your qualities, think about what type of qualities employers might be interested in. Engineering often involves working in teams, so it can be useful to let the employer know if you are good in teams, but you should make sure that this is relevant to the specific position that you are considering. If the work will involve working in teams, but this is not something that you are particularly good at, then you should think carefully about whether you should pursue your application for this job. You should avoid claiming that you are good at teamwork if you are not as you may get the job but later wish you had not if you fail to fit in or do not enjoy the work.

Application letter

CV

Curriculum vitae – Latin for 'course of life'; it is a record of significant events and achievements that have occurred during your life and that are relevant for potential employment

The letter of application is your opportunity to develop the information provided on your CV and to address those aspects of the job for which you feel you are particularly well qualified. It provides you with the opportunity to highlight the skills, qualifications and experience that you feel are most relevant.

Requests vary between businesses as to whether the letter should be written by hand or typed. The important thing is to produce the letter in the form that has been requested and to ensure that it is neat, legible and without errors. Handwritten letters should always be produced using black ink unless you are specifically requested to do otherwise.

Key rules for your letter of application are as follows:
- show your full address on the top right-hand corner
- place the date directly underneath
- place the name and address of the person to whom you are directing the letter on the left hand of the letter directly above the salutation
- avoid reference to sir or madam in your salutation but ensure that you have the correct name and spelling; it is important to include the correct title (e.g. Dr, Mrs, Miss, Ms)
- give your letter a heading
- make sure that each paragraph in your letter deals with a separate point, each of which is directly relevant to issues raised in the job details
- letters that begin with a salutation to a named person should end with 'yours sincerely'
- you should always print your name clearly under your signature, including reference to your title, if appropriate
- do not waffle – keep the letter to the point.

There are some who advise job applicants that it is important to do something that will make them stand out from the crowd. This does not mean using red, green or luminous ink, nor does it mean using scented paper; it means thinking of how you can best present yourself as being the best person for the job.

Never add a postscript (PS) as this indicates that you have not thought out the letter very carefully. If you think of something after you have finished the letter and feel it absolutely must go in, then rewrite the letter. Always keep a copy of your letter.

Activity 11.20

Select two engineering jobs from a local paper. Write an application letter for each job.

Application form

An application form is usually anything from two to four sides of A4. Businesses may ask for this alone, or with a letter of application and/or a CV.

The application form may have to be typewritten but is usually completed in handwriting. Black ink should always be used unless you are specifically required to do otherwise.

Always read through all the questions before beginning to answer any; you will sometimes find a number of questions that ask for broadly similar answers, in which case you need to plan how you are going to answer each one. You are given limited space in an application form, and it is best to aim to use the space to your best advantage; this will not be achieved if you simply repeat what you have said in an earlier question.

Take note of what you are asked and how you are asked to record your answers. If you are asked to type your answers do not hand write them as your application form will be thrown out at the first level, because you have shown that you are unable to follow simple instructions.

Photocopy the blank form and use this as your practice copy. In this way you can find out how much text you can fit in the appropriate spaces. It is always a good idea to try to fill the spaces allowed for a particular answer if you can, but not if it means providing irrelevant information. The application form needs to be as neat as possible, and directly relevant to the post for which you are applying.

Always keep a copy of your completed application form, with the contact details.

Some people will be tempted to exaggerate or simply make up information to put on their application forms. Even if this gets them to interview it is not likely to get them the job as they are likely to get caught out in the interview. If they actually manage to get the job they may find themselves wishing they had not. If you have to make up information in order to get the job it is highly unlikely that you will be able to do the job, in which case you could end up as a danger not only to yourself, but also to colleagues and workmates.

Curriculum vitae

A curriculum vitae (CV) is a summary of those aspects of you and your life that are relevant to the employment market.

Your CV should include the following.

APPLICATION FORM
Please complete the following details using black ink.

Job Title: .... Ref: ....

PERSONAL DETAILS
Title: ....
Surname: ....
First Names: ....
Address: ....
....
Postcode: ....
Date of Birth: ....
Daytime Tel: ....
Evening Tel: ....

Do you hold a current driving licence: YES/NO

EMPLOYMENT - please start with the most recent

Date (from - to)	Position Title and Description	Employer Name & Address

EDUCATION AND QUALIFICATIONS

Date (from - to)	School/college etc.	Qualification	Grade

Figure 11.15 Example of an application form

Careers in engineering

Please give a general statement about the way in which you consider your skills, knowledge and experience to be relevant to this position and your reasons for applying.

REFERENCES
Please give the names and addresses of two people prepared to provide references.

Name:	Name: ....
Position:	Position: ....
Company:	Company: ....
Address:	Address: ....
....	....
....	....
....	....
Postcode:	Postcode: ....
Telephone:	Telephone: ....

Figure 11.15 Example of an application form (continued)

Personal details
- name
- current address
- telephone number
- date of birth (do not include age unless you want to update your CV every year)
- marital status.

Education
- Education establishments attended, with the appropriate dates; the best order is to put the one you attended most recently first on the list, e.g. college before secondary school.

Qualifications
- List your qualifications with year passed and grades (those passed most recently should be top of the list), e.g. GNVQs, GCSEs, others. Do not forget to include qualifications for which you are currently studying.

Employment:
Provide a summary of any employment you have had, whether part-time or full-time, permanent or temporary work; information should include the name of the organisation for which you worked, the dates of employment, job title and a summary of the type of work undertaken, e.g.

April–September 1994
Plastique Ltd, Production Engineering Department Trainee
Produced documents for company accreditation to BS 5750
Assembled and piloted production schedules for vacuum formed products
Designed and implemented total quality procedures

- if you have never been employed, include work experience, work shadowing or work visits that you have undertaken in order to show that you do have some experience of the world of work.
- do not dismiss part-time jobs, particularly if you have held a job for a reasonable period of time.

Other
- Include details of other skills, experience, qualifications or interests that you may have that you think could be relevant or that help to describe you as an individual, for example the fact that you can drive or that you play football for a local team.
- Skills gained during the course of your GNVQ such as teamwork skills, management of time and resources, working to deadlines, communication, numeracy and information technology skills are all of value and should be included if you feel this is appropriate to the positions for which you are applying.

References
- Include the names and contact details of two people who have agreed to provide you with references; if possible, one of these should be an employer or someone from your education establishment.

Careers in engineering

Activity 11.21

Draw up a CV for Pamela Wilson (see page 437) or Colin Tompkins (see page 438).

A CV is probably the first piece of information that the prospective employer will receive about you. It is not only the information printed on the page that is important. The presentation of the information itself will also tell the employer something about you. For example, coloured or grubby paper, typing errors, spelling mistakes and grammatical errors will all provide the employer with information about you, although it might not be the information you would like them to receive. Unfortunately, however well qualified and well suited to the job you may be, a poorly presented CV could cancel this out and may even prevent you from getting an interview.

Points to bear in mind when drawing up your CV:
- make sure that your CV is typewritten, not handwritten
- produce your CV on white or cream paper
- clearly set out the different sections (personal, education, employment, other, references)
- make sure you have the permission of the referees to be approached
- ask someone else to read through and check the spelling, typing, punctuation, etc.
- if possible send your CV in a large envelope so that it does not need to be folded
- a CV should be clear and concise, additional information should be provided on a letter of application or application form
- your CV may be one of hundreds that someone has to read through; there will probably be a set of requirements that have to be met to get through to the next stage and this will continue until there is a reasonable number of candidates to be assessed in terms of being invited to interview
- keep spare copies.

Activity 11.22

Draw up your own curriculum vitae. Present it in a suitable form.

National record of achievement (NRA)

The national record of achievement was introduced by the government in 1991. The aim was to introduce a standardised system of recording and presenting personal achievements that could be maintained throughout the course of the individual's life.

The key difference between this and a CV is that it should be a continuously maintained document showing not only the actual achievements, but also the target setting and action planning that led to those achievements.

For employers it is a useful document that provides evidence of your ability to set targets and to plan your way towards meeting such targets. It can be seen as evidence of your ability to take responsibility for your own learning, not only in full-time education, but throughout your adult life.

Engineering in society and the environment

NAME	David Gray
HOME ADDRESS	28 Murray Road, Oldbury OL14 6PX
TELEPHONE	01234 56789
DATE OF BIRTH	18 February 1970
MARITAL STATUS	Single NATIONALITY Irish

EDUCATION AND QUALIFICATIONS

1984-89	**St Mary's College, St Mary's Road, Galway, Ireland.** Intermediate Certificate : 1 grade A & 3 grade B Leaving Certificate: 1 grade B & 2 grade C
1989-91	**Regional Technical College, Dublin Road, Galway , Ireland** Higher National Certificate in Industrial Engineering (Merit)
1991-1996	**Brunel University** Manufacturing Systems Engineering with Management and Business Studies BEng (Hons)
1996 - Present	**Thames Valley University** Certified Postgraduate Diploma in Accounting and Finance A.C.C.A. accredited
Present	**I.E.E.** Associate Member of the Institute of Electrical Engineers

WORK EXPERIENCE

Present	**John Crane U.K. Ltd, Slough, Middlesex.** Production Engineer, Manufacturing Engineering Office. ○ Formulated and implemented raw materials rationalisation plan. ○ Justified capital expenditure.
April 95-Sep 95 (Work Placement)	**Ford Motor Company, Basildon Radiator Plant, Basildon, Essex.** Undergraduate Trainee, Simultaneous Engineering Office. ○ Designed and developed automatic visual inspection system. ○ Researched and recommended vision system equipment purchases. ○ Designed and implemented bar-code component recognition system.
April 94-Sep 94 (Work Placement)	**Plastique Ltd.** Undergraduate Trainee, Production Engineering Department. ○ Produced documents for company accreditation to B.S.5750 ○ Assembled and piloted production schedules for vacuum formed products. ○ Designed and implemented total quality procedures.
Apr 93-Sep 93 (Work Placement)	**Stannah Stairlifts Ltd.** Undergraduate Trainee, Project Engineering. ○ Redesigned PCB cleaning processes to comply with European Environmental Engineering Standards. ○ Sourced and justified vertical milling machines for manufacture of new generation products. ○ Analysed sales order processing system for integration to company I.T. Network.
Apr 92-Sep 92 (Work Placement)	**Stannah Stairlifts Ltd.** Undergraduate Trainee, Production Engineering. ○ Designed flexible manufacturing cell utilising robotic technology to replace manual assembly operations. ○ Prepared and presented financial analysis of manufacturing system.
Jun 90-Sep 90 Jun 91-Sep 91	**Digital Equipment Corporation Ltd. Galway, Ireland.** Engineering Apprentice, Software Processing Unit. ○ Conducted work study investigations and designed layouts to meet proposed reduction in W.I.P. and order lead times. ○ Devised and implemented solutions to a broad range of day-to-day production problems.

Figure 11.16 A sample curriculum vitae

Careers in engineering

SKILLS AND APPOINTMENTS

Information Technology

I have a working understanding of several common computer languages. I am also familiar with most Windows and DOS based software and have a good working understanding of Internet technology.

Languages

I have a moderate working understanding of French.

1991- 1993 Officer Training Corp, Territorial Army, Handel St., London.
Cadet Officer of the Territorial Army, in R.E.M.E.
Completed basic training and gained M.o.D. Class II driving licence.

1991 - 1993 Brunel University Irish Society
Executive committee member and vice-president of the Brunel Irish society. Brunel Irish society is one of Brunel's largest societies, organising social events and outings for its members of all ethnic backgrounds.

1989 - 1991 Regional Technical College Karate Club, Galway, Ireland.
Club treasurer.

ACTIVITIES AND INTERESTS

I enjoy competing in a great number of sports, particularly Jitsu, Karate, Horse-riding, Squash and Swimming. I am a member of Brunel University Jitsu Club and I am approaching my first grading.

Since starting at Brunel I have been an active member and committee member of the University's Irish Society. The society organises an extremely broad range of activities that bring together students from all backgrounds. My involvement in this society has enable me to improve my personal skills and meet many good friends.

I have a inherent interest in mechanical devices and machines which extends to servicing and repairing my own car and agricultural machinery on my uncle's farm.

REFEREES

Mr Marconi	Dr M.Faraday
Ford Motor Company	Brunel University
Christopher Martin Road	Cleveland Road
Basildon	Uxbridge .
Essex	Middlesex

Figure 11.16 A sample curriculum vitae (continued)

Engineering in society and the environment

National record of achievement

A system of recording personal achievements

If you started to develop your NRA in school, it is worth keeping it up to date, reviewing your targets and your plans for the future against your achievements.

Many employers will be looking for employees who are capable of flexibility, and adapting to new situations. The rapid changes in technology require engineers to be able to update their engineering knowledge and skills in relation to new products and processes. This means that those entering the engineering field must accept the need for life-long learning. The NRA is therefore, a useful tool.

Activity 11.23

Check your own national record of achievement. Ensure it is up to date. Add any skills that you have acquired as a result of your studies on your GNVQ course. Consider your future plans. Add your current action plan to your NRA.

Interviews

An interview generally represents the selection by the prospective employers of a small number of the original applicants for further discussion. The employers will usually already have a copy of the applicant's CV and job application form or letter. They may or may not have already checked the applicant's references, although the referees are not normally contacted until after the job offer has been made.

The interview is usually held at the site where the job will be based.

Interviews can involve any number of people. It may take place on a one-to-one basis or there may be a panel of people involved in the interview.

Interviews can take many forms. It may be include a series of questions and answers, a practical test or an intelligence test.

Remember that an interview works both ways. In the same way that you should be aiming to impress the interviewer(s), answer all the questions and make sure that they offer you the job, the interviewer should be prepared to answer all your questions, and for you to leave impressed with the organisation and wanting to work there.

Preparation for an interview

Confirm your attendance at the interview in plenty of time. If you are unable to attend on the date or at the time specified contact the employer to let him or her know, and find out if it is possible to arrange an alternative date or time. If you decide not to attend, write to let the organisation know. Remember, it may have other jobs that may be of interest to you in the future.

As soon as you are invited to interview, make sure you know where the interview is and how to get there. Ask for directions or a map if you do not know where it is or if this is not automatically provided. Ensure that you know how long it will take you to get there and how long you should allow for possible delays, for example with traffic or train connections.

Make sure that you know where to go and what to do when you get there, for example should you go the reception or a specific room number. Also make sure that you know the name of the person running the interview.

Think about the type of person you think the business will want to employ, and consider very carefully whether you think this is you. If you are confident that it is you, then think about how you can convince the interviewer you are the person they should appoint.

A useful rule is: **Make an effort to show that you have made an effort!**

Think about how you are going to look at your interview. The most important issue is to dress appropriately. Clean, smart and suitable appearance can help provide a good first opinion. It can also help you to feel more confident.

Something else that can help you feel more confident is to make sure you are clear about what the job will entail. A small amount of research before the interview into the nature of the business in general, and the specific job for which you have applied in particular, can help to provide you with an edge over other applicants. This may simply entail making sure that you read any paperwork the business sends to you before the interview, or you might want to find out if there is any additional information you can read up on, such as the annual report or general marketing information.

Write a list of the questions that you think that they are likely to ask you, and think about how you will answer each one. When you have thought through each of your answers, find someone to ask you the questions so that you can practise going through the answers. Speaking the answers out loud is very different from, and harder than, going through them in your mind, so this practice can be very useful.

Write a list of the questions you want the answers to. You should not worry about asking any questions you want to, although you should be aware that questions about how many holidays or tea breaks you are entitled to might not give the impression you want to give. Try to think of questions you can ask that will give the right impression; for example, if you are ambitious ask questions relating to how you can get ahead in the company.

You should not worry about memorising these questions; after all if you take out a list of questions to ask at interview, it proves you have prepared yourself for the interview. Neither should you worry about writing down the answers given by the interviewers, as it shows that you are organised.

When you are preparing for the interview it is also a good idea to think about the type of documentation, such as certificates, that it might be appropriate to take to interview, or evidence of work that you have produced, such as designs or models, that may be of interest to interviewers. Obviously, the work must be of relevance to the job for which you have applied, and must show a high standard of work, for it to be worth showing. If you have models that you feel would support you in your interview but which are bulky, consider taking photographs or designs instead. If you have a large amount of material that you feel could be appropriate, think carefully about what points you feel it could help you make. Your interview is of limited time, and you do not want to overwhelm the interviewer(s) with material, especially if it is all evidence of the same skills or qualities. Think about whether you could put together a portfolio that you would be prepared to leave with the interviewer(s) if required. Whatever you take with you, make sure that it is well presented and easy to retrieve from your bag or folder.

Much of the preparation for interviews is common sense. Unfortunately, because of the nature of interviews, common sense often disappears when you find yourself sitting on one side of a table looking at a group of people all waiting to ask you questions.

When you are invited into the interview room, sit in the chair that is indicated to you. Make sure you are comfortable, but do not slouch, and make sure that you are organised, with any documentation or evidence of work within easy reach. Sit leaning slightly forward in your chair and pay attention to the ques-

Engineering in society and the environment

tions being asked. If there is more than one interviewer, try to look at each of them when answering your questions.

Stay positive. If you are unable to answer a question, just be honest. Ask if you can think about it and come back to it later.

Do not give up if you do not get the job on your first interview. The fact that you were able to get through to interview is an important step, and practice at interviews is important.

Nerves can get in the way of interviews. Take a couple of deep breaths before you go in. Remember that these are people who have also had to go through interviews to get to their present job. They are interested in you and your ability to do the job. Remember that the fact that you have made it to interview shows that someone thinks you can.

Remember that an interview is just as important and can be just as stressful for the interviewer. It may even be their first interview. An interviewer typically has 20–30 minutes to find the right person to fill a particular position. If he or she chooses the wrong person this can cost the business hundreds, if not thousands, of pounds.

Some sample interview questions are listed below.
- What interested you about this position?
- What do you think you can offer this firm?
- What abilities and skills are you able to bring to this job?
- Why do you want to work for this company?
- Where do you want to be in 5 years' time?

Activity 11.24

You are the personnel manager for a local engineering company. You are getting ready to interview candidates for a job vacancy. Write a list of the questions you would ask to help you decide who to appoint. Think about how the answers to these questions will affect your decision and the answers you might expect to hear.

Figure 11.17 Interviewing panel

Careers in engineering

Activity 11.25

Identify an engineering job that you would like to apply for. Set up two interviews, one that is arranged on a one-to-one basis and one that involves a panel of interviewers. In each case the interviewer(s) should be fellow students in role play. For this to be effective and realistic you must be able to research the job and obtain background information about the industry, or a specific business, that would offer this job.

Progress check

1. Briefly explain the difference between a CV and a national record of achievement.
2. List five things that should be considered before going to interview.
3. Describe the nature of an application form.
4. List five things that should be listed on a CV.
5. What is the difference between an application form and a letter of application?

Assignment 11
Investigate employment options in engineering

This assignment provides evidence for:
Element 4.2: Explore career options and pathways in engineering
and the following key skills:
Communication 2.1: Take part in discussions
Communication 2.2: Produce written material
Communication 2.3: Use images
Communication 2.4: Read and respond to written materials
Information technology 2.1: Prepare information
Information technology 2.2: Process information
Information technology 2.3: Present information

Your tasks

The following tasks should be prepared and presented using information technology. Diagrams, photographs and other visual evidence should be employed to support your work, as appropriate.

1. List the main engineering sectors, and identify two employers in each.

2. Identify the field of engineering in which you plan to make your career. Describe how you intend to enter this field and the route you hope to follow within your career. Within this description you should highlight the skills, knowledge and understanding and experience you intend to acquire in order to enable your intended progression.

3. Select a job vacancy advertised in your local area. Write a letter of application for that job and supporting documentation, such as a copy of your CV and national record of achievement.

4. Working with a group of students on your course, organise and conduct a mock interview in which you are interviewed for the job vacancy identified in task 3. Evaluate your performance, highlighting areas of strength and areas which could benefit from improvement.

Appendix

Chartered engineers and incorporated engineers

The Engineering Council recommends that posts involving one or more of the duties and responsibilities listed below should be occupied by those registered as chartered engineers or incorporated engineers.

Chartered and incorporated engineers are expected to apply their codes of professional conduct, to undertake work within their expertise and to exercise a responsible attitude to society with regard to the ethical, economic and environmental impact of technical need and change.

The following lists are arranged in an approximately descending order of responsibility level, although clearly the relative importance of the items may vary somewhat from one situation to another.

Design
- Managerial responsibility for an engineering design function or group.
- Supervising preparation of designs.
- Engineering design outside the scope of established procedures, standards and codes of practice to a competitive level of cost, safety, quality and reliability.
- Promotion of advanced designs and design methods. Continual development of standards and codes internationally.
- Failure analysis and value engineering.
- Design work involving established procedures and the use of engineering standards and codes of practice to a competitive level of cost, safety, quality and reliability and appearance.

Research and development
- Leading research and development effort in engineering, resulting in the design, development and manufacture of products, equipment and processes to a competitive level of cost, safety, quality, reliability and appearance.
- Managing engineering research and development groups, planning and execution of research and development programmes, carrying out research and development assignments.
- Evaluation of test results and interpretation of data. Preparing reports and recommendations.

Engineering practice
- The exercise of independent technical judgement and the application of engineering principles.
- Application of theoretical knowledge to the marketing operation and maintenance of products and services.
- Development and application of new technologies.
- Monitoring progress on a worldwide basis, assimilation of such information and independent contributions to the development of engineering science and its applications.
- Work involving the need to understand and apply analytical and technical skills and judgement and the use of a range of equipment, techniques and methods for measurement, control, operation, fault diagnosis, maintenance and for protection of the environment.

Manufacture, installation, construction
- Managerial responsibility for a production, installation, construction or dismantling function.
- Organisation of cost effective manufacturing functions.
- The introduction of new and more efficient production techniques and of installation and construction concepts.
- Organisation of quality-driven manufacture, installation and construction functions.
- Day-to-day organisation and supervision of manufacturing, installation, and construction functions from raw material input to finished product.

Operation and maintenance
- Managerial responsibility for an operation or maintenance function or group.
- Providing specifications of operational maintainability standards to be achieved in design and production.
- Determining operational maintenance requirements in terms of tasks to be performed and the time intervals between tasks.
- Managing the quality of the output of operational maintenance activities.
- Developing and specifying diagnostic techniques and procedures.
- Developing and specifying repair and rectification methods.
- Assessing the actual and expected effect on performance of deterioration in service.

Health, safety, reliability
- Making the appropriate provision in engineering projects to ensure safety and the required standards of reliability, not only with employees and customers in mind but in the general public interest.
- Responsibility for health, safety, reliability in situations involving engineering plants, systems, processes or activities.
- Accident investigation.
- Supervision of inspection and test procedures.

Management planning
- Overall company/commercial responsibility as a director with engineering activities and functions.
- Management of the development and implementation of new technologies with estimation of the cost/benefit of the financial, social and political decisions taken.
- Pioneering of new engineering services and management methods.
- Effective direction of advanced existing technology involving high risk and capital intensive projects.
- Direct responsibility for the management or guidance of technical staff and other resources.
- Supervision of engineering staff and resources and the legal, financial and economic practice at a level commensurate with the scale of the activity and size of organisation within the constraints of the relevant environment.
- Short range planning of engineering activities and functions.

Engineering aspects of marketing
- Management responsibility for a technical marketing function.
- Top-level customer and contract negotiations.
- Setting marketing objectives and policies.
- Territorial or market planning forecasts and targets.

- Management responsibility for the dissemination of accurate technical information.
- Customer technical advisory service.
- Market analysis, contract negotiations.
- Non-standard customer requirements.
- Sales operations, efficient market coverage.
- Preparing cost estimates and proposals.

Engineering technicians

The Engineering Council recommends that posts involving one or more of the following duties and responsibilities should be occupied by those registered as engineering technicians. These duties are normally undertaken under the general supervision of a chartered or incorporated engineer. It is recognised that there will be some overlap with duties towards the end of the previous lists for chartered and incorporated engineers.

There is no order of responsibility level of the items under each heading.

Design
- Preparation of plans and designs of limited scope and complexity.
- Costs estimates and checking designs and products against specifications.
- Use of standard codes and specifications within established practice.

Research and development
- Developing and constructing test equipment, conducting and assisting with experiments and investigations.
- Recording and calculating results as scheduled.

Engineering practice
- Detailed knowledge of appropriate established technology.
- Assistance in manufacturing schedules and production control.
- Fault investigation and correction.

Manufacture, installation, construction
- Installation, commissioning and dismantling of specified plant and processes.
- Operational responsibility for manufacturing and construction systems.
- Inspection and fault correction. Work study. Assessing processes for production.
- Improving existing plant and processes.

Operation and maintenance
- Preparing programmes to implement operation and maintenance requirements.
- Planning the provision of resources to support the maintenance programme.
- Implementing and controlling the execution of operation and maintenance programmes.
- Ensuring continuity of supply and services as scheduled.

Health, safety, quality
- Ensuring efficient day-to-day cleanliness and safety of equipment and services.
- Responsibility for detailed compliance with health and safety requirements.
- Operation and development of detailed procedures for quality assurance and control.

Management and planning
- Planning manufacturing layouts, installation and maintenance schedules ad procedures.
- Supervision of shop floor and similar activities.
- Technical guidance and organisation of routine tasks.
- Trouble shooting of technical activities.

Marketing
- Cost estimates and proposals to suit customers' requirements for standard engineering plant and products.
- Efficient processing of enquiries, orders and arrangements for delivery.
- After-sales service and technical advice to customers.

Teaching, training and career development
- Technical support functions in laboratory, design office and project work.
- General assistance to academic staff and students in teaching establishments.
- Training of those engaged in craft, maintenance and associated functions.

Useful addresses

Mechanical Engineers, Schools Liaison Service, Northgate Avenue, Bury St Edmunds, Suffolk, IP32 6BN. Tel: 01284 763277.

Institution of Electrical Engineers, Michael Faraday House, Six Hills Way, Stevenage, Hertfordshire, SG1 2AY. Tel: 01438 313311.

Institution of Chemical Engineers, The Education Liaison Unit, Institution of Chemical Engineers, Davis Building, 165–189 Railway Terrace, Rugby, CV21 3HQ. Tel: 01788 578214.

The Engineering Training Authority (EnTra) Vector House, 41 Clarendon Road, Watford, Herts, WD1 1HS. Freephone 0800 282167.

Sample unit test for Unit 4

1. Automation in the home has led to:
   a Labour-saving devices
   b Higher levels of waste in the home
   c Mass production
   d Greater levels of repetitive tasks.

2. Catalytic converters for cars have led to:
   a Higher car maintenance
   b Lower car pollution
   c Lower use of unleaded petrol
   d Lower mileage.

3. Engineering technology refers to:
   a Automation
   b Information technology
   c Materials
   d All of the above.

4. The use of automated systems in high-volume industrial production is most likely to lead to increases in:
   a Labour costs
   b Production time
   c Productivity
   d Numbers of staff.

5. A key material in modern computer technology is:
   a Plastic
   b Silicon
   c Stainless steel
   d Paper.

6. Bar codes are used to help:
   a Stock control
   b Sales
   c Advertising
   d Production.

7 Ambulatory biomonitors enable:
   a Ambulances to reach scenes of accidents faster
   b Patients to be monitored in ambulances en route to hospital
   c Medical staff to be monitored while they move around
   d Patients to be monitored while they move around.

8 Computerised axial tomography scans produce images of slices of the body through:
   a X-ray technology
   b Surgery
   c Amputation
   d Donation of organs.

9 Which of the following does not refer to the tapping of natural resources?
   a Mining and quarrying
   b Assembly and processing
   c Oil and gas extraction
   d Agriculture and forestry.

10 Which of the following is not a classification of waste?
   a Special
   b By-product
   c Toxic
   d Non-degradable.

11 An example of solid waste is:
   a Plastic
   b Oil
   c Methane
   d Sewerage.

12 The Environment Agency was set up in:
   a 1980
   b 1986
   c 1990
   d 1996.

13 The Environment Agency is responsible for the environmental protection of:
   a Air
   b Land
   c Water
   d All of the above

14 An engineer involved in the development and manufacture of plastics is most likely to be working in which industry?
   a Aeronautical
   b Chemical
   c Communications
   d Power

15 An engineer involved in the design and manufacture of a motor vehicle is most likely to involved in which area of engineering?
 a Mechanical
 b Electrical
 c Chemical
 d Civil

16 The best source of career information tailored to your specific needs is from:
 a Advertisements
 b Local telephone directories
 c General-interest magazines
 d Radio

17 The best place to find advertisements for vacancies for engineers is:
 a Trade press
 b Local telephone directories
 c General-interest magazines
 d Radio

18 An engineer at craftperson level needs the following:
 a To be over 25
 b An NVQl evel 4
 c Three GCSEs
 d An honours degree

19 Which of the following is the highest classification for an engineer?
 a Chartered engineer
 b Craftsperson
 c Incorporated engineer
 d Engineering technician

20 A non-written method of presenting data to prospective employers is:
 a National record of achievement
 b Interview
 c Application form
 d Application letter

21 You have written a letter of application to Mr Jardin of Jardin Engineering. Which of the following should not appear in your letter?
 a Date
 b Yours faithfully
 c Yours sincerely
 d Your address

22 Information which need not be included in your curriculum vitae is:
 a Name
 b Reference
 c Age
 d Date of birth

Answers to the sample unit tests

Unit 1	Unit 2	Unit 3	Unit 4
1 b	1 b	1 c	1 a
2 a	2 a	2 d	2 b
3 c	3 d	3 b	3 d
4 d	4 c	4 b	4 c
5 c	5 a	5 a	5 b
6 b	6 a	6 b	6 a
7 c	7 d	7 a	7 d
8 a	8 a	8 b	8 a
9 d	9 a	9 a	9 b
10 c	10 b	10 c	10 b
11 a	11 d	11 d	11 a
12 c	12 c	12 b	12 d
13 a	13 b	13 c	13 d
14 a	14 d	14 b	14 b
15 d	15 a	15 a	15 a
16 a	16 c	16 b	16 c
17 a	17 b	17 c	17 a
18 b	18 a	18 d	18 c
19 c	19 a	19 a	19 a
20 a	20 b	20 c	20 b
21 c	21 c	21 b	21 b
22 a	22 c	21 a	22 c
23 c	23 d	22 a	
24 a	24 a	23 d	
25 b	25 b	24 c	
26 d	26 b	25 c	
27 a	27 a	26 a	
28 d	28 d	27 b	
29 d	29 c		
30 b	30 b		
31 a			
32 d			

Index

Abbreviations 204, 213
Absolute temperatures 289
Academic qualifications 434–5
Acceleration 311, 312, 315–18, 353–4
Accident reporting 89, 90, 91
Acid rain 407
Adhesive bonding 77–9
Advertisements, job 429–30
Aeronautical industry 419
Aeroplanes 393–5
Agriculture 406
Air bags 393
Alarms, responding to 91
Allen keys 63, 64
Alloys 3, 5, 6, 9
Alternating current 326–9
Aluminium 7, 89
Aluminium alloys 9
Aluminium bronze 8, 9
Amorphous ceramics 14–15
Amplitude 327
Amps (amperes) 319
Analogue models 130
Annealing 80–1
Application forms 440, 441
Application letters 440–1
Apprenticeships 433–4
Arbors 60
Arc welding 74–6, 102, 103
Arm protection 94
Artificial intelligence 382
Artist's impressions 131–3
Assembly drawings 136–8, 161
Assembly processes 63–79, 101–3
Automation 382, 385–6, 397
Auxiliary projection 192–3

Bakelite 12–13
Bar charts 153, 154, 155
Bar codes 399
Base units 286, 287
Bearings 247, 296
Bicycles 392
Block diagrams 112, 140–4, 161
Bluing 89
Bolts 18–19, 63, 216, 218
Bonded ceramics 15, 16
Books 391
Brass 7–8

Brinell test 25
Bronze 8
Butyl rubber 14
By-products 409

Calibration 333
Calorising 89
Capacitors 31–3, 38–9, 251
Carbon 3, 4, 5
Carbon dioxide 406
Carburising 83
Careers advisers 427–8
Cars 392–3, 422
Case hardening 83
CAT scans 401
Catalytic converters 393
Catalytically drying paints 87–8
Catchplates 48–9
Celsius scale 288, 289
Cements 15
Centre lathes 45–6
Ceramic composites 16–17
Ceramics 14–15, 27, 31, 80
Cermets 16–17
CFCs 406–7
Chartered engineers 425, 453–5
Chemical cleaning 84
Chemical engineers 418
Chemical industry 420
Chemical treatment 84–5, 103–4
Chromium plating 85
Chuck keys 54, 99–100
Chucks 46, 47, 54, 99
Circlips 220, 223
Circuit breakers 250, 251
Circuit diagrams 138–40, 238–40, 241, 249–57, 262–4
Circuits, electrical 65–6, 319–20, 322–3
Civil engineers 419
Clean Air Act 1993 415
Cleaning and washing 388–9
Clothing, protective 93
Communications 395–6, 420
Compasses 175, 176
Components 244
 conventions 204–30
 electrical and electronic 30–41, 249–57
 mechanical 17–25, 245–9

representation of physical situations 258–66
Composite materials 16–17
Computer numerical control (CNC) 51, 382, 397
Computer-based drawings 182–8
Computers 347–8, 399
Conceptualising 160–1
Concrete 17
Connecting wires and cables 30–1, 36–7
Contact makers/breakers 250, 252
Control mechanisms 255, 256
Control of Substances Hazardous to Health Regulations (COSHHR) 1988 103, 414–15
Control valves 255, 257
Conventions, drawing 204–30, 244–69
Cookers 387
Coolants 51, 62, 116
Copper 6
Copper wire 30
Corrosion 27
Costs 163–5
 graphical methods 168–9
 materials and components 28
Couplings 248
Courtney washers 65, 66
Craftspersons 424
Crimped teminations 65, 66
Crystalline ceramics 15
Cupro-nickel 8, 9
Current, electric 319–20, 322–3, 326–9, 341, 351–3
Curves (drawing) 176, 177
Curves (graphs) 357–60, 362–3
Cutters, milling 57–8, 60–1
Cutting speeds 50–1, 56, 62
Cutting tools (turning) 44, 49, 50
CVs (curriculum vitae) 441–5, 446–7
Cyanoacrylate adhesives 79
Cylinders 255

Dampers 248–9
Datums 110, 209–10
Dead weight scales 340
Denominators 363–4, 365
Dependent variables 344

Index

Design 44, 160–1, 172, 426
Detail drawings 133–5, 161
Detail information 159
Digital multimeter (DMM) 340–2
Dimensions and dimensioning 110, 204–17
 measurement devices 335–7
Diodes 33–5, 40, 253
Direct current (DC) 326
Directories, local 430
Dishwashers 390
Displacement 311
Distance–time graphs 312–13
Dividers 175, 176
Doppler effect 332
Dowel pins 220, 223
Draughting machines 178, 179
Drawing boards 174
Drawing instruments 174–8
Drawing sheets 175, 233, 234, 235
Drawings 109–10, 172–3
 computer-based 182–8
 conventions 204–30, 244–57
 information from 269–73
 manual 173–82
 projections and sectioning 189–203
 representation of physical situations 258–66
 standards 231
 see also Graphical methods
Drilling 52–6, 99
Ductility of materials 26
Dynamic systems 311–18

Ear protection 94
Elastomers 13, 79
Electrical and electronic components 30–6, 65–6
 drawings 249–57, 260, 262–4
 selection 36–41
Electrical and electronic engineers 418
Electrical measurement devices 340–3
Electrical systems 319–29
Electroforming 85
Electromotive force (EMF) 250, 319
Electrons 15–16, 319
Electroplating 84–5
Emergency equipment 97–9
Emergency services 402
Emissions 413–14
Employment agencies 428–9
Energy 286, 296
Engineering Council 424, 453
Engineers 417–27, 431
Entertainment, home 390–1
Environment Protection Act (EPA) (1990) 415
Environmental issues 403–15

Environmental Protection (Prescribed Processes and Substances) Regulations 1991 415
Epoxy resin 13
Eraser shields 178
Ergonomics 426
Estimating 163–4
Etching 84
Exercise 392
Expansion, linear 293–6
Eye protection 95

Face masks and respirators 96
Faceplates 47–8
Fahrenheit scale 288, 289
Ferrous metals 3–6
Fibreoptics 381
Finishes, surface
 processes 86–9, 104
 specifications 110–11
 symbols 111, 228
Fire 97–9
First-aid equipment 97
Flame spraying 88, 104
Flexible curves 176, 177
Flow charts 112, 113, 145–8, 162
Fluid power systems 305–11
 drawings 240–1, 252–7, 261, 264–5
Fluxes 68, 71, 74
Food preparation 386–7
Footwear, protective 93–4
Force 296, 299–304, 317, 353–4
 measurement 338–40, 345
Forestry 405–6
Formica 13
Formulae transposition 368–70
Fractions 363–5
French curves 176, 177
Frequency 326–7
Fuses 250, 251

Gantt charts 163
Gas extraction 405
Gears 222–5, 245, 246
Global warming 406
GNVQs 432
Gradients 351–2, 354, 356, 357–8
Graphical methods 130
 for different purposes 159–66
 selection criteria 167–70
 types 131–56
 types of information 157–9
 see also Drawings
Graphite 4, 5
Graphs 153, 312–14, 346–7, 350–63
Grey cast iron 4–5
Grinding 86, 104
Guards, machine 92, 100–1

Hand protection 94
Hand tools maintenance 117
Hard soldering 70–1, 101, 102
Hardness of materials 25, 112
Head protection 94–5
Health and medicine 400–2, 421
Health and safety 89–90, 414–15
 equipment 92–9
 procedures 90–1
 specific processes 99–104
Health and Safety at Work Act 1974 89, 90
Heat treatment 80–3, 103
Hexagonal keys 63, 64
Hidden details 191, 193
High-speed steel 5
Home activities 385–91
Hooke's law 297
Horizontal mills 57–8, 60, 61
Hot dipping 88, 104
Hot spraying 88, 104
Hydraulic systems 252, 305–6, 307
Hydrostatic pressure 307–11
Hygiene 91
Hysteresis 335

Iconic models 130
Imperial unit system 288
Incorporated engineers 424, 453–5
Indentation tests 25
Independent variables 344, 345
Inductors 33, 39–40, 251
Information technology (IT) 347–8, 380, 385, 398–9
Information types 157–9
Installation 166, 270–2
Insulators and insulating materials 31, 37
Integrated circuits 131, 249, 253, 380
Interviews, job 448–50
Iron 3, 27

Job centres 429
Joining processes 63–79, 101–3
Kelvin scale 289
Keys 217, 221, 222, 248

Kilogram units 287

Lacquers 87
Lamps 251, 252
Lasers 381
Latent heat 289–93
Lathes 44–51, 100, 115
Layouts 258, 265–6
LEDs (light-emitting diodes) 33–5, 40, 253
Leisure activities 390–2
Lettering 182
Line types 179–81
Linear expansion 293–6

Index

Linear relationships 354–7
Locking devices 21–3

Machine screws *see* Screws
Machine tools maintenance 115–16
Macro information 158
Magnetic resonance imaging (MRI) 401
Maintenance 115–18, 272–3
Malleability 26
Manufacturing 397–9, 422, 427
    application of engineering 397–9
    and drawings 269–70
    environmental issues 406
    and graphical methods 166
Marketing 166–7
Material removal 44–62, 99–101
Materials 382–4, 386, 397–8
    for mechanical products 3–16
    selection 25–8
    specifications 109–10
Measurements and measuring 331–4
    devices 253, 335–43, 348
    recording correctly 344–8
    types of error 334–5
Mechanical components 17–25
    drawings 245–9, 258–9, 261–2
Mechanical engineers 417–18
Mechanical properties of materials 25–6
Medicine 400–2
Melamine 13
Mercury thermometers 337–8
Metal products industry 422
Metals 3–9, 27
Metric system 286–7
Micrometers 336
Microwaves 386, 387
Milling 57–62, 99
Millscale 27, 84
Mining 404–5
Moments, principles of 302–4
Motion *see* Dynamic systems
Motor vehicle industry 392–3, 422
MRI (magnetic resonance imaging) 401
Multicomponent drawings 261–6

n-type semiconductors 16, 33, 35
Natural resources 404–6
Negative numbers 365–7
Neoprene 14
Newton's laws 300, 317
Non-ferrous metals 6–9, 27
Non-linear relationships 357–8
Normalising 81–2
NRA (national record of achievement) 445–8
Numerators 363–4, 365

Nuts 18–19, 20–1, 22, 63
NVQs 424, 433
Nylon 11, 12

Ohm's law 322–3
Oil extraction 405
Oil paints and varnishes 87
Open dimensions 110
Operators 424
Orthographic projections 134, 189–92
Oscilloscopes 253, 342–3
Oxidising 89, 104
Oxy-acetylene welding 72–4, 102
Ozone layer 406

p-type semiconductors 16, 33, 35
Painting 87, 104
Parkerizing 89
Pencils 173–4
Pens 174
Perspex 11, 12
PERT (programme review and evaluation technique) 163
Phosphating 89
Photodiodes 33, 34, 40, 253
Photographs 151–3
Pickling 84
Pictorial projections 134, 196–7
Pie charts 153, 155
Pillar drills 52, 53
Plain-carbon steels 3–4, 82
Planning 162–3
Plans, process 112–14
Plasticisers 87
Plastics 10–13, 383
Plating 84–5, 88–9
Pneumatic systems 252, 305, 306
Polishing 86, 104
Pollution 405, 406–7, 409–10
Polyester resin 13
Polymers 10, 11, 12, 13, 27
Polypropene 10, 11
Polystyrene 11
Polythene 10, 11
Pop rivets 23, 24, 67
Positive numbers 365–7
Potentiometer 38, 251
Powder bonding 89, 104
Power 291, 320
Power industry 423
Power tools maintenance 115–16
Pressure 305–11, 346
Process plans 112–14
Product specifications 109–12
Production processes 44–89, 406, 429
    selection 109–14
Professional qualifications 435
Progression, career 436–9
Projections 134, 189–93, 196–7

Proportional relationships 351–4
Prototypes 161
Protractors 176, 177
PTFE 11, 12, 30
PVC 10–11

Qualifications 432–9
Qualitative information 157
Quality assurance 426–7
Quantitative information 158
Quarrying 404–5
Quench hardening 82, 103

Radio 396
Raster-based graphics 184–5, 186
Rates of change 358–60
Reamers 55
Recrystallisation 80–1
Recycled materials 383–4
Refrigeration 292–3
Relays 250, 252
Representations, drawing 204–30, 244–69
Research and development 425
Resistance, electrical 319, 322–5
    measurement 341–2, 346
Resistance to attack (materials) 27–8
Resistors 31, 37–8, 251
Resolution (measuring devices) 335
Respirators 96
Rivets 23–4, 66–7, 217, 219, 220, 221
rms (root means square) current 328
Rubbers 13–14, 79

Safe working load (SWL) 101
Safety *see* Health and safety
Scale drawings *see* Drawings
Scales, electronic 339–40
Schematic drawings *see* Drawings
Screw threads 18, 206, 214–15, 245–7
Screwdrivers 63
Screwed fastenings 18–21, 63–6, 214–17
Screws 19–20, 63, 216, 218
Sectioning 193–6
Self-tapping screws 21
Semiconductor devices 251, 253
Semiconductor materials 15–16
Sensitive bench drills 52
Serrations 248
Set squares 176, 177
Setscrews *see* Screws
Sheradising 89
Shock absorbers 248–9
SI system of units 286–7
Silicon chips 380, 398
Silicone rubber 14, 30

463

Index

Single-component drawings 258–60
Sketches 131–3
Slings, lifting 101
Soldering 249
    hard 70–1, 101, 102
    soft 67–70, 101, 102
Solvents 27–8
Spanners 63, 64
Specifications
    process 112–14
    product 109–12
Splines 248
Sport 391–2
Spreadsheets 153–6, 165, 347–8
Spring balances 338–9
Springs 226, 245, 296–9
Stainless steel 5
Standard form (numbers) 367–8
Standards, drawing 231
Static equilibrium 299–304
Static systems 296–304
Steam boilers 291–2
Steel 5, 27, 34, 82, 384
Steel rules 336
Stencils 178
Stoving paints and enamels 87
Straight line relationships 350–7
Strain gauges 339–40
Studs 20–1, 216, 218
Styrene rubber 13–14
Subassembly drawings 136–8
Super glues 79
Surface finishes *see* Finishes, surface
SVQs 424
Swarf 101
Switches 250, 252

SWL (safe working load) 101
Symbols *see* Conventions

Tables 344–6
Tailstocks 45, 50
Technicians 424, 455–6
Telephones 395–6
Television 391
Temperatures 80
    and electrical resistance 324–5
    measurement devices 337–8
    scales 288–9
Tempering 82
Templates 178
Terminals, electrical 65–6, 249–50
Terylene 11, 12
Test readings 112
Texture *see* Finishes, surface
Thermal conductivity 27
Thermal expansivity 27, 293–6
Thermal systems 288–96
Thermometers 337–8
Thermoplastic resins 78
Thermoplastics 10–12, 31, 80
Thermosetting plastics 12–13, 16, 80
Thermosetting resins 78–9, 87
Tightening order (screwed fastenings) 65
Tin 7
Tinplating 88
Title blocks 232, 236
Toasters 387
Tolerances 110, 207–9, 210–11
Toolposts 49
Torque wrenches 63, 65
Touch-screen technology 400
Toughness of materials 26

Toxic waste 407
Transistors 35–6, 40–1, 253
Transport 392–5
Transposition of formulae 368–70
Turning 44–51, 99
Twist drills 52, 54, 55, 56

Ultimate tensile strength (UTS) 25

Valves 255, 257
Vaporisation 289–93
Vector graphics 184, 185, 186
Velocity 311
Velocity–time graphs 313–14, 315, 355–7
Vernier callipers 336–7
Vertical mills 57, 58, 59, 60, 61
Vices 53–4, 57
Vickers pyramid test 25, 112
Vocational qualifications 432–4
Voltage 319, 322–3, 327–8, 341, 351–3
Volts 319

Washers 19, 21, 22
Washing and cleaning 388–9
Washing machines 389
Waste 407–8, 412–13, 415
Waveforms 327, 328, 329
Welding 71–6, 101–3
    symbols 228–30
Wood composites 17
Work 296
Work areas 115
Work experience 435–6

X-rays 401

Zinc 6, 88, 89